Mathematics in Programming

Xinyu Liu

Mathematics
in Programming

 Springer

Xinyu Liu
Amazon.com Inc
Beijing, China

ISBN 978-981-97-2431-4   ISBN 978-981-97-2432-1   (eBook)
https://doi.org/10.1007/978-981-97-2432-1

Jointly published with China Machine Press Co., Ltd., Beijing, China
The print edition is not for sale in the mainland of China. Customers from the mainland of China please order the print book from: China Machine Press, Beijing, China

This Springer imprint is published by the registered company Springer Nature Singapore Pte Ltd.
The registered company address is: 152 Beach Road, #21-01/04 Gateway East, Singapore 189721, Singapore

If disposing of this product, please recycle the paper.

Preface

Martin Gardner gives an interesting story in his popular book *aha! Insight*. In a country fair, there was a game called "Fifteen" on the carnival midway. Mr. Carny, the carnival operator, explained to people the rules: "We just take turns putting down coins on a line of numbers from 1 to 9. It doesn't matter who goes first. You put on nickles, I put on silver dollars. Whoever is the first to cover three different numbers that add to 15 gets all money on the table."

$$1 \mid 2 \mid 3 \mid 4 \mid 5 \mid 6 \mid 7 \mid 8 \mid 9$$

A lady joined this game. She went first by putting a nickle on 7. Because 7 was covered, it couldn't covered again by either player. And it's the same for other numbers. Mr. Carny then put a dollar on 8.

1	2	3	4	5	6	7	8	9
						nickle	dollar	

The lady next put a nickle on 2, so that one more nickle on 6 would make 15 and win the game for her. But the man blocked her with a dollar on 6. Now he could win by covering 1 on his next turn.

1	2	3	4	5	6	7	8	9
	nickle				dollar	nickle	dollar	

Seeing this threat, the lady put a nickle on 1 to block his win.

1	2	3	4	5	6	7	8	9
nickle	nickle				dollar	nickle	dollar	

The carnival man then put a dollar on 4. He would win by covering 5 next. The lady had to block him again. She put a nickle on 5.

1	2	3	4	5	6	7	8	9
nickle	nickle		dollar	nickle	dollar	nickle	dollar	

But the carnival man placed a dollar on 3. He won because $3 + 4 + 8 = 15$. The poor lady lost all nickles.

1	2	3	4	5	6	7	8	9
nickle	nickle	dollar	dollar	nickle	dollar	nickle	dollar	

Many people joined to play the game. The town's Mayor was fascinated by the game, too. After watching it for a long time, he decided that the carnival man had a secret that made him never to lose the game except he wanted to. The Mayor was awake all night trying to figure out the answer.

The key to the secret can be traced back to 650BC. There was a legend about *Lo Shu* in ancient China around the time of huge flood. A turtle emerged from the river with a curious pattern on its shell: a 3×3 grid in which circular dots of numbers were arranged.

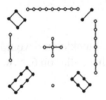

4	9	2
3	5	7
8	1	6

(a) Lo Shu Square   (b) Magic square of order 3

This is known as Lo Shu square, a magic square of order 3. The sum of the numbers in each row, column, and diagonal is the same: 15. For example, the sum of the first row $4 + 9 + 2 = 15$; the sum of the third column $2 + 7 + 6 = 15$; and the sum of the diagonal from left up to right bottom $4 + 5 + 6 = 15$. The insight to the carnival fifteen game is exactly the magic square. All three numbers sum to

15, form the rows, columns, and diagonals in that square. If the carnival man held a secret Lo Shu map, he was essentially playing the tick-tack-toe game with other people.

4	9	2
3	5	7
8	1	6

(a) magic square   (b) tick-tack-toe game

The game between the lady and Mr. Carny is equivalent to such a tick-tack-toe game. The man trapped the lady in step three. He could line up both a column and a diagonal. If the lady put on 3, then the man could win the game by playing on 5. If you know a bit about game theory or programming, one will never lose the tick-tack-toe game if plays carefully. The carnival man with the secret Lo Shu square map does have the advantage over other people. As the fifteen game proceeds, the carnival operator mentally plays a corresponding tick-tack-toe game on his secret map. This makes it easy for the operator to set up traps of winning position.

This interesting story reflects an important mathematical idea, isomorphism. A difficult problem can be transformed to an isomorphic one, which is mathematical equivalent and easy to solve. A line of 9 numbers corresponds to a 3 × 3 grid; the target sum of fifteen corresponds to one of the rows, columns, and diagonals; Lo Shu pattern corresponds to the magic square of order 3. This is what this book intents to tell: programming is isomorphic to mathematics. Just like in art and music, there are interesting stories and mathematicians behind the great minds.

(a) magic square

(b) Step 1, the lady puts on 7, the man puts on 8.

(c) Step 2, the lady puts on 2, the man puts on 6.

(d) Step 3, the lady puts on 1, the man puts on 4.

(e) Step 4, the lady puts on 5, the man wins on 3.

There is another idea behind this story: under the surface of the problem, there hides the theoretical essence, which is abstract and needs to understand. With the rapid development of artificial intelligence and machine learning, can we keep moving forward with a little cleverness and engineering practice? Can we open the mysterious black box to find the map to the future?

Beijing, China   Xinyu Liu
May 2019

Exercise 1

1.1 It's a classic exercise to program a tick-tack-toe game. It's trivial to test if three numbers sum up to 15. Use this point to implement a simplified tick-tack-toe program that never loses.[1]

Answer: page 303

---

[1] Refer to appendix A for the answers to all exercises.

Contents

Chapter 1
Numbers

Numbers are the highest degree of knowledge. It is knowledge itself.

Plato

1.1 The History of Number

Numbers emerged with human evolution. Some people believe that human languages were inspired by numbers. Our ancestors learned the numbers from gathering and hunting activities, for example, to count the gathered fruits. As trading developed, people needed numeral tools to handle bigger numbers than previously encountered.

We found in the regions of Iran, people made clay tokens for recordkeeping around 4000 BC. They baked two round tokens with "+" sign to represent "two sheep." As it would be impractical to represent a hundred sheep with many tokens, they invented different tokens for 10 sheep, 20 sheep, and so on. In order to avoid the record being altered, people further made a clay envelope in which tokens were placed, sealed, and baked. If anybody disputed the number, they could break the envelope and do a recount. They also pressed signs outside the envelope before baking; these signs became the first written language for numbers [1]. Figure 1.1 shows the ancient clay tokens and envelopes found in Uruk period.

As the number increasing, the tokens and envelopes were gradually replaced by more powerful numerals. About 3500 BC, the Sumerians in Mesopotamia used round stylus in flat clay tablets to carve pictographs representing various tokens. Each sign represented both the things being counted and the quantity.

The next leap happened around 3100 BC. Abstract numbers were dissociated from the things being counted. We found from the clay tablets the things were indicated by pictographs carved with a sharp stylus next to the round-stylus numerals. These abstracted numerals later evolved to Babylonian cuneiform characters.

© China Machine Press 2024
X. Liu, *Mathematics in Programming*,
https://doi.org/10.1007/978-981-97-2432-1_1

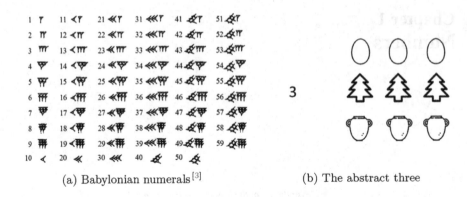

(a) Babylonian numerals[3]   (b) The abstract three

Fig. 1.1 The envelope of tokens in Uruk period from Susa. Louvre Museum

The abstract numbers emerged from intelligent mind. People realized the abstract number 3 could represent 3 eggs, 3 trees, 3 jars, etc. It's a powerful tool. People can manipulate the pure numbers and apply the result to the concrete things. When adding 1 to 3 to get 4, we know gathering another egg after 3 eggs gives 4 eggs; we also know baking another jar after 3 jars gives 4 jars. We resolve a whole kind of problems instead of one by one.

Starting from numbers, people developed addition and subtraction and then the more powerful methods of multiplication and division. When measuring the length, angle, area, and volume, we connected the number to geometric quantity. People from different places found the inner relationship and laws for numbers and shapes.

Ancient Egyptian, Greek, and Chinese found the Pythagoras theorem independently and applied it to the amazing works like building the Great Pyramid. Tracing back from the modern civilization, we find the natural number is the source of mathematics and science. "God made the integers;[1] all else is the work of man." said Kronecker.

1.2 Peano Axioms

Euclid's *Elements* is the classic example of the axiomatic method. From five "postulates" (now known as axioms[2]), Euclid developed the propositions one by one elaborately. Every one is only based on the axioms and the propositions proved previously. In this way, he built a great system including geometry (plane and three-dimensional), ratio and proportions, incommensurables (irrational numbers), number theory, method of exhaustion, etc. Such achievement happened in ancient time more than 2000 years ago. One may surprise that there was no axiomatic system for natural numbers until the late nineteenth century. People considered natural numbers were straightforward and obvious. The Italian mathematician Giuseppe Peano (Fig. 1.2) established the axioms for natural numbers in 1889, known as Peano axioms nowadays. Interestingly, there are also five axioms.

1. 0 is a natural number: $0 \in \mathbb{N}$;
2. For every natural number n, there is a successor n', written as $n' = succ(n)$;

It seemed that we could define infinitely many natural numbers only with these two axioms. From 0, the next is 1, then 2, 3, ..., n, and $n + 1$, ... However, here is a counterexample: a set of two numbers $\{0, 1\}$. Define 1 as the successor of 0, while 0 as the successor of 1. It satisfies the above two axioms well although unexpected. In order to avoid this situation, we need the third axiom.

3. 0 is not the successor of any natural number, i.e., $\forall n \in \mathbb{N} : n' \neq 0$.

Is it enough? Here's another counterexample: a set of $\{0, 1, 2\}$. Define 1 as the successor of 0, 2 as the successor of 1, and 2 as the successor of 2 again. It satisfies all the three axioms. Therefore, we need the fourth.

4. The successors are distinct. If m and n have the same successor, then they are identical, i.e., $m = n$ if and only if $m' = n'$.

However, it is still insufficient. For another counterexample: set $\{0, 0.5, 1, 1.5, 2, 2.5, ...\}$. Define 1 as the successor of 0, 2 as the successor of 1, 1.5 as the successor

---

[1] Strictly speaking, integers contain the negate of natural numbers.

[2] The two terms are interchangeable nowadays. Postulates were specific to geometry in Euclid's time.

Fig. 1.2 Giuseppe Peano
(1858–1932)

of 0.5, and 2.5 as the successor of 1.5, but 0.5 is not the successor of any number. In order to exclude such "unreachable" elements, we need the last axiom.

5. If some subset S of natural numbers contains 0, and every element in S has a successor, then $S = \mathbb{N}$.

Why does the fifth axiom exclude the counterexample of $\{0, 0.5, 1, 1.5, 2, 2.5, ...\}$? Consider its strict subset of $\{0, 1, 2, ...\}$. 0 belongs to it, and every element has a successor. But it is not identical to the original set. As 1.5, 2.5, ... are not in this subset, it does not satisfy the fifth axiom. The last axiom is also known as the *axiom of induction*. It's often stated equivalently as below.

5. For any proposition to natural numbers, if it holds for 0, and assuming it holds for some number n, we can prove it also holds for n', then the proposition holds for all natural numbers. (This axiom ensures the correctness of mathematical induction.)

These are the five Peano axioms. They establish the first-order arithmetic, also known as Peano arithmetic.[3]

Peano was a mathematician, logician, and linguist. He was born and raised on a farm at Spinetta, a hamlet now belonging to Cuneo, Italy. He entered the University of Turin in 1876, graduating in 1880 with high honors, after which the university employed him to teach calculus course. In 1887, he married Carola Crosio. In 1886, he began teaching concurrently at the Royal Military Academy. From 1880s, Peano started to study mathematical logic. He published Peano axioms, a formal foundation for the natural numbers in 1889. He started the Formulario Project. It was the "Encyclopedia of Mathematics," containing all known formulas and theorems. In 1900, the Second International Congress of Mathematicians was held in Paris. At the conference Peano met Bertrand Russell and gave him a copy of Formulario. Russell

---

[3] Some people use 1 but not 0 as the first natural number. The order is different from the original works published by Peano, where the fifth axiom of induction was listed as the third.

was so struck by Peano's innovative logical symbols that he left the conference and returned home to study Peano's text [3].

When Russell and Whitehead wrote *Principia Mathematica*, they were deeply influenced by Peano. Peano played a key role in the axiomatization of mathematics and was a leading pioneer in the development of mathematical logic and set theory. As part of this effort, he made key contributions to the modern rigorous and systematic treatment of the method of mathematical induction. He spent most of his career teaching mathematics at the University of Turin. He also wrote an international auxiliary language, Latino sine flexione ("Latin without inflections," later called Interlingua), which is a simplified version of classical Latin. Most of his books and papers are in Latino sine flexione. Although Peano put a lot of effort to rewrite his works in the new language, few people read it. On the other hand, his early works in French influenced many mathematicians, especially to Bourbaki group, which came out many top mathematicians like André Weil and Henri Cartan. Giuseppe Peano died on April 20, 1932.

1.3 Natural Numbers and Programming

People didn't establish the axioms for programming before making amazing achievement with modern computer systems. Along the great success of computer application, the foundation of computer science is gradually formalized. The similar things happened several times in history. For example, calculus was developed by Newton and Leibniz in the seventeenth century, then applied to a wide range of areas, including fluid dynamics, astronomy, and so on. However, it was not formalized until the nineteenth century by Cauchy, Weierstrass, etc. [3].

With Peano axioms, we can define natural numbers for computer programs. In a bare computer system, define natural numbers as below:[4]

```
data Nat = zero | succ Nat
```

It says a natural number is either 0 or the successor of some natural number. Symbol "|" is mutually exclusive, implicating the axiom that 0 is not the successor of any natural number. We can further define addition:

```
a + zero = a
a + (succ b) = succ (a + b)
```

There are two rules for addition. First, adding any natural number a to 0 gives a itself; second, adding a natural number a to a successor of some natural number b

---

[4] We use Haskell programming language with minor modification to illustrate programs.

equals to the successor of $a + b$. Here is the compact form:

$$a + 0 = a$$
$$a + b' = (a + b)'$$
(1.1)

Taking 2+3, for example, 2 is `succ(succ zero)`, and 3 is `succ(succ(succ zero))`. According to the definition:

```
  succ(succ zero) + succ(succ(succ zero))
= succ(succ(succ zero) + succ(succ zero))
= succ(succ(succ(succ zero) + succ zero))
= succ(succ(succ(succ(succ zero) + zero)))
= succ(succ(succ(succ(succ zero))))
```

The result is the fifth successor of 0. It's not practical to apply *succ* function again and again for big numbers like 100. Let's introduce a simplified notation for natural number n.

$$n = foldn(succ, zero, n)$$
(1.2)

It applies *succ* function to 0 for n times. Function *foldn* is realized as below:

$$foldn(f, z, 0) = z$$
$$foldn(f, z, n') = f(foldn(f, z, n))$$
(1.3)

Function *foldn* defines some operation on natural number. When z is *zero*, and f is the *succ* function, then it applies the succeed operation multiple times to obtain a specific natural number, for example:

```
foldn(succ, zero, 0) = zero
foldn(succ, zero, 1) = succ(foldn(succ, zero, 0)) = succ zero
foldn(succ, zero, 2) = succ(foldn(succ, zero, 1)) = succ(succ zero)
...
```

Multiplication for natural numbers can be defined on top of addition.

```
a . zero = zero
a . (succ b) = a . b + a
```

or in compact form:

$$a \cdot 0 = 0$$
$$a \cdot b' = a \cdot b + a$$
(1.4)

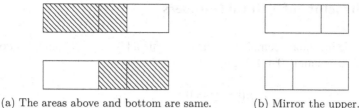

(a) The areas above and bottom are same.   (b) Mirror the upper.

Fig. 1.3 Geometric explanation for associativity and commutativity of +

It turns out that neither the associativity nor the commutativity of addition or multiplication is axiom. Each can be proved with Peano axioms and the definition. Let's prove the associativity of addition (Fig. 1.3), for example. To prove $(a + b) + c = a + (b + c)$, we firstly show it holds when $c = 0$. According to the first rule of addition:

$$(a + b) + 0 = a + b$$
$$= a + (b + 0)$$

For induction, assuming $(a + b) + c = a + (b + c)$ holds, we are going to show $(a + b) + c' = a + (b + c')$.

$$
\begin{aligned}
(a + b) + c' &= (a + b + c)' && \text{2nd rule of + (converse)} \\
&= (a + (b + c))' && \text{induction assumption} \\
&= a + (b + c)' && \text{2nd rule of +} \\
&= a + (b + c') && \text{2nd rule of + (converse)} \quad \square
\end{aligned}
$$

However, it is a bit complex to prove the commutativity of addition; see the proof in Appendix B.

Exercise 1.2

1.2.1 Define 1 as the successor of 0, prove $a \cdot 1 = a$ holds for all natural numbers.
1.2.2 Prove the distribution of multiplication over addition.
1.2.3 Prove the associativity and commutativity of multiplication.
1.2.4 How to verify $3 + 147 = 150$ with Peano axioms?
1.2.5 Give the geometric explanation for associativity, commutativity, and distribution of multiplication.

Answer: page 306

1.4   Structure of Natural Numbers

We can define more complex operations with addition and multiplication, for example, summation: $0 + 1 + 2 + ...$

$$sum(0) \quad = 0$$
$$sum(n + 1) = (n + 1) + sum(n) \tag{1.5}$$

and the factorial $n!$

$$fact(0) \quad = 1$$
$$fact(n + 1) = (n + 1) \times fact(n) \tag{1.6}$$

They are similar. Although artificial intelligence achieves incredible results today, the machine can't jump out of the box, as intelligent mind, to abstract at a higher level. This is one of the most complex, powerful, and mysterious parts in the human brain [4]. We observe two patterns: (1) Corresponding to 0 in natural numbers, both summation and factorial have a start value. Summation starts from 0; factorial starts from 1. (2) For recursion, both apply some operation to a number n and its successor n'. For summation, it's $n' + sum(n)$; for factorial, it's $n' \times fact(n)$. If the start value as c and the operation as h are abstracted, then we get the same form for both.

$$f(0) \quad = c$$
$$f(n + 1) = h(f(n)) \tag{1.7}$$

We call this scheme the *structural* recursion over natural numbers. Below examples show how it behaves over the first several numbers.

n	$f(n)$
0	c
1	$f(1) = h(f(0)) = h(c)$
2	$f(2) = h(f(1)) = h(h(c))$
3	$f(3) = h(f(2)) = h(h(h(c)))$
...	...
n	$f(n) = h^n(c)$

where $h^n(c)$ means applying the operation h over c for n times. It's an instance of the more general *primitive* recursion [5, p5]. It is further related to the *foldn* defined in Eq. (1.3).

$$f = foldn(h, c) \tag{1.8}$$

However, there are three variables in the original *foldn* definition—why only two appear? We can actually write it as $f(n) = foldn(h, c, n)$. When binding the first two variables to *foldn*, it turns to be a new function accepting one argument. We can consider it as $foldn(h, c)(n)$.

We call *foldn* the *fold* operation on natural numbers. When c is *zero* and h is *succ*, we obtain a sequence of natural numbers:

$$zero, succ(zero), succ(succ(zero)), \cdots, succ^n(zero), \cdots$$

When c and h are other things than *zero* or *succ*, then $foldn(h, c)$ defines something isomorphic[5] to natural numbers. Here are some examples:

$$(+m) = foldn(succ, m)$$

This operation increases a number by m. When applying to natural numbers, it generates a sequence of $m, m + 1, m + 2, ..., n + m, ...$

$$(\times m) = foldn((+m), 0)$$

This operation multiplies a number by m. When applying to the natural numbers, it generates a sequence of $0, m, 2m, 3m, ..., nm, ...$

$$m^{()} = foldn((\times m), 1) \tag{1.9}$$

This operation takes the power for a number m. When applying to natural numbers, it generates a sequence of $1, m, m^2, m^3, ..., m^n, ...$ All the three sequences are isomorphic to natural numbers. Can we use the abstract tool *foldn* to define the summation and factorial? Observe below table.

n	0	1	2	3	...	n'
$sum(n)$	0	$1 + 0 = 1$	$2 + 1 = 3$	$3 + 3 = 6$	...	$n' + sum(n)$
$n!$	1	$1 \times 1 = 1$	$2 \times 1 = 2$	$3 \times 2 = 6$	...	$n' \times (n!)$

We know that h needs to be a binary operation as it manipulates n' and $f(n)$. To solve it, we define c as a pair (a, b).[6] Then define some kind of "succ" operation on the pair. We also need functions to extract a and b out of the pair.

$$1st(a, b) = a$$
$$2nd(a, b) = b \tag{1.10}$$

---

[5] See Sect. 3.2.3 for isomorphism. It means the thing has the same structure as natural numbers (equivalent).

[6] Also known as *tuple* in computer programming.

With this setup, we define summation as:

$$c = (0, 0) \qquad\qquad \text{starting pair}$$

$$h(m, n) = (m', m' + n) \qquad\qquad \text{succeed the 1st, then add to the 2nd}$$

$$sum = 2nd \circ foldn(h, c)$$

Here we use the symbol "$\circ$" to "connect" the $2nd$ function and the $foldn(h, c)$ function. We call it *function composition*. $f \circ g$ means firstly apply g to the variable, then apply f on top of the result. That is, $(f \circ g)(x) = f(g(x))$. Starting from $(0, 0)$, below table gives the steps of summation.

(a, b)	$(a', b') = h(a, b)$	b'
$(0, 0)$	$(0 + 1 = 1, 1 + 0 = 1) = (1, 1)$	1
$(1, 1)$	$(1 + 1 = 2, 2 + 1 = 3) = (2, 3)$	3
$(2, 3)$	$(2 + 1 = 3, 3 + 3 = 6) = (3, 6)$	6
...	...	...
$(m, sum(m))$	$(m + 1, m + 1 + sum(m))$	$sum(m + 1)$

Similarly, we define factorial with *foldn*.

$$c = (0, 1) \qquad\qquad \text{starting pair for } n!$$

$$h(m, n) = (m', m'n) \qquad\qquad \text{iteration for } n!$$

$$fact = 2nd \circ foldn(h, c)$$

Let's see another example powered by this abstract tool, Fibonacci sequence. It's named after the medieval mathematician Leonardo Pisano (Fig. 1.4). Fibonacci originates from "filius Bonacci" in Latin. It means the son of (the) Bonacci. Fibonacci's father was a wealthy Italian merchant, who often did trading around North Africa and the Mediterranean coast. Fibonacci traveled with him as a young

Fig. 1.4 Leonardo Pisano, Fibonacci (1175–1250)

boy. It was in Bugia (now Béjaïa, Algeria) that he learned about the Hindu-Arabic numeral system. Fibonacci realized many advantage of this numeral system. He introduced it to Europe through his book, the *Liber Abaci* (Book of Abacus or Book of Calculation, 1202). European people were using Roman numeral system before that. We can still see Roman numbers in clock today. The Roman number for year 2018 is MMXVIII, where M stands for 1000, so the two M letters mean 2000, X stands for 10, V stands for 5, and the three I letters mean 3. Summing them up, we get 2018. The Hindu-Arabic numeral system is a positional decimal numeral system. It uses the zero invented by Indian mathematicians. Numbers at different position mean different value. This advanced numeral system was widely used in business, for example, converting different currencies and calculating profit and interest, which were important to the growing banking industry. It influenced the mathematics renaissance in Europe greatly and is used almost everywhere today.

Fibonacci numbers, a well-known problem described in the *Liber Abaci*, can be traced back around 200 BC in India. Assume a newly born pair of rabbits, one male and one female, are put in a field. Rabbits are able to mate at the age of 1 month so that at the end of the second month a female will produce another pair of rabbits. Rabbits never die and a mating pair always produces a new pair (a male, a female) every month from the second month on. How many pairs will be there in 1 year?

At the start, there is a pair in the first month. In the second month, there is a newborn pair, in total two pairs. In the third month, the matured pair produces another pair, while the newborn in the previous month are still young. In total, there are $2 + 1 = 3$ pairs. In the fourth month, the two pairs of matured rabbits produce another two pairs of baby. In addition to the three pairs in the third month, there are in total $3 + 2 = 5$ pairs. Repeating it gives a sequence of numbers (Fig. 1.5).

1, 1, 2, 3, 5, 8, 13, 21, 34, 55, 89, 144, ...

It's easy to find the pattern of this sequence. From the third number, every number is the sum of the previous two. We can understand the reason behind it like this. Let there be m pairs of rabbits in the previous month and n pairs in this month. As the

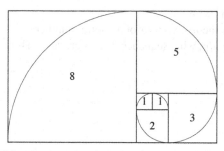

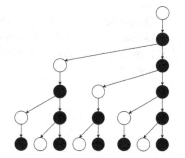

(a) The length of the squares give a Fibonacci sequence

(b) black: mature, white: new born

Fig. 1.5 Expansion of Fibonacci numbers

new additional $n - m$ pairs are all newborn, the rest m pairs are mature. In the next
month, the $n - m$ pairs grow mature, while the m pairs of big rabbits produce another
m pairs of baby rabbits. The total pairs in the next month is the sum of big and baby
rabbits, which is $(n - m) + m + m = n + m$. This gives the recursive definition of
Fibonacci numbers:

$$\begin{aligned} F_0 &= 0 \\ F_1 &= 1 \\ F_{n+2} &= F_n + F_{n+1} \end{aligned} \qquad (1.11)$$

The starting numbers are defined as 0 and 1 by convention.[7] As Fibonacci
numbers start from a *pair* of natural numbers and the recursion also uses a pair
of numbers, we can apply the abstract tool *foldn* to define Fibonacci numbers.

$$F = 1st \circ foldn(h, (0, 1))$$
$$h(m, n) = (n, m + n) \qquad (1.12)$$

Can this definition be realized in real-world computer program? Is it too
idealistic? Below is a real piece of code that implements Fibonacci numbers in
Haskell.[8] Running command `fib 10` outputs the tenth Fibonacci number, 55.[9]

```
foldn _ z 0 = z
foldn f z n = f (foldn f z (n - 1))

fib = fst o foldn h (0, 1) where
    h (m, n) = (n, m + n)
```

Exercise 1.3

1.3.1 Define square of natural number $()^2$ with *foldn*.
1.3.2 Define $()^m$ with *foldn*, which gives the m-th power of a natural number.
1.3.3 Define sum of odd numbers with *foldn*; what sequence does it produce? Hint:
     Fig. 1.6.

---

[7] If starts from 1 and 3, it produces the Lucas sequence 1, 3, 4, 7, 11, 18, ...
[8] Haskell does not support n+k pattern matching since 2010, or we could use:
    `foldn f z (n+1) = f (foldn f z n)`.
[9] One line of code produces the first 100 Fibonacci numbers:
    `take 100 $ map fst $ iterate (λ(m, n) → (n, m + n)) (0, 1)`.

Fig. 1.6 Cover of PWW
(proof without words), partial

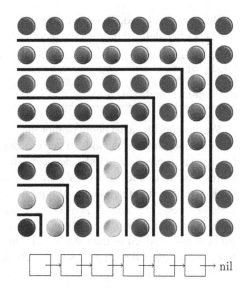

Fig. 1.7 Linked list

1.3.4 There is a line of holes (infinitely many) in the forest. A fox hides in a hole. It moves to the next hole every day. If we can only check one hole a day, is there a way to catch the fox? Prove this method works. What if the fox moves more than one hole a day [6]?

Answer: page 308

1.5 List

We see the examples that the set of natural numbers is isomorphic to its subsets, like odd and even numbers, squares, and Fibonacci numbers. Natural numbers can also be isomorphic to other things. One interesting example is the list in computer programming. Here is its definition.

```
data List A = nil | cons(A, List A)
```

As a data structure, a list of type A is either empty ($\varnothing$), represented as nil, or contains two parts: a node with data of type A and the rest sub-list. Function cons links an element of type A to another list.[10] Figure 1.7 shows a list of six nodes.

---

[10] The name cons comes from the Lisp convention.

Because every node links to the next one or nil, the list is also called as "linked list." It's often defined through the record data structure[11] traditionally, for example:

```
Node of A:
    key: A
    next: Node of A
```

We can consider a list being isomorphic to natural numbers. According to the first Peano axiom, nil corresponds to zero. Based on the second Peano axiom, for any list, we can apply cons to link a new element of type A to the left. We can treat cons corresponding to succ to natural numbers. There are two different things: first, the list is augmented with elements of type A—cons(1, cons(2, cons(3, nil))), cons(2, cons(1, cons(3, nil))), cons(1, cons(4, cons(9, nil))), and cons('a', cons('b', cons('c', nil))) are all different lists; second, the new element is not added to the right at tail, but is added to the left on head. Different from the intuition, the list grows to the left but not to the right.

It's inconvenient to represent a long list with nested cons. We simplify cons(1, cons(2, cons(3, nil))) to [1, 2, 3] and use symbol ":" for cons. This list can also be written as 1:[2, 3] or 1:(2:(3:nil)). When the type A is a character, we write the list as string in quotes. For example, "hello" is the simplified form of ['h', 'e', 'l', 'l', 'o']. For empty, we also use [ ], "" (for string), or $\varnothing$ interchangeably. Similar to addition of natural numbers, we define the concatenation for lists as the following:

$$nil + \!\!\!+ \, y = y$$
$$cons(a, x) + \!\!\!+ \, y = cons(a, x + \!\!\!+ \, y)$$

(1.13)

There are two rules: first, concatenating empty with any list produces the same list; second, when concatenating the "successor" of a list to another one, it equals to firstly concatenate the two lists, then take the successor. Comparing to "+" of natural numbers, list concatenation is symmetric (mirrored).

$$\begin{array}{c|c} y = & nil + \!\!\!+ \, y \\ cons(a, x + \!\!\!+ \, y) = & cons(a, x) + \!\!\!+ \, y \end{array} \quad \begin{array}{c} a + 0 = a \\ a + succ(b) = succ(a + b) \end{array}$$

This symmetry gives us hint to the associativity of list concatenation. We prove $(x + \!\!\!+ \, y) + \!\!\!+ \, z = x + \!\!\!+ \, (y + \!\!\!+ \, z)$ by induction on x. We first show it holds when $x = nil$.

$$\begin{array}{ll} (nil + \!\!\!+ \, y) + \!\!\!+ \, z = y + \!\!\!+ \, z & \text{1st rule of } + \!\!\!+ \\ = nil + \!\!\!+ \, (y + \!\!\!+ \, z) & \text{1st rule of } + \!\!\!+ \text{ (converse)} \end{array}$$

---

[11] In most cases, the data stored in a list have the same type. However, there is also a heterogeneous list, like in Lisp.

For induction, assuming $(x + y) + z = x + (y + z)$ holds, we are going to show $((a:x) + y) + z = (a:x) + (y + z)$.

$$
\begin{aligned}
((a:x) + y) + z &= (a:(x + y)) + z & \text{2nd rule of } + \\
&= a:((x + y) + z) & \text{2nd rule of } + \\
&= a:(x + (y + z)) & \text{induction assumption} \\
&= (a:x) + (y + z) & \text{2nd rule of } + \text{ (converse)} \quad \square
\end{aligned}
$$

Different from the natural numbers, however, list concatenation is not commutative.[12] For example, $[2, 3, 5] + [7, 11] = [2, 3, 5, 7, 11]$, while $[7, 11] + [2, 3, 5] = [7, 11, 2, 3, 5]$, a different result.

Considering the similarity to the natural numbers, we also define the abstract folding operation for lists. Corresponding to the abstract start value c and the binary operation h, define the recursive scheme as below.

$$
\begin{aligned}
f(nil) &= c \\
f(cons(a, x)) &= h(a, f(x))
\end{aligned} \tag{1.14}
$$

Next, let $f = foldr(h, c)$ abstract list folding. We name it as *foldr* to call out the folding that starts from right to left.

$$
\begin{aligned}
foldr(h, c, nil) &= c \\
foldr(h, c, cons(a, x)) &= h(a, foldr(h, c, x))
\end{aligned} \tag{1.15}
$$

We can define varies of list manipulation with *foldr*, for example, to sum and multiply a list of elements:

$$
\begin{aligned}
sum &= foldr(+, 0) \\
product &= foldr(\times, 1)
\end{aligned} \tag{1.16}
$$

Let's expand *sum* with examples. For empty list, $sum(nil) = foldr(+, 0, nil) = 0$. For list of $[1, 3, 5, 7]$:

$$
\begin{aligned}
sum([1, 3, 5, 7]) &= foldr(+, 0, 1 : [3, 5, 7]) \\
&= 1 + foldr(+, 0, 3 : [5, 7]) \\
&= 1 + (3 + foldr(+, 0, 5 : [7])) \\
&= 1 + (3 + (5 + foldr(+, 0, cons(7, nil))))
\end{aligned}
$$

---

[12] This is the reason why we avoid using symbol + for concatenation, although many programming languages adopt the + sign for string. It causes potential issues in practice.

$$= 1 + (3 + (5 + (7 + foldr(+, 0, nil))))$$
$$= 1 + (3 + (5 + (7 + 0)))$$
$$= 16$$

We can count the length of a list with *foldr*, essentially mapping the list to a natural number.

$$length = foldr(cnt, 0)$$
$$cnt(a, n) = n + 1$$
(1.17)

where function *cnt* increases the counter n by 1 for whatever element a. We can use $|x| = length(x)$ to represent the length of a list. The next example defines list concatenation with *foldr*.

$$(+\!\!\!+\ y) = foldr(cons, y)$$
(1.18)

It corresponds to $(+m)$ to a natural number. Further corresponding to $\times$, we define "multiplication" of lists, i.e., to concatenate a list of lists.

$$concat = foldr(+\!\!\!+, nil)$$
(1.19)

For example, $concat([[1, 1], [2, 3, 5], [8]])$ "flattens" to $[1, 1, 2, 3, 5, 8]$. To wrap up this section, we define two important list operations with *foldr*: filtering and mapping.[13] Filtering is to compose a new list from the elements that satisfy a given predication. In order to realize it, we need to introduce the conditional expression,[14] written as $(p \mapsto f, g)$. Given a variable x, if the predication $p(x)$ holds, then the expression result is $f(x)$, else it's $g(x)$. We also use the form "if $p(x)$, then $f(x)$, else $g(x)$" interchangeably. Below is the definition of filtering.

$$filter(p) = foldr((p \circ 1st \mapsto cons, 2nd), nil)$$
(1.20)

Let's use an example to understand how it works. We want to select all even numbers from a list $filter(even, [1, 4, 9, 16, 25])$. It expands to $h(1, h(4, h(9, ...)))$ till the right end $cons(25, nil)$. According to the definition of *foldr*, the result is c when the list is *nil*. So the next step is to compute $h(25, nil)$, where h is the conditional expression. When applying $even \circ 1st$ to the pair $(25, nil)$, function $1st$ picks 25. As it's odd, the predication *even* does not hold. Hence the conditional expression $2nd$ is evaluated and gives the result *nil*. Then we enter the upper level to compute $h(16, nil)$. Function $1st$ extracts 16 out. As it is even, the predicate *even*

---

[13] We actually define the *for each* map of a list, but not a map from a list to something else.

[14] Also known as McCarthy conditional form. It was introduced in 1960 by computer scientist John McCarthy, the inventor of Lisp.

holds, so the conditional expression sends to $cons(16, nil)$, which produces the list [16]. Then we enter one more upper level to compute $h(9, [16])$; the conditional expression sends to $2nd$, which again produces [16]. The computation enters to $h(4, [16])$ next. The conditional expression sends to $cons(4, [16])$, which produces the list [4, 16]. The computation finally reaches to the top level $h(1, [4, 16])$. The conditional expression sends to $2nd$, which produces the final result [4, 16].

Mapping is to transform every element x in a list to value $f(x)$ through a function f and form a new list. That is, $map(f, [x_1, x_2, \cdots, x_n]) = [f(x_1), f(x_2), \cdots, f(x_n)]$. Define it with $foldr$ as below.

$$map(f) = foldr(h, nil)$$
$$h(x, c) = cons(f(x), c) \tag{1.21}$$

Define $first$ that applies a function f to the first value in a pair, as $first(f, (x, y)) = (f(x), y)$. Then we can simplify the definition to $map(f) = foldr(cons \circ first(f), nil)$.

Exercise 1.4

1.4.1 What does the expression $foldr(cons, nil)$ define?

1.4.2 Read in a sequence of digits (string of digit numbers); convert it to decimal with $foldr$. How to handle hexadecimal digit and number? How to handle the decimal point?

1.4.3 Jon Bentley gives the maximum sum of sub-vector puzzle in *Programming Pearls*. For integer list $[x_1, x_2, \cdots, x_n]$, find the range i, j that maximizes the sum of $x_i + x_{i+1} + \cdots + x_j$. Solve it with $foldr$.

1.4.4 The longest sub-string without repeated characters. Given a string, find the longest sub-string without any repeated characters in it. For example, the answer for string "abcabcbb" is "abc." Solve it with $foldr$.

Answer: page 310

1.6 Form and Structure

The same idea can be reflected in different forms, for example, the arithmetic laws in equation and in Fig. 1.3, Fibonacci numbers and Fibonacci spiral in Fig. 1.5, Euclid axioms to geometry and Peano axioms to arithmetic, natural numbers and lists. We want to express the beauty of correspondence. Aristotle said, "The chief forms of beauty are order and symmetry and definiteness, which the mathematical sciences demonstrate in a special degree." (*Metaphysic*) When Raphael, the supreme High Renaissance artist, created the world famous fresco *School of Athens* in Fig. 1.8, he used the same approach by inventing a system of iconography. Many great

Fig. 1.8 Raphael, School of Athens (partial)

Greek philosophers in ancient time were illustrated with the figures in Raphael's time of Renaissance. The center figures are Plato and his student Aristotle. Of which Plato is depicted in the image of Leonardo da Vinci; Aristotle is in the image of Giuliano da Sangallo. Both were great artists in Renaissance. The elder Plato is walking alongside Aristotle with his right hand finger pointed up, while Aristotle is stretching his hand forward. It is popularly thought that their gestures indicate central aspects of their philosophies, for Plato, his Theory of Forms, and for Aristotle, his empiricist views, with an emphasis on concrete particulars. Plato argues a sense of timelessness, while Aristotle looks into the physicality of life and the present realm. Below the steps in the middle, the great philosopher Heraclitus leans the box and meditates. He was famous for the thoughts about simple dialectics and materialism. The image of Heraclitus is by another great artist Michelangelo in Renaissance. The front left is centered on the great mathematician Pythagoras. He is writing something. On the right side of Pythagoras is a blond young man in a white cloak, considered to be Francesco Maria della Rovere. He was Urbino's future Grand Duke. The center of the bottom right is the great mathematician Euclid with a compass in hand (or Archimedes in other opinion); he is surrounded by Ptolemy, the great astronomer with the celestial sphere in hand. The opposite is the painter Raphael's fellow villager, the architect Bramante. The one who wears a white hat on the right is the painter Sodom; the young man next to him with half head and a hat, is the painter, Raphael himself. This reminds us of the great musician, Johann Sebastian Bach, who spelled out his name B-A-C-H, as B-flat, A, C, and B-nature, which is an H in German conventions, in his work *The Art of Fugue*. The School of Athens reflects the classic art and philosophy in ancient Greece with the figures in Italian Renaissance, with multiple levels of correspondence between the form and content, the structure and thoughts. It is seen as "Raphael's masterpiece and the perfect embodiment of the classical spirit of the Renaissance".

Exercise 1.5

1.5.1 In the definition of Fibonacci numbers with *foldn*, the successor is computed as $(m', n') = (n, m + n)$. It is essentially the matrix multiplication:

$$\begin{pmatrix} m' \\ n' \end{pmatrix} = \begin{pmatrix} 0 & 1 \\ 1 & 1 \end{pmatrix} \begin{pmatrix} m \\ n \end{pmatrix}$$

where it starts from $(0, 1)^T$. Then Fibonacci numbers is isomorphic to natural numbers under the matrix multiplication:

$$\begin{pmatrix} F_n \\ F_{n+1} \end{pmatrix} = \begin{pmatrix} 0 & 1 \\ 1 & 1 \end{pmatrix}^n \begin{pmatrix} 0 \\ 1 \end{pmatrix}$$

Write a program to compute the power of 2×2 square matrix, and use it to give the n-th Fibonacci number.

Answer: page 314

Chapter 2
Recursion

GNU means GNU's Not Unix

Richard Stallman

People explore the world with numbers. Chapter 1 shows examples that are isomorphic to natural numbers under Peano axioms, like the list data structure in programming. Natural number is a basic tool. However, we accept the recursive definition without proof of its correctness, for example the factorial.

$$fact(0) = 1$$
$$fact(n + 1) = (n + 1)fact(n)$$

Why does it work? What is the theory behind recursion? Can we formalize it? We try to answer these questions in this chapter.

Pythagoras (about 570 BC–490 BC)

© China Machine Press 2024

X. Liu, *Mathematics in Programming*,

https://doi.org/10.1007/978-981-97-2432-1_2

2.1   Everything Is Number

Pythagoras was the first mathematician and philosopher who studied the universe with numbers. He is famous all over the world in the theory named after him. He was born in the island of Samos, off the coast of modern Turkey. Pythagoras might have learned from Thales of Miletus. With Thales suggestion, he went to Egypt to learn more about mathematics in about 535 BC. He spent 10 years in Egypt till the Persian Empire conquered Egypt in 525 BC. He was taken prisoner to Babylon with the army. Pythagoras learned mathematics and astronomy from the Babylonians. He might also arrive in India. Wherever he went, Pythagoras learned from the local scholars to enrich his knowledge [7].

Pythagoras returned his hometown Samos after a long journey abroad in around 520 BC. He founded a school and began giving lectures. Two years later, he left Samos, possibly because he disagreed with the local tyranny. He arrived in the Greek colony of Croton (southern Italy today), where he earned trust and admiration from people. He founded the philosophical and religious school that had many followers. Several prominent members of his school were women. The school was devoted to study astronomy, geometry, number theory, and music. They were called quadrivium and influenced European education for more than 2000 years [9]. Quadrivium reflected the Pythagoreans' philosophy that everything is number. They believed the planetary motion corresponded to geometry, while geometry was built on top of numbers. Pythagoras discovered the pattern of octave in mathematics. He was revered as the founder of mathematics and music. There are different sayings about Pythagoras's death. In around 497 BC, the Pythagorean Society at Croton was attacked due to political conflict. It's said Pythagoras escaped to Metapontum and died there [8], while some said he was killed.

The Pythagoreans, who were the first to take up mathematics, not only advanced this subject, but saturated with it. They fancied that the principles of mathematics were the principles of all things, said Aristotle in *Metaphysics*. They studied the numbers and their relations to nature. They developed the early number theory, one of the most important areas in mathematics. The Pythagoreans classified natural numbers and defined many important concepts including even and odd numbers, prime and composite numbers, and so on. They found some numbers were the sum of their proper positive divisors[1] and named them as perfect numbers. For example $6 = 1 + 2 + 3$, where 1, 2, 3 are the all three proper divisors of 6. Pythagoreans found the first two perfect numbers[2] 6 and 28 ($28 = 1 + 2 + 4 + 7 + 14$).

---

[1] Proper means the divisor is less than the number.

[2] Euclid proved a formation rule (*Element*, Book IX, proposition 35) whereby $q(q + 1)/2$ is an even perfect number whenever q is a prime of the form $2^p - 1$ for prime p, called a Mersenne prime nowadays. Much later, Euler proved that all even perfect numbers were of this form. This is known as the Euclid–Euler theorem.

Fig. 2.1 Triangular number

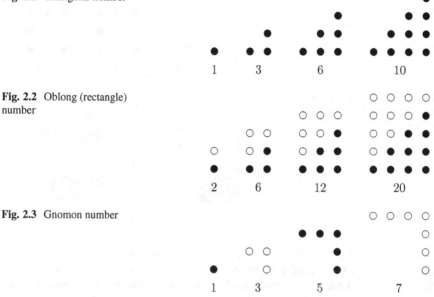

Fig. 2.2 Oblong (rectangle) number

Fig. 2.3 Gnomon number

The Pythagoreans found a class of figurate numbers when they formed geometric figure with small pebbles.[3] Figures 2.1 and 2.2 show the triangular numbers and oblong (rectangle) numbers. It's easy to see that the oblong number is two times of the corresponding triangle number, while the triangle number is the sum of the first n positive integers. In this way, the Pythagoreans found the below formula.

$$1 + 2 + 3 + ... + n = \frac{1}{2}n(n + 1)$$

The Pythagoreans found the odd number could be represented in gnomon[4] shape as shown in Fig. 2.3. And the first n gnomon shapes form a square, as in Fig. 2.4. In this way, they found the formula to sum n-odd numbers.

$$1 + 3 + 5 + ... + (2n - 1) = n^2$$

This is the answer to Exercise 1.3.3. More things could be explained in numbers. Given two strings under the same tension, Pythagoras found the tune was harmonic if the ratio of their lengths is 2:1, 3:2, 4:3, or 9:8. He developed the earliest

---

[3] In Latin, the word calculus meant "pebble" [7].

[4] The word "gnomon" originated in Babylonia, which probably meant upright stick whose shadow was used to tell time. In Pythagoras's time it meant a carpenter's square. It also meant what was left over from a square when a smaller square was cut out of one corner. Later Euclid extended from square to parallelogram [8, p31].

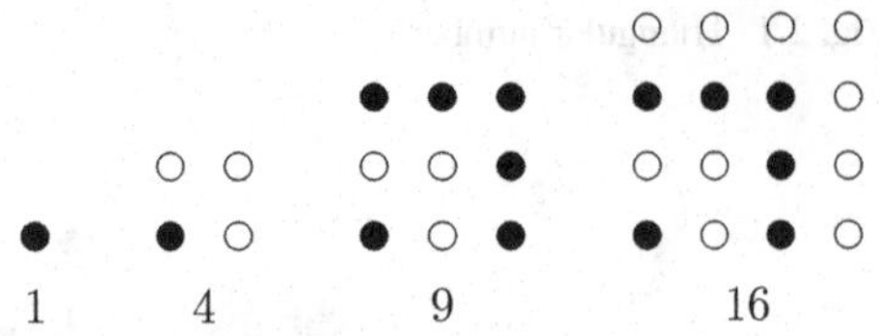

Fig. 2.4 Square and gnomon numbers

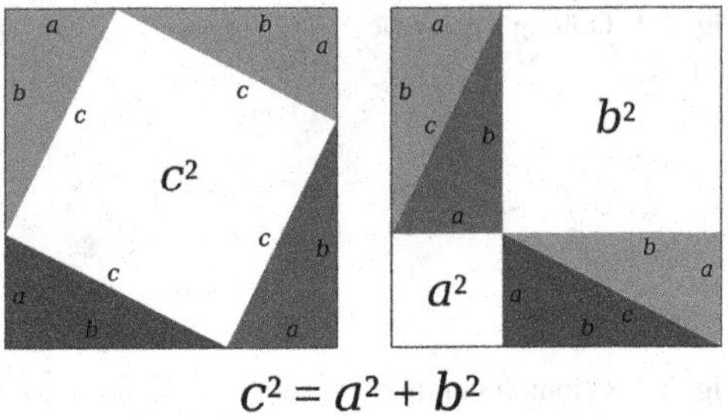

1   4   9   16

Fig. 2.5 Pythagoras theorem and a proof (equal the white areas)

$$c^2 = a^2 + b^2$$

music theory based on this. It seemed that music and mathematics were totally different things, while finally Pythagoras concluded that music was mathematics. Such unexpected relationship influenced Pythagoras greatly. He conjectured that all things could be explained with numbers or the ratio of numbers[5] The Pythagoreans started to find more and more things related to numbers; they believed the meaning of the whole universe was the harmonic of numbers and developed the philosophy based on number. This led to the attempt to build the geometry and the overall mathematics on top of numbers.

The Pythagoreans' most famous achievement is the Pythagoras theorem as shown in Fig. 2.5. However, this theorem led to a recursive circle and gave a counterexample against "everything is number." In order to understand this, we need the concept of commensurable and Euclidean algorithm. To build geometry with numbers, the Pythagoreans defined how to measure a line segment with another. If segment A can be represented by duplicating segment V finite times, one says V measures A. It means the length of a segment is the integer times of the other. There can be variations of measurements for a given line. When one can use the same segment to measure different lines, it has to be the *common measure*. That is to say if and only if segment V can measure both A and B, it is the common measure of them. The Pythagoreans believed for any two segments, there must be a common measure. If this was true, then the whole geometry can be built on top of numbers.

2.2 Euclidean Algorithm

When there are multiple common measures, we can define the greatest common measure. If segment V is the common measure of A and B, and V is greater than any other common measures, we say V is the *greatest common measure* of A and

---

[5] Actually means integers by the Pythagoreans.

B, abbreviated as "gcm." Given two segments, how to find the gcm? There is a famous Euclidean algorithm[6] that solves this problem. It is named after Euclid. This algorithm is defined in proposition 3, in Book X[7] of Euclid's Elements [10].

2.2.1 Euclid's Elements

Euclid of Alexandria was the most prominent ancient Greek mathematician, often referred to as "the father of geometry." His *Elements* is one of the most influential works in the history. However, little is known of Euclid's life except that he taught at Alexandria in Egypt. The year and place of his birth and death are unknown. Proclus, the last major Greek philosopher who lived around 450 AD, introduced Euclid briefly in his *Commentary on the Elements*. He mentioned an interesting story. When Ptolemy I of Alexandria (King of Egypt 323 BC–283 BC) grew frustrated at the degree of effort required to master geometry via Euclid's *Elements*, he asked if there was a shorter path; Euclid replied there was no royal road to geometry. This becomes the learning maxim of eternal. Another story told by Stobaeus said someone who had begun to learn geometry with Euclid, when he had learned the first theorem, asked "What shall I get by learning these things?" Euclid said "Give him three pence since he must make gain out of what he learns." Euclid disagreed with the narrow practical view of learning [10].

Euclid, about 300 BC

---

[6] The Euclidean algorithm was also developed independently in ancient India and China. The Indian mathematician Aryabhata used this method to solve the Diophantine equation around the end of the fifth century. The Euclidean algorithm was treated as a special case of the Chinese remainder theorem in *Sunzi Suanjing*. Qin Jiushao gave the detailed algorithm in his *Mathematical Treatise in Nine Sections* (数书九章) in 1247.

[7] The same algorithm for integers is in proposition 1, book VII. However, the algorithm for segments covers the integer case.

From ancient time to the late nineteenth century, people considered the *Elements* as a perfect example of correct reasoning. Although many results in *Elements* originated with earlier mathematicians, one of Euclid's accomplishments was to present them in a single, logically coherent framework, making it easy to use and reference, including a system of rigorous mathematical proofs that remains the basis of mathematics 2300 years later. More than a thousand editions have been published, making it one of the most popular books after the Bible. Even today, *Elements* is still widely taught in school[8] as one of the basic ways to train logic reasoning [7].

2.2.2 Euclidean Algorithm

Proposition 2.2.1 (Euclidean Algorithm) *To find the greatest common measure of two given commensurable magnitudes.*

The solution given by Euclid only uses recursion and subtraction. It means the gcm can be solved only with edge and compass essentially. This algorithm is defined as below.[9]

$$\text{gcm}(a, b) = \begin{cases} a = b: & a \\ b < a: & \text{gcm}(a - b, b) \\ a < b: & \text{gcm}(a, b - a) \end{cases} \tag{2.1}$$

Suppose segment a and b are commensurable. If they are equal, then either one is the greatest common measure, we return a as the result. If a is longer than b, use compass to intercept b from a repeatedly (through recursion), then find the gcm for the intercepted segment a' and b; otherwise if b is longer than a, we intercept a from b repeatedly and find the gcm for a and b'. Figure 2.6 lists the steps for example. We can also apply the algorithm to integers, for example, $a = 42$, $b = 30$, in below table.

gcm(a, b)	a	b
gcm(42, 30)	42	30
gcm(12, 30)	12	30
gcm(12, 18)	12	18
gcm(12, 6)	12	6
gcm(6, 6)	6	6

---

[8] The most popular version is edited by the French mathematician Legendre (1752–1833).

[9] For integers, we often use "gcd" as the abbreviation of the greatest common **divisor**.

Fig. 2.6 Example steps

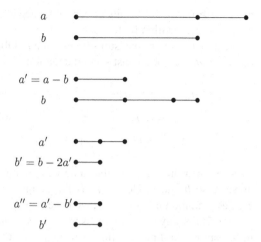

Because $\text{gcm}(a, b) = \text{gcm}(b, a)$, we can swap a and b to ensure $b \le a$ always holds. When $b \ne 0$, by repeatedly subtracting b from a to get a', we essentially perform division with remainder: $a' = a - \lfloor a/b \rfloor b$, or with the notion of modulo $a' = (a \bmod b)$. As $(a \bmod b)$ must be less than b, it means $\text{gcm}(a, b) = \text{gcm}(a \bmod b, b) = \text{gcm}(b, a \bmod b)$. If a is some multiple of b, i.e., $a = nb$ for some integer n, then the gcm is b, and the remainder $(a \bmod b) = 0$. We define $\text{gcm}(0, b) = \text{gcm}(b, 0) = b$. This simplifies Euclidean algorithm as below:

$$\text{gcm}(a, b) = \begin{cases} b = 0 : & a \\ \text{otherwise} : & \text{gcm}(b, a \bmod b) \end{cases} \tag{2.2}$$

Why does this algorithm give the greatest common measure? We prove it in two steps. *Step 1*: We show this algorithm gives the common measure. Suppose $b \le a$, apply division to get $a = bq_0 + r_0$, where q_0, r_0 are the quotient and remainder. Unless the remainder is 0, we repeat the division:

$$a = bq_0 + r_0$$
$$b = r_0q_1 + r_1$$
$$r_0 = r_1q_2 + r_2$$
$$r_1 = r_2q_3 + r_3$$
$$\dots$$

As far as a and b are commensurable, the above list is finite. This is because we always intercept the segment with compass of integer times. Hence, the quotient is integer, and every remainder is strict less than its divisor. We have $b > r_0 > r_1 >$

$r_2 > \cdots \geq 0$. As the remainder cannot be negative, and b is finite, we must reach to $r_{n-1} = r_n q_{n+1}$ within finite steps.

Next is to show r_n measures both a and b. Obviously, r_n measures r_{n-1} because $r_{n-1} = r_n q_{n+1}$. For the last second equation:

$$r_{n-2} = r_{n-1} q_n + r_n$$

$$\qquad = r_n q_{n+1} q_n + r_n \qquad\qquad \text{substitute } r_{n-1} = r_n q_{n+1}$$

$$\qquad = r_n (q_{n+1} q_n + 1)$$

So r_n measures r_{n-2}. In the same way, r_n measures every left hand step by step upwards to b and a. Hence, r_n is the common measure of a and b. Suppose the greatest common measure is g; we have $r_n \leq g$.

Step 2: For any common measure c, let $a = mc$, $b = nc$ for some integer m and n. Substitute the first equation $a = bq_0 + r_0$ to $mc = ncq_0 + r_0$, then change to $r_0 = (m - nq_0)c$. It means c measures r_0. In the same way, c measures $r_1, r_2, \cdots$, r_n one by one. Hence, it proves any common measure also measures r_n, so does the greatest one g. It means $g \leq r_n$.

Combine the result of *Steps 1* and *2* that $r_n \leq g$ and $g \leq r_n$, the gcm $g = r_n$. It does not only prove Euclidean algorithm, but also tells g is the gcm of every pair:

$$g = \text{gcm}(a, b) = \text{gcm}(b, r_0) = \ldots = \text{gcm}(r_{n-1}, r_n) = r_n \qquad (2.3)$$

2.2.3   Extended Euclidean Algorithm

For given a and b, in addition to computing their gcm g, the extended Euclidean algorithm also finds two integers x and y, satisfying $ax + by = g$, called Bézout's identity.[10] Why does Bézout's identity always hold? Here is a proof. Consider a set of all the positive linear combinations of a and b.

$$S = \{ax + by > 0 \mid x, y \in \mathbb{Z}\}$$

For line segments, S must not be empty, as it contains at least a (where $x = 1, y = 0$) and b (where $x = 0, y = 1$). Since all elements in S are positive, as far as a, b are commensurable, we are able to choose the smallest one[11] as $g = as + bt$.

---

[10] French mathematician Méziriac discovered and proved this identity for integers. Bézout proved it for polynomials. Bézout identity can be extended to any principle ideal domain (PID).

[11] The smallest one is obviously well defined when a, b are natural numbers. It's well defined if and only if a, b are commensurable.

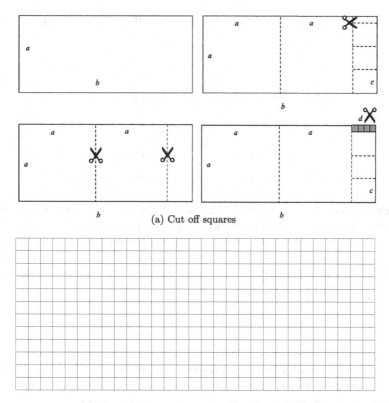

(a) Cut off squares

(b) Tile with the small squares of length $gcm(a, b)$

A geometric explanation of Euclidean algorithm

(a) Étienne Bézout, 1730 - 1783   (b) Claude Gaspard Bachet de Méziriac,
1581–1638

We'll next show that g is the gcm of a and b. Represent a as the quotient and remainder of g.

$$a = qg + r \tag{2.4}$$

where $0 \le r < g$. It is either 0 or belongs to S because:

$$
\begin{aligned}
r &= a - qg & &\text{Eq. (2.4)} \\
&= a - q(as + bt) & &\text{substitute } g \\
&= a(1 - qs) - bqt & &\text{combination of } a \text{ and } b
\end{aligned}
$$

Hence, r is a linear combination of a and b. If it's not zero, then it must be in S. However, this is impossible because we chose g as the smallest element in S, but $r < g$. This forces $r = 0$, so $a = gq$ (Eq. (2.4)), and g measures a. In the same way, g also measures b. Therefore, g is the common measure of a, b. We next show that g is the greatest. For any common measure c, let $a = mc$ and $b = nc$ for some integers m, n. Then g can be expressed as:

$$g = as + bt = mcs + nct = c(ms + nt)$$

c measures g and implies $c \le g$. Hence, g is the gcm. Summarizing the above, we prove Bézout's identity. There exist integers, such that $ax + by = g$ holds. Besides, we know the greatest common measure is the minimum positive value among all the linear combinations. $\qquad\square$

Euclidean algorithm can be extended with Bézout's identity.

$$
\begin{aligned}
ax + by &= \text{gcm}(a, b) & &\text{Bézout's idendity} \\
&= \text{gcm}(b, r_0) & &\text{Eq. (2.3)} \\
&= bx' + r_0 y' & &\text{Bézout's idendity for } b \text{ and } r_0 \\
&= bx' + (a - bq_0)y' & &\text{by } a = bq_0 + r_0 \\
&= ay' + b(x' - y'q_0) & &\text{as linear combination of } a \text{ and } b \\
&= ay' + b(x' - y'\lfloor a/b \rfloor) & &q_0 \text{ is the quotient of } a \text{ and } b
\end{aligned}
$$

This gives the recursive relation:

$$
\begin{cases}
x = y' \\
y = x' - y'\lfloor a/b \rfloor
\end{cases}
$$

The edge case happens when $b = 0$, we have $\text{gcm}(a, 0) = 1a + 0b$. Combined together, we define the extended Euclidean algorithm as below.

$$\text{gcm}_{ex}(a, b) = \begin{cases} b = 0: & (a, 1, 0) \\ b \neq 0: & (g, y', x' - y'\lfloor a/b \rfloor) \text{ where } (g, x', y') \\ & = \text{gcm}_{ex}(b, a \bmod b) \end{cases} \quad (2.5)$$

Here is an example application of the extended Euclidean algorithm [12]. Given two water jugs, 9 and 4 liters, how to get 6 liters of water from the river? This puzzle has history back to ancient Greece. A story said French mathematician Poisson solved this puzzle when he was a child. This puzzle has some variants. The goal and capacities can be other numbers, like to get 2 liters of water from two jugs of 899 liters and 1147 liters.

Let the small jug be A of capacity a and the big jug be B of capacity b. There are 6 operations each time: (1) Fill jug A. (2) Fill jug B. (3) Empty jug A. (4) Empty jug B. (5) Pour from jug A to B. (6) Pour from jug B to A. The below table lists a series of operations (assume $a < b < 2a$).

A	B	Operation
0	0	Start
a	0	Fill A
0	a	Pour A to B
a	a	Fill A
2a - b	b	Pour A to B
2a - b	0	Empty B
0	2a - b	Pour A to B
a	2a - b	Fill A
3a - 2b	b	Pour A to B
...	...	...

Whatever operations, the water in each jug must be $xa + yb$ for some integers x and y. It is some linear combination of a and b. From the proof of Bézout's identity, we know the smallest positive number of this linear combination is exactly the greatest common measure g. It is possible to get c liters of water if and only if g measures c.[12] We assume c is not greater than the capacity of the bigger jug.

For example, we can't get 5 liters of water with two jugs of 4 and 6 liters. This is because the gcm of 4 and 6 is 2, which can't measure 5. (In other words, we can't get odd liters of water with two jugs of even liters capacities.) If a and b are coprime, i.e., $\text{gcm}(a, b) = 1$, then we are sure to be able to get any natural number c liters of water.

---

[12] For integer capacities, if and only if the greatest common divisor g divides c.

Although we can tell if the puzzle is solvable when g measures c, we don't know the detailed steps. Actually, we can design the steps from the two integers x and y that satisfy $ax + by = c$. Assume $x > 0$, $y < 0$; we fill jug A for x times and empty jug B for y times. For example, the small jug $a = 3$ liters, the big jug $b = 5$ liters, and the goal is to get $g = 4$ liters of water. Because $4 = 3 \times 3 - 5$, so $x = -1$, $y = 3$. We design below operations:

We fill jug A for 3 times, empty jug B for 1 time. The next problem is how to find x and y satisfying $ax + by = c$. With the extended Euclidean algorithm, we can find a solution to $ax_0 + by_0 = g$. If g measures c, let $m = \frac{c}{g}$, we make a solution by multiplying x_0 and y_0 with m.

$$\begin{cases} x_1 = x_0 \dfrac{c}{g} \\[2mm] y_1 = y_0 \dfrac{c}{g} \end{cases}$$

From this solution, we can generate all the integer solutions to the Diophantine equation[13] $ax + by = c$.

$$\begin{cases} x = x_1 - k\dfrac{b}{g} \\[2mm] y = y_1 + k\dfrac{a}{g} \end{cases} \tag{2.6}$$

where k is some integer. We get all the solutions to the water jug puzzle. Further, we can find a special k that minimizes $|x| + |y|$; it gives the fastest steps.[14] Below example program in Haskell solves this puzzle.

```haskell
import Data.List
import Data.Function (on)

-- Extended Euclidean Algorithm
gcmex a 0 = (a, 1, 0)
gcmex a b = (g, y', x' - y' * (a `div` b)) where
  (g, x', y') = gcmex b (a `mod` b)

-- Solve the linear Diophantine equation ax + by = c
solve a b c | c `mod` g /= 0 = (0, 0, 0, 0)  -- no solution
            | otherwise = (x1, u, y1, v)
```

---

[13] Named after ancient Greek mathematician, Diophantus of Alexandria (about 200–284 AD). In his book *Arithmetica*, he made important advances in mathematical notation, becoming the first person known to use algebraic notation and symbolism. Diophantus is often called "the father of algebra" [11].

[14] One way to get this special k is to represent the solution as two lines in a Cartesian plane. When taking the absolute value, it flips the lower part of the x-axis. Then we can find the k minimizes $|x| + |y|$ from the figure.

```
where
  (g, x0, y0) = gcmex a b
  (x1, y1) = (x0 * c `div` g, y0 * c `div` g)
  (u, v) = (b `div` g, a `div` g)

— Optimal by minimizing |x| + |y|
jugs a b c = (x, y) where
  (x1, u, y1, v) = solve a b c
  x = x1 - k * u
  y = y1 + k * v
  k = minimumBy (compare `on` (λi → abs (x1 - i * u) +
                                    abs (y1 + i * v))) [-m..m]
  m = max (abs x1 `div` u) (abs y1 `div` v)
```

With x and y, we can populate the steps as in Sect. 2.7.

2.2.4 Commensurable

In the spirit of all things relating to numbers, Euclidean algorithm perfectly finds the greatest common divisors for numbers (integers). However, Euclid applied it to the geometric line segments. We see the separation of geometry from numbers.[15] Geometry was not built on top of numbers, but developed independently to solve the generic problems not limited to numbers. The ancient Greek had a tradition that even for a proposition about numbers, one needed to give proof in terms of geometry. People followed it even in the sixteenth century. For example, the Italian mathematician Cardano still used geometric cubic filling method when solving the cubic and quartic equations in his book *Ars Magna* in 1545 [11].

Euclidean algorithm was the first famous recursive algorithm. "The structure of the whole number theory is based on the same foundation, which is the greatest common divisor algorithm." commented by Dirichlet, the founder of analytic number theory. The modern RSA cryptosystem[16] utilizes the extended Euclidean algorithm directly.

Euclid algorithm is a double-edged sword. The powerful recursive method attacks back to the idea that all things relate to numbers. The Pythagoreans believed any two numbers were commensurable, i.e., there must be a common measure, because all things are essentially the ratio of numbers. About 500 BC, Hippasus, a member of Pythagorean school, attempted to find the common measure for the side and diagonal of a square. However, no matter how short the segment being used, he could not measure them. It surprised other members and led to a crisis for the foundation of Pythagoreans. In another legend, Hippasus was inspired by the

---

[15] This is the reason we use distinct names of gcm and gcd.

[16] The world's first public-key asymmetric crypto algorithm was developed by Ron Rivest, Adi Shamir, and Leonard Adelman in MIT, 1977, abbreviated as RSA.

Hippasus of Metapontum, 500 BC

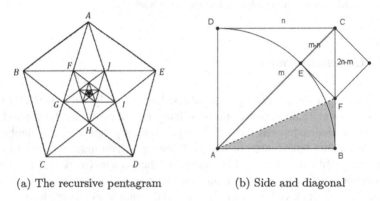

(a) The recursive pentagram   (b) Side and diagonal

Fig. 2.7 Irrational magnitude

mysterious pentagram logo of Pythagorean school. The Pythagoreans use pentagram as the school's badge and liaison symbol. There was a story about a school member who met difficulty in a foreign land, poor and sick. The landlord helped take care of him. He drew a pentagram on the door before he died. A few years later, someone in Pythagorean school saw the sign. He asked about the past, paid the landlord a lot of money, then left [7]. In the Walt Disney's film *Donald in Mathmagic Land* in 1959, Duck Donald met Pythagoras and his friends; they discovered the mathematical principles of music together. After shaking hands with Pythagoras, who then vanishes, Donald found on his hand a pentagram, the symbol of the secret Pythagorean society. As shown in Fig. 2.7a, Hippasus might found segment AC and AG were incommensurable.

Scottish mathematician George Chrystal reconstructed Hippasus's proof in the nineteenth century. By reduction to absurdity, assume there exists some segment c that measures both the side and diagonal of a square. Let the side be nc and the diagonal be mc for some positive integers m, n. As shown in Fig. 2.7b, taking the side as radius, draw an arc around A as the center, which intersects the diagonal AC at point E. Then draw a line from E that is perpendicular to the diagonal and intersects side BC at point F.

For arc, the length of AE equals to the square side. Therefore, $|EC| = |AC| - |AB| = (m - n)c$. Because $EF \perp AC$ and $\angle ECF = 45°$, $\triangle ECF$ is the isosceles right triangle. The two sides $|EC| = |EF|$ (isosceles). Observe the two right-angled triangles $\triangle AEF$ and $\triangle ABF$: side $AE = AB$ (arc), AF is shared. They are congruent, implying $|EF| = |FB|$. So the three segments $|EC| = |EF| = |FB|$ are of the equal length of $(m - n)c$, while segment $|CF| = |BC| - |FB| = nc - (m - n)c = (2n - m)c$. We list them side by side as below.

$$\begin{cases} |AC| = mc \\ |AB| = nc \end{cases} \qquad \begin{cases} |CF| = (2n - m)c \\ |CE| = (m - n)c \end{cases}$$

big square | small square

As both m, n are integers, c measures both diagonal CF and side CE of the small square. In the same way, we can draw another even smaller square. Repeating it will give smaller and smaller infinitely many squares. But c measures both the diagonal and the side for every square. It conflicts with the fact that m, n are finite positive integers, i.e., we can't endlessly repeat this process. Therefore, our assumption is not true; there is no common measure for the diagonal and side of a square. □

It is a loophole in Pythagoreans' philosophy. There are segments that can't be represented by any ratio of natural numbers. Hippasus discovered irrational number essentially. It was said the Pythagoreans didn't want this secret to be disclosed; they drowned Hippasus at sea. The discovery of irrational number greatly boosts mathematics. The ancient Greek philosophers and mathematicians, including Eudoxus, Aristotle, and Euclid, treated this problem seriously. They finally defined the concept of incommensurable and incorporate it into mathematics through geometry.

Proposition 2.2.2 (Euclid's Elements, Book X, Proposition 2) *If, when the less of two unequal magnitudes is continually subtracted in turn from the greater that which is left never measures the one before it, then the two magnitudes are incommensurable.*

It's interesting that incommensurable is defined by checking whether the Euclidean algorithm terminates or not. Euclidean algorithm is recursive; it means the condition is essentially whether the recursion terminates or not. It brings our attention to the nature of recursion.

Exercise 2.1

2.1.1 Euclidean algorithm in Eq. (2.2) is recursive. Eliminate recursion and implement it and the extended Euclidean algorithm only with loops.

2.1.2 Most programming environments require integers for modulo operation. However, the length of segment isn't necessarily integer. Implement modulo operation that manipulates segments. What's about its efficiency?

2.1.3 In the proof of Euclidean algorithm, we mentioned $b > r_0 > r_1 > r_2 > \cdots \geq 0$. As the remainder cannot be negative and b is finite, we must reach to $r_{n-1} = r_n q_{n+1}$ within finite steps. Can r_n infinitely approximate zero, but not be zero? Does the algorithm always terminate? What does the precondition that a and b are commensurable ensure?

2.1.4 For the linear Diophantine equation $ax + by = c$, let x_1, y_1 and x_2, y_2 be two pairs of integral solution. Proof that the minimum of $|x_1 - x_2|$ is $b/\gcd(a, b)$, and the minimum of $|y_1 - y_2|$ is $a/\gcd(a, b)$

2.1.5 For the regular pentagon with side of 1, how long is the diagonal? Show that in the pentagram in Fig. 2.7a, segment AC and AG are incommensurable. What's their ratio in real number?

Answer: page 315

2.3 The λ Calculus

As human beings, we perform recursive computation well. Mechanically, one enters the next level of computation when recursion is met, then returns back to the upper level after that. However, it matters when building a machine to compute recursively. Several computation models were developed in the 1930s independently. The most famous ones were the Turing machine (1935 by Turing) and λ-calculus (1932–1941 by Church, and 1935 by Stephen Kleene. The Greek letter λ is pronounced as lambda) (Fig. 2.8).

(a) Alan Mathison Turing, 1912 - 1954   (b) Alonzo Church, 1903 - 1995

Fig. 2.8 Turing and Church

Turing was a British mathematician and logician. He was highly influential in the development of theoretical computer science, providing a formalization of the concepts of algorithm and computation with the Turing machine, which can be considered a model of a general-purpose computer. Turing is widely considered to be the father of theoretical computer science and artificial intelligence [13].

During the Second World War, Turing worked for the government at Bletchley Park, the British code breaking center that produced ultra intelligence. He devised a number of techniques for speeding the breaking of German ciphers, including improvements to the pre-war Polish bombe method, an electro-mechanical machine that could find settings for the Enigma cipher machine. Turing played a pivotal role in cracking intercepted coded messages that enabled the Allies to defeat the Nazis in many crucial engagements. It has been estimated that this work shortened the war in Europe by more than 2 years and saved millions lives.

After the war, Turing worked on the Automatic Computing Engine (ACE) in the 1950s, which was one of the first designs for a stored-program computer. It influenced a number of later computers around the world. From 1950, Turing worked in "Computing Machinery and Intelligence." He addressed the problem of artificial intelligence and proposed an experiment that became known as Turing test, an attempt to define a standard for a machine to be called "intelligent." The idea was that a computer could be said to "think" if a human interrogator could not tell it apart, through conversation, from a human being. Turing was elected a fellow of the Royal Society (FRS) in 1951. Turing died in 1954. Since 1966, the Turing Award has been given annually by the Association for Computing Machinery (ACM) for technical or theoretical contributions to the computing community.

The formalization of computation itself is called *metamathematics*. Besides Turing machine, this attempt led to another great work, the λ-calculus. There's an anecdote about the name. When considering the computation itself, people realized we should distinguish function and its evaluation result. For example, if we say "if x is odd, then $x^2 + 1$ is even," we mean the evaluated value of the function, while if we say "$x^2 + 1$ is even symmetric (unchanged under negation $(-x)^2 + 1 = x^2 + 1$)," we mean the function itself. To differentiate the two concepts, we write function as $x \mapsto x^2 + 1$, but not barely $x^2 + 1$. The "$\mapsto$" symbol was introduced by Nicolas Bourbaki[17] around 1930. But before that Russel and Whitehead had already used the notation $\hat{x}(x^2 + 1)$ in their influential book *Principia Mathematica* (first published in three volumes between 1910 and 1913). Church wanted to use the same $\hat{x}$ notation in the 1930s. However, his publisher didn't know how to print the "hat" symbol above x. Alternatively, he printed the uppercase Greek letter Λ (lambda) before x and later changed to lowercase λ. That is why we see it's in the form of $\lambda x \,.\, x^2 + 1$ today [14]. Although the mapping notation $x \mapsto x^2 + 1$ is widely accepted nowadays, people tend to use Church's notation particularly in logic and computer science and name it as the "λ-calculus."

---

[17] Nicolas Bourbaki is the collective pseudonym of Bourbaki group.

2.3.1 Expression Reduction

We start from some examples to understand how λ-calculus formalizes the computation and algorithm. We treat the basic arithmetic operations of $+$, $-$, $\times$, and $/$, as functions as well. For instance $1 + 2$ can be considered as applying function "$+$" to two arguments 1 and 2. Following the tradition to write function name first, this expression can be written as $(+\ 1\ 2)$. We can present the process of expression evaluation as a series of reduction steps, for example:

$$(+\ (\times\ 2\ 3)\ (\times\ 4\ 5))$$

$$\rightarrow (+\ 6\ (\times\ 4\ 5))$$

$$\rightarrow (+\ 6\ 20)$$

$$\rightarrow 26$$

The arrow $\rightarrow$ is read as "reduce to." Note that when applying f to variable x, we don't write it as $f(x)$, but $f\ x$. For multi-variable functions, like $f(x, y)$, we don't write as $(f\ (x, y))$, but use the uniformed way as $((f\ x)\ y)$. For example, three plus four is written as $((+\ 3)\ 4)$. Under this convention, expression $(+\ 3)$ actually represents a function; it adds 3 to any argument passed in. The whole expression $((+\ 3)\ 4)$ means "apply function $+$ to a variable that equals to 3, it gives a function as result. Then apply this function to another variable that equals to 4." In this way, we treat every function takes only one argument. This method was first introduced by Schönfinkel (1889–1942) in 1924, then widely used by Haskell Curry from 1958, known as *Currying* [15]. We call a function fed with a single argument as the *Curried* form. There might be too many parentheses in Curried form. To make it concise, we may omit some parentheses without causing ambiguity. For example, simplify $((f\ ((+\ 3)\ 4))\ (g\ x))$ to $(f\ (+\ 3\ 4)\ (g\ x))$.

We need to define the basic components for expression reduction. We can define arithmetic for numbers with Peano axioms in theory. For example, define $+$ as Eq. (1.1) and $\times$ as Eq. (1.4), then define $-$ and $/$ as their reversed operations. For the numbers as operands, define them with zero and the successor. In practice, arithmetic operators and numbers are built-in realized for performance. So as the logic and ($\wedge$), or ($\vee$), not ($\neg$) operators, Boolean constant values of true and false are built-in. The conditional expression is realized in McCarthy form as $(p \mapsto f, g)$, or in if form as below:

$$\textbf{if } \textit{true} \quad \textbf{then } e_t \textbf{ else } e_f \mapsto e_t$$
$$\textbf{if } \textit{false} \quad \textbf{then } e_t \textbf{ else } e_f \mapsto e_f \tag{2.7}$$

where both e_t, e_f are expressions. For the compound data structure defined with *cons* in Sect. 1.5, the following functions extract every component:

$$head\ (cons\ a\ b) \mapsto a$$
$$tail\ (cons\ a\ b) \mapsto b \tag{2.8}$$

2.3.2 λ Abstraction

When defining a function, like $f(x) = x+1$, we actually give it a name f. However, λ abstraction defines an unnamed function (anonymous), for example:

$$(\lambda x\ .\ +\ x\ 1)$$

A λ abstraction contains four parts: (1) The λ symbol starts the function definition. (2) Next is a variable, called formal parameter. It's x in this example. (3) A dot separates the formal parameter and the rest. (4) Function body that extends to the right most. It's "$+\ x\ 1$" in this example. We may add parentheses to avoid ambiguity about the right boundary. For easy memory, we correspond the four parts of λ abstraction to natural language as below.

$$
\begin{array}{cccc}
(\lambda & x & . & +\,x\,1) \\
\uparrow & \uparrow & \uparrow & \uparrow \\
\text{That function of} & x & \text{which} & \text{add } x \text{ to } 1
\end{array}
$$

We also use the equivalent arrow form $x \mapsto x + 1$ interchangeably. Note that λ abstraction is *not* equivalent to λ expression, as shown below; λ abstraction is only one out of the four types of λ expression:

$$
\begin{array}{llll}
\text{<exp>} & = & \text{<constant>} & \text{built-in numbers, Boolean values, etc.} \\
& | & \text{<variable>} & \text{variable names} \\
& | & \text{<exp> <exp>} & \text{applications} \\
& | & \lambda\ \text{<variable> . <exp>} & \text{λ abstraction}
\end{array}
$$

2.3.3 λ Conversion Rules

To evaluate expression $(\lambda x.+\ x\ y)$, we need to know the value of the global variable y, but needn't for x. Because x appears as the formal parameter and is expected to give a value later, we say x is *bound* to λx. When applying this λ abstraction to some argument, for example, $(\lambda x. + x\ y)\ 2$, we replace x with 2. On the contrary, as y is not bound by λ, we say y is *free*. So this expression is evaluated to $2+y$. Overall, the expression value is determined by the unbound free variables. A variable is either bound or free. Here is another example:

$$\lambda x. + ((\lambda y. + y\ z)\ 3)\ x$$

Or make it clear with the arrow notation:

$$x \mapsto ((y \mapsto y + z)\ 3) + x$$

Both x and y are bound, while z is free. In a more complex expression, the same variable name may be bound in some part, while free in other part, for example:

$$+ x\ ((\lambda x. + x\ 1)\ 2)$$

or in the arrow form:

$$
\begin{array}{ccc}
x + ((x \mapsto x + 1)\ 2) \\
\uparrow \qquad \uparrow \\
\text{free} \qquad \text{bound}
\end{array}
\tag{2.9}
$$

x is free in its first occurrence, but is bound in the second occurrence. It may cause confusion to have the same name representing different variables in a complex λ expression. The first λ conversion rule helps to avoid name conflict. It's called α-conversion, where α (alpha) is the first Greek letter. This rule renames a variable, for example:

$$\lambda x. + x\ 1 \overset{\alpha}{\longleftrightarrow} \lambda y. + y\ 1$$

or in arrow form:

$$x \mapsto x + 1 \overset{\alpha}{\longleftrightarrow} y \mapsto y + 1$$

Let's apply the *alpha* conversion to eliminate name conflict in Eq. (2.9):

$$x + ((x \mapsto x + 1)\ 2) \overset{\alpha}{\longleftrightarrow} y + ((x \mapsto x + 1)\ 2)$$

λ abstraction constructs an unnamed function. How to apply the constructed function to a given parameter? The second rule, β-conversion, does this job. It replaces all the free occurrences of the formal parameter in function body to its value. For example, $(x \mapsto x + 1)\ 2$ applies the λ abstraction $x \mapsto x + 1$ to 2. The formal parameter is x; there is only one its free occurrence in $x + 1$. Replace it with 2 gives $2 + 1$.

$$(x \mapsto x + 1)\, 2 \xrightarrow{\beta} 2 + 1$$

We call the left direct conversion β-reduction, the right direct β-abstraction when using conversely. For more examples, the first is about multiple occurrences of the formal parameter.

$$(x \mapsto (x + 3) \times (x - 1))\, 2 \xrightarrow{\beta} (2 + 3) \times (2 - 1)$$
$$\longrightarrow 5 \times 1$$
$$\longrightarrow 5$$

Here's another example that the formal parameter does not occur at all.

$$(x \mapsto 1)\, 2 \xrightarrow{\beta} 1$$

This example is known as constant projection. It ignores any input, i.e., projects any value to a constant. The next example looks like applying a function to two parameters, but actually it is through Currying with two β-conversions.

$$(x \mapsto (y \mapsto y - x))\, 2\, 4 \xrightarrow{\beta} (y \mapsto y - 2)\, 4 \qquad \text{Currying}$$
$$\xrightarrow{\beta} 4 - 2 \qquad \text{inner reduction}$$
$$\longrightarrow 2 \qquad \text{built-in arithmetic}$$

The Currying process performs reduction from inner to outer. When following this order, we omit the parentheses to simply the expression as below.

$$(\lambda x.(\lambda y.E)) \Rightarrow (\lambda x.\lambda y.E)$$

or in arrow form:

$$(x \mapsto (y \mapsto E)) \Rightarrow (x \mapsto y \mapsto E)$$

where E represents the function body. When applying a function through β-reduction, the parameter can be another function, for example:

$$(f \mapsto f\, 5)\, (x \mapsto x + 1) \xrightarrow{\beta} (x \mapsto x + 1)\, 5$$
$$\xrightarrow{\beta} 5 + 1$$
$$\longrightarrow 6$$

The last rule we give here is the η-conversion. Below is its definition:

$$(\lambda x.F\ x) \xleftrightarrow{\eta} F$$

or in arrow form:

$$x \mapsto F\ x \xleftrightarrow{\eta} F$$

where F is a function and x is *not* the free variable in F, for example:

$$(\lambda x. +\ 1\ x) \xleftrightarrow{\eta} (+\ 1)$$

The λ-expressions on both sides behave same. When applied to a parameter, the effect is adding 1 to it. The reason why x must not be free in F is to avoid wrongly converting expression like $(\lambda x. +\ x\ x)$ to $(+\ x)$, where x is the free variable in $(+\ x)$. It's also necessary to limit F to function; otherwise it could convert 1 to $(\lambda x.1\ x)$, which does not make sense. We call the conversion direction from left to right as η-reduction.

As a summary, here are the three conversion rules for λ expression:

1. α-Conversion to change the name of formal parameter
2. β-Reduction to realize function application
3. η-Reduction to eliminate redundant λ abstraction

Besides these three rules, we call the built-in functions, like arithmetic operations, logic and, or, and not, as δ-conversion. To easily memorize, when performing β-reduction for expression $(\lambda x.E)\ M$, we use M to replace x in E, written as $E[M/x]$. The below table lists the three rules with this simplified notation:

Conversion	λ form			Arrow form		
α	$(\lambda x.E)$	$\xleftrightarrow{\alpha}$	$\lambda y.E[y/x]$	$x \mapsto E$	$\xleftrightarrow{\alpha}$	$y \mapsto E[y/x]$
β	$(\lambda x.E)\ M$	$\xleftrightarrow{\beta}$	$E[M/x]$	$(x \mapsto E)\ M$	$\xleftrightarrow{\beta}$	$E[M/x]$
η	$(\lambda x.E\ x)$	$\xleftrightarrow{\eta}$	E	$x \mapsto E\ x$	$\xleftrightarrow{\eta}$	E

All these conversions are bidirectional, left or right. There raises two questions by nature: first, will every reduction terminate? second, if terminates, do different reduction steps lead to the same result? For the first question, the answer is not deterministic; the reduction process is not sure to terminate.[18] For example, this

---

[18] Note the answer is not "no," but nondeterministic. It's essentially as same as the Turing halting problem. There is no determined process that can tell if a given reduction process terminates; see Sect. 7.1.

Fig. 2.9 Church-Rosser
confluence

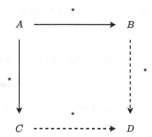

expression loops endlessly: $(D\ D)$, where D is defined as $\lambda x.x\ x$, or in arrow form:
$x \mapsto x\ x$. When evaluated, it leads to the following result:

$$(D\ D) \rightarrow (x \mapsto x\ x)\ (x \mapsto x\ x)$$

$$\xrightarrow{\alpha} (y \mapsto y\ y)\ (x \mapsto x\ x) \qquad \text{rename } x \text{ to } y \text{ in the first half}$$

$$\xrightarrow{\beta} (x \mapsto x\ x)\ (x \mapsto x\ x) \qquad \text{substitute } y \text{ with } (x \mapsto x\ x)$$

$$\cdots \qquad\qquad\qquad\qquad\qquad\quad \text{repeat}$$

For a more interesting example: $(\lambda x.1)\ (D\ D)$, if firstly reduce the $(\lambda x.1)$ part,
it terminates with the result of 1. But if firstly reduce $(D\ D)$ part, it loops endlessly
as above. Church and Rosser[19] proved a pair of theorems that completely answered
the second question.

Theorem 2.3.1 (Church-Rosser Theorem 1) *If $E_1 \leftrightarrow E_2$, then there exists E that
$E_1 \rightarrow E$ and $E_2 \rightarrow E$.*

This theorem says, if the reduction process terminates, then the results confluence. Various reduction steps give the same result, as shown in Fig. 2.9. Church
and Rosser further proved the second theorem with the concept of the *normal form*.
The normal form, also known as β normal form, is an expression that can't be β-
reduced anymore. It means all functions have already been applied. A more strict
normal form is the $\beta - \eta$ normal form, in which neither β-reduction nor η-reduction
can be done. For example, $(x \mapsto x + 1)\ y$ is not a normal form, because one can

---

[19] John Barkley Rosser Sr. (1907–1989) was an American mathematician and logician. Besides
Church-Rosser confluence theory, he also founded the Kleene-Rosser paradox with Kleene. In
number theory, he developed what is now called "Rosser sieve" and proved Rosser theorem that the
n-th prime number $p_n > n \ln n$. Rosser gave a stronger form for the Gödel's first incompleteness
theorem. He improved the nondeterministic proposition to "For any proof to this proposition, there
exists a shorter one for the negated one."

apply β-reduction to change it to $y + 1$. Below is the recursive definition of the normal form.

$$normal((\lambda x.y)\, z) = \textbf{false} \qquad \text{can do } \beta\text{-reduction}$$

$$normal(\lambda x.(f\ x)) = \textbf{false} \qquad \text{can do } \eta\text{-reduction}$$

$$normal(f\ x) = normal(f) \land normal(x) \qquad \text{application: both } f \text{ and } x$$
$$\text{are normal}$$

$$normal(x) = \textbf{true} \qquad \text{others}$$

Theorem 2.3.2 (Church-Rosser Theorem 2) *If $E_1 \to E_2$, and E_2 is normal form, then there exists normal order to convert from E_1 to E_2.*

Theorem 2 requires the reduction process terminates. The *normal order* is the order to reduce from left to right, outer to inner.

2.4   Recursion

λ abstraction defines unnamed simple function. But how to define recursive function? For example, factorial is often recursively defined as below:

$$fact = n \mapsto \textbf{if } n = 0 \textbf{ then } 1 \textbf{ else } n \times fact(n - 1)$$

But this is *not* a valid λ expression. λ abstraction can only define anonymous functions, while we don't know how to name a function. The recursive factorial definition has the pattern like:

$$fact = n \mapsto (...fact...)$$

Conversely using β-reduction (i.e., β-abstraction), we get:

$$fact = (f \mapsto (n \mapsto (...f...)))\, fact$$

Then further abstract to:

$$fact = H\, fact \qquad\qquad (2.10)$$

where

$$H = f \mapsto (n \mapsto (...f...))$$

H is not recursive anymore after above conversion. It is a valid λ expression. Observe Eq. (2.10); it represents recursion. It is in equation form, a functional equation. The differential equation, we learned in calculus, is also a functional equation. Solving equation $y' = sin(x)$ gives a function $y = a - cos(x)$. If we solve the equation $F = H\ F$, then we are able to define factorial with λ-calculus. This equation says when applying H to F, the result is still F. It is called *fixed point* in mathematics, i.e., F is the fixed point of H. For example, the fixed points for λ expression $x \mapsto x \times x$ are 0 and 1; this is because we have $(x \mapsto x \times x)\ 0 = 0$ and $(x \mapsto x \times x)\ 1 = 1$.

We want to figure out the fixed point of H. Obviously, the fixed point only depends on H. We introduce a function Y; it accepts a function, then returns its fixed point. Y behaves like this:

$$Y\ H = H\ (Y\ H) \tag{2.11}$$

Y is called the fixed-point combinator, also known as the Y combinator. With Y, we define the solution to Eq. (2.10) as:

$$fact = Y\ H \tag{2.12}$$

This is a non-recursive definition; let us verify this solution:

$$
\begin{aligned}
fact &= Y\ H && \text{by Eq. (2.12)}\\
&= H\ (Y\ H) && \text{by Eq. (2.11)}\\
&= H\ fact && \text{converse of Eq. (2.12)}
\end{aligned}
$$

Y is so powerful that it can describe any recursive functions. However, it is still a black box so far. We define it in λ abstraction:

$$Y = \lambda h.(\lambda x.h\ (x\ x))\ (\lambda x.h\ (x\ x)) \tag{2.13}$$

Or in arrow form:

$$Y = h \mapsto (x \mapsto h\ (x\ x))(x \mapsto h\ (x\ x)) \tag{2.14}$$

We are opening a magic box, let's verify if Y behaves as expected: $Y\ H = H\ (Y\ H)$.

Proof

$$
\begin{aligned}
Y\ H &= (h \mapsto (x \mapsto h\ (x\ x))(x \mapsto h\ (x\ x)))\ H && \text{definition of } Y\\
&\overset{\beta}{\leftrightarrow} (x \mapsto H\ (x\ x))\ (x \mapsto H\ (x\ x)) && \text{substitute } h \text{ with } H
\end{aligned}
$$

$$\overset{\alpha}{\leftrightarrow} (y \mapsto H (y\,y))\,(x \mapsto H (x\,x)) \qquad \text{rename } x \text{ to } y \text{ in the}$$
$$\text{first half}$$

$$\overset{\beta}{\leftrightarrow} H ((x \mapsto H (x\,x))\,(x \mapsto H (x\,x))) \qquad \text{substitute } y \text{ with}$$
$$(x \mapsto H (x\,x))$$

$$\overset{\beta}{\leftrightarrow} H (h \mapsto (x \mapsto h (x\,x))\,(x \mapsto h (x\,x))\,H) \quad \text{extract } H \text{ out as a}$$
$$= H (Y\,H) \qquad\qquad\qquad\qquad\qquad\qquad \text{parameter definition of } Y$$

$$\square$$

Finally, let us define factorial with Y:

$$Y (f \mapsto (n \mapsto \textbf{if } n = 0 \textbf{ then } 1 \textbf{ else } n \times f (n - 1)))$$

It is more significant in theory than in practice to define Y in λ abstraction. Y is often implemented as a built-in function in real environment, which directly converts $Y\,H$ to $H (Y\,H)$.

2.5   Compound Data

The λ calculus is significant to model the complex computation with a set of simple rules. We can describe Euclidean algorithm in λ abstraction and Y combinator, then perform β-reduction to evaluate the gcm of given a and b. It's feasible although it looks complex for implementation. Not limiting to Euclidean algorithm, Church expected to represent any computable functions in *lambda* calculus. People later proved that λ calculus and Turing machine are equivalent. λ calculus originates from the function concept in mathematics and conversion rules. People attempt to formalized metamathematics with it in the 1930s. However, Kleene and Rosser proved that the original lambda calculus was inconsistent in 1935 [16]. Subsequently, in 1936 Church isolated and published just the portion relevant to computation, what is now called the untyped lambda calculus. In 1940, he also introduced a computationally weaker, but logically consistent system, known as the simply typed lambda calculus.

Besides the simple and recursive functions, arithmetic and logic operations can also be represented with λ calculus (see Exercises 2.2.2 and 2.2.3). We are yet to cover the compound data structures. The building blocks are *cons*, *head*, and *tail*, where *cons a b* constructs a compound data from *a* and *b*, and others extract the component, respectively:

$$cons = (\lambda a.\lambda b.\lambda f.f\ a\ b)$$

$$head = (\lambda c.c\ (\lambda a.\lambda b.a))$$

$$tail = (\lambda c.c\ (\lambda a.\lambda b.b))$$

or in arrow form:

$$cons = a \mapsto b \mapsto f \mapsto f\ a\ b$$

$$head = c \mapsto c\ (a \mapsto b \mapsto a)$$

$$tail = c \mapsto c\ (a \mapsto b \mapsto b)$$

Let's verify *head* (*cons p q*) = *p* holds.

$$head\ (cons\ p\ q) = (c \mapsto c\ (a \mapsto b \mapsto a))\ (cons\ p\ q)$$

$$\overset{\beta}{\to} (cons\ p\ q)\ (a \mapsto b \mapsto a)$$

$$= ((a \mapsto b \mapsto f \mapsto f\ a\ b)\ p\ q)\ (a \mapsto b \mapsto a)$$

$$\overset{\beta}{\to} ((b \mapsto f \mapsto f\ p\ b)\ q)\ (a \mapsto b \mapsto a)$$

$$\overset{\beta}{\to} (f \mapsto f\ p\ q)\ (a \mapsto b \mapsto a)$$

$$\overset{\beta}{\to} (a \mapsto b \mapsto a)\ p\ q$$

$$\overset{\beta}{\to} (b \mapsto p)\ q$$

$$\overset{\beta}{\to} p \qquad\qquad\qquad\qquad \square$$

This example composites data with λ expression; hence, the compound data structure needn't necessarily be built-in. We can build recursive list with *cons* (Sect. 1.5), traverse it with *tail*, and manipulate any element with *head*. For example, given a list F of Fibonacci numbers, *head* (*tail* (*tail F*)) gives the 3rd number, 2. When extended from a pair to n components, we could define n-tuple that holds arbitrary number of data,[20] for example, $(\lambda a.\lambda b.\lambda c.\lambda f.f\ a\ b\ c)$ defines a 3-tuple.

---

[20] We do not limit the type of *a*, *b*, and *cons*. Actually we ignore the type limitation in all λ expressions in this section. However, we assume the list contains the same type of data; otherwise, we can use tuple instead.

Exercise 2.2

2.2.1 Use λ conversion rules to verify $tail\ (cons\ p\ q) = q$.

2.2.2 We can define numbers with λ calculus. The following are called Church numbers:

0	$\lambda f.\lambda x.x$
1	$\lambda f.\lambda x.f\ x$
2	$\lambda f.\lambda x.f\ (f\ x)$
3	$\lambda f.\lambda x.f\ (f\ (f\ x))$
$\cdots$	$\cdots$

Define $+$ and $\times$ for Church numbers.

2.2.3 Below are Church Boolean values and the relative logic operators:

true	$\lambda x.\lambda y.x$
false	$\lambda x.\lambda y.y$
and	$\lambda p.\lambda q.p\ q\ p$
or	$\lambda p.\lambda q.p\ p\ q$
not	$\lambda p.p$ **false true**

where **false** is defined as same as Church number 0. Use the λ conversion rules to (1) prove: **and true false** = **false** and (2) define "if ... then ... else ..." expression.

Answer: page 320

2.6   More Recursive Structures

With recursive function and compound data being supported by λ calculus, we define more complex data structures like binary trees in this section. We return to the normal function and variable from the low level λ notations.

```
data Tree A = nil | node (Tree A, A, Tree A)
```

A binary tree of type A is either empty (denoted as $\varnothing$ or nil), or a branch node consists of three parts: two sub-trees of type A and an element of A. We often call the element as key and label the sub-trees as left and right (Fig. 2.10). This definition is recursive: the sub-tree is again a binary tree. When compared with list, we see list has one recursive component, while binary tree has two.

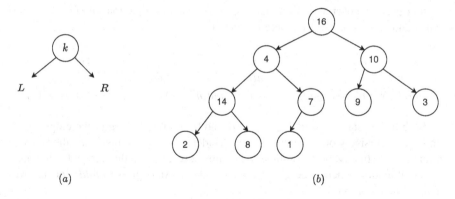

Fig. 2.10 Binary tree

List A = nil |	Tree A = nil |
cons(A, List A)	node(Tree A, A, Tree A)
1 sub-list	2 sub-trees
nil or *cons*(*x*, *xs*)	*nil* or *node*(*l*, *x*, *r*)

A is the type parameter, for example, let $A = \mathbb{N}$, *Tree* $\mathbb{N}$ be the type of natural number tree; node(nil, 0, node(nil, 1, nil)) is such a tree. Analogous to $map(f)$ (Eq. (1.21)), let's define for-each map for binary tree. For function f that maps x of type A to y of type B, then the function type is $f : A \to B$. $mapt(f)$ applies f to every key x in a binary tree; generate a new tree of the same structure, while all keys are replaced by $y = f(x)$. For example, $mapt(x \mapsto -x, t)$ negates every element in a tree.

$$
\begin{aligned}
mapt(f, nil) &= nil \\
mapt(f, node(l, x, r)) &= node(mapt(f, l),\ f(x),\ mapt(f, r))
\end{aligned}
\tag{2.15}
$$

If the tree is empty, the new tree is also empty; otherwise, it recursively maps the left $l' = mapt(f, l)$, the right $r' = mapt(f, r)$; map the key to $y = f(x)$; finally, construct a new tree $node(l', y, r')$.

For another example, let's define $sizet(t)$ to count how many elements in a tree. This is analogous to the *length* of list (Eq. (1.17)). The size of empty tree is definitely 0; for $node(l, x, r)$, recursively count the size of left $s' = sizet(l)$ and right $s'' = sizet(r)$, count x as 1, then the total size is $s' + 1 + s''$.

$$
\begin{aligned}
sizet(nil) &= 0 \\
sizet(node(l, x, r)) &= sizet(l) + 1 + sizet(r)
\end{aligned}
\tag{2.16}
$$

It counts whatever x as 1 and is exactly a constant projection: $one(x) = 1$. Put *mapt* and *sizet* side by side; there is a pattern.

	mapt	*sizet*
nil	*nil*	0
$node(l, x, r)$	$node(mapt(f, l), f(x), mapt(f, r))$	$sizet(l) + 1 + sizet(r)$

Both has a "starting value" z (e.g., *nil* and 0), corresponding to the empty tree; both run recursively on sub-trees; if we consider 1 as $one(x)$, both map the key via a function f; finally, both combine the recursive results and the mapped value. Let the combination function be g (e.g., *node* and $+$). Analogous to *foldr* for list, we abstract *foldt* for tree.

$$
\begin{aligned}
foldt(f, g, z, nil) &= z \\
foldt(f, g, z, node(l, x, r)) &= g(foldt(f, g, z, l), f(x), foldt(f, g, z, r))
\end{aligned}
\tag{2.17}
$$

Let's see how it works. The result is z for empty tree; otherwise, it recursively folds the l and r separately with the same starting value z. Denote the two sub-folding result as a and b, and let $c = f(x)$, then the final fold result is the combination of $g(a, c, b)$.

As a side word, the type of binary function $z = f(x, y)$ is $(X \times Y) \to Z$, where $x \in X$, $y \in Y$, and $z \in Z$, respectively. If only pass x, in other words, it only binds x, then one gets a Curried function, $y \mapsto f(x, y)$ of type $Y \to Z$. Explicitly, we view $f(x) = y \mapsto f(x, y)$ as a Curried unary function of type $Y \to Z$. If it sends every x to a unary function $f(x)$, then one gets a function of another function: $x \mapsto (y \mapsto f(x, y))$ of type $X \to (Y \to Z)$. We omit the parentheses to get the final Curried form: $f : X \to Y \to Z$. Extend from binary to n variables $y = f(x_1, x_2, \cdots, x_n)$ of type $(X_1 \times X_2 \cdots X_n) \to Y$. Bind x_1 to get the Curried function $f(x_1)$ of type $(X_2 \times X_3 \cdots X_n) \to Y$; next bind x_2 to get the Curried function $f(x_1, x_2)$ of type $(X_3 \cdots X_n) \to Y$; ... to the Curried function $f(x_1, x_2, \cdots, x_{n-1})$ of type $X_n \to Y$ (unary). Translate to λ abstraction (arrow form), function:

$$
x_1 \mapsto (x_2 \mapsto (\cdots x_n \mapsto f(x_1, x_2, \cdots x_n)) \cdots)
$$

has the type

$$
X_1 \to (X_2 \to (\cdots \to (X_n \to Y)) \cdots)
$$

Omit the parentheses to get the final Curried form: $f : X_1 \to X_2 \cdots X_n \to Y$.

The type of the Curried function $foldt(f, g, z)$ is $Tree\ A \to B$, where the starting value z is of type B. The type of the combination function g is $(B \times B \times B) \to B$, or in Curried form $g : B \to B \to B \to B$. It takes three inputs of B:

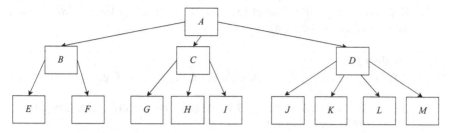

Fig. 2.11 Multi-trees

the recursive sub-folding results a, b, and the mapped key c; then combine them to a result of B. As the application of *foldt*, let's redefine *mapt* and *sizet*:

$$
\begin{aligned}
mapt(f) &= foldt(f, node, nil) \\
sizet &= foldt(one, sum, 0)
\end{aligned}
\tag{2.18}
$$

where $sum(a, b, c) = a + b + c$, is a ternary function.

List has a recursive component; binary tree has two. Can we extend along this way of $1 \to 2 \to n$, i.e., some data structure with n recursive components? We use a list to hold multiple sub-trees, call it *multi-trees*.

```
data MTree A = nil | node (A, List (MTree A))
```

A multi-tree of type A is either empty or a node containing an element of type A together with a list of multiple sub-trees (Fig. 2.11). This time, we directly define the abstract tree folding operation; inside it, we recursively call *foldr*.

$$
\begin{aligned}
foldm(f, g, c, nil) &= c \\
foldm(f, g, c, node(x, ts)) &= foldr(g(f(x), c), h, ts) \\
h(t, z) &= foldm(f, g, z, t)
\end{aligned}
\tag{2.19}
$$

Analogous to *sizet*, we count the size of a multi-tree as: $sizem = foldm(one, +, 0)$.

Exercise 2.3

2.3.1 Define *depth*, which counts for the maximum depth of a binary tree.

2.3.2 Define *toList*, which converts a binary tree to list. It's also known as the "flatten" operation.

2.3.3 Someone thought the abstract fold operation for binary tree, $foldt$, should be defined as the following:

$$foldt(f, g, c, nil) = c$$
$$foldt(f, g, c, node(l, x, r)) = foldt(f, g, g(foldt(f, g, c, l), f(x)), r)$$

That is to say $g : (B \times B) \to B$ is a binary operation like add. Can we use this $foldt$ to define $mapt$?

2.3.4 The binary search tree (BST) is a special tree that the type A is comparable. For any nonempty $node(l, k, r)$, all elements in the left sub-tree l are less than k, and all elements in the right sub-tree r are greater than k. Define function $insert(x, t) : (A \times Tree\ A) \to Tree\ A$ that inserts an element into the tree.

2.3.5 Can we define mapping for multi-trees with folding? If not, how should we modify $foldm$?

Answer: page 321

2.7 The Recursive Pattern

Recursion exists from ancient Euclidean algorithm to modern computer systems. Besides, the fascinating recursive patterns appear in various of arts in human civilizations. Figure 2.12 shows a piece of Islamic mosaic art with recursive patterns, big and small, polygon and strip, through colorful tiles. Figure 2.13a is a sketch by Leonardo da Vinci, the famous Renaissance artist. It also has recursive patterns. Using the same radius, he drew six interlaced circles along the centered one, forming a snowflake-style figure. Moreover, they recursively form the same pattern in a bigger scope. Figure 2.13b shows Chinese handmade katydid cages with a similar pattern. The mesh of the cage is hexagonal, while the overall shape of the cage is also hexagonal (prism).

Fig. 2.12 Zellige terracotta tiles in Marrakech (a city of the Kingdom of Morocco)

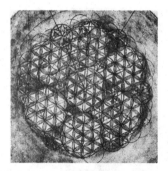   

(a) A sketch of Leonardo da Vinci   (b) Chinese katydid cages

Fig. 2.13 The recursive pattern in art and artifact

Fig. 2.14 Minuet of Haydn's String Quartet in D Minor, Op. 76, No. 2

Recursion also appears in music. The polyphonic music, for example, canon and fugue, can have recursive musical texture. Canon is a contrapuntal music that employs a melody with one or more imitations played after a given duration. The initial melody is called the leader, while the imitative melody, which is played in a different voice, is called the follower. The follower must imitate the leader, either as an exact replication of its rhythms and intervals or some transformation thereof. The weaving of the leader and followers results in a continuous effect. Each part imitates the theme but contains various changes, such as raising or lowering the pitch, retrograde overlapping, faster (diminution) or slowing (augmentation), melody reflection, and so on. A fugue is like a canon, in that it is usually based on one theme which gets played in different voices and keys and occasionally at different speeds or upside down or backwards [4] (Fig. 2.14).

Recursion can be infinite. Figure 2.15 is a lithograph by Dutch artist Escher. Two hands are mutual recursively drawing each other. The upper hand is drawing the lower hand with a pencil, so as the lower, at the same time. The recursion loops endlessly.

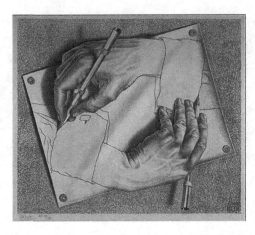

Fig. 2.15 M.C. Escher, Drawing Hands, 1948

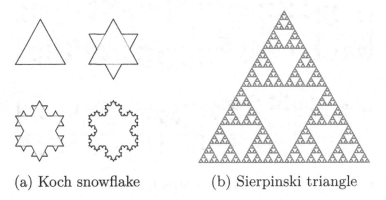

(a) Koch snowflake   (b) Sierpinski triangle

Fig. 2.16 Recursive fractal patterns

Fractal, generated by infinite mathematical recursion, looks like fine art. For example, here is the generation rule of Koch snowflake: divided every segment into 3 equal parts, draw a equilateral triangle based on the middle section, then erase the bottom side. Figure 2.16a shows the first three rounds of recursion, gradually forming a curve like snowflake. Figure 2.16b shows another fractal pattern called Sierpinski triangle (the first four recursive rounds). The generation rule is to connect the three middle points of each side recursively.

We show another two fractal patterns to close this chapter: Fig. 2.17a is Julia set, developed by humans; Fig. 2.17b is a plant by nature.

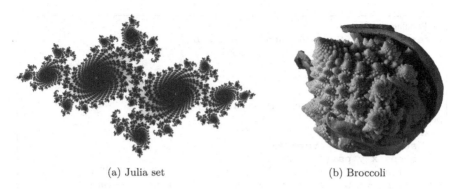

(a) Julia set   (b) Broccoli

Fig. 2.17 Fractal patterns

Exercise 2.4

2.4.1 What are the area and circumference of the Koch snowflake?

Answer: page 55

Appendix: Example Program

After figuring out the integer solutions for the puzzle of two water jugs, the below program outputs the steps in format like in Table 2.1.

Table 2.1 Steps to get 4

A	B	Operation
0	0	Start
3	0	Fill A
0	3	Pour A to B
3	3	Fill A
1	5	Pour A to B
1	0	Empty B
0	1	Pour A to B
3	1	Fill A
0	4	Pour A to B

```
— Populate the steps
water a b c = if x > 0 then pour a x b y
                else map swap $ pour b y a x
  where
    (x, y) = jugs a b c

— Pour from a to b, fill a for x times, and empty b for y times.
pour a x b y = steps x y [(0, 0)]
  where
    steps 0 0 ps = reverse ps
    steps x y ps@((a', b'):_)
      | a' == 0 = steps (x - 1) y ((a, b'):ps)    — fill a
      | b' == b = steps x (y + 1) ((a', 0):ps)    — empty b
      | otherwise = steps x y ((max (a' + b' - b) 0,
                                min (a' + b') b):ps) — a to b
```

Run this program, enter `water 9 4 6`; the best pour steps are as below:

`[(0,0),(9,0),(5,4),(5,0),(1,4),(1,0),(0,1),(9,1),(6,4),(6,0)]`

Chapter 3
Symmetry

One must be able to say at all times–instead of points, straight lines, and planes–tables, chairs, and beer mugs.

David Hilbert

Symmetry appears everywhere. It is often related to a sense of harmonious, order, pattern, beautiful proportion, and balance. The human body is symmetric, left and right of bilateral halves against the sagittal plane. The *Vitruvian Man* by Leonardo da Vinci describes the symmetry of a human body. Butterfly, fish, and birds represent the symmetry in biology. People created symmetric artifacts, arts, and buildings in civilization. The ancient clay jars were rotational symmetric; the Arabic carpets and rags have translational and glide symmetric patterns; the fine Chinese-style windowpane is radial symmetric. The great buildings like Taj Mahal and the Forbidden City are reflectional symmetric. We are surprised about the symmetry in snowflakes; every one is unique but follows the same hexagon symmetric pattern. When stepping into the garden in spring, we see all kinds of the beautiful flowers with radial symmetry, while in the woods in autumn, the matured fruits, spikes, and the colorful leaves demonstrate an amazing canvas in the language of symmetry.

The huge spiral arms of the galaxy rotate symmetrically; the symmetric crystal lattice reflects light from the tiny mystical particles. A palindrome poem leads to deep thinking; a music in rando form moves our feeling. Our minds work with symmetric concepts, connotation and denotation, abstraction versus material. Symmetry appears in mathematics, in geometry, in algebra, in equations, in curves, so as in programming, stack push and pop, computation scheduling, allocation and release.

© China Machine Press 2024
X. Liu, *Mathematics in Programming*,
https://doi.org/10.1007/978-981-97-2432-1_3

3.1   What Is Symmetry?

What is symmetry? How to accurately define or measure symmetry? Use *Vitruvian Man*, for example; we get the same figure when it is reflected against the middle vertical axis. Put a figure in Cartesian plane; let the symmetric axis be *y*-axis. When negating $x \mapsto -x$ for all points in the figure, the new figure is congruent with the original one if it is reflective symmetric, expressed as:

$$f(x) = f(-x)$$

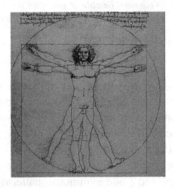

Leonardo da Vinci, Vitruvian Man, 1490

Abstract the reflective symmetric figure as a segment with two ends of 1 and 2. The segment is congruent when exchanging the ends from 1, 2 to 2, 1 as in Fig. 3.1. We call the transform $(1, 2) \mapsto (2, 1)$ a permutation.

For a set of two things $\{a, b\}$, like two end points of a segment, two coordinates, two persons, two computer programs, and whatever, let 1 denote the first thing, 2

Fig. 3.1 Reflective
symmetry

denote the second. There are two permutations: $(1, 2) \mapsto (2, 1)$ and $(1, 2) \mapsto (1, 2)$. We call the first "transpose," denoted as s, the second "identity," denoted as e:

$$s = \begin{pmatrix} 1 & 2 \\ 2 & 1 \end{pmatrix}, \; e = \begin{pmatrix} 1 & 2 \\ 1 & 2 \end{pmatrix}$$

Taking s, for example, it permutes 1 2 in the first row to 2 1 in the second row. We can combine and apply multiple permutations. $s \cdot s$ applies transpose twice: $(1, 2) \xrightarrow{s} (2, 1) \xrightarrow{s} (1, 2)$, and restores the elements finally; it equals to identity, i.e., $s \cdot s = e$. In other words, s reverses itself: $s^{-1} = s$. We say apply one permutation after another a *composition*, denoted as "$\cdot$." There are four compositions among s and e: (1) $s \cdot e = s$, (2) $e \cdot s = s$, (3) $s \cdot s = e$, and (4) $e \cdot e = e$, summarized in below table:

$\cdot$	s	e
s	e	s
e	s	e

Such table is called Cayley table, named after British mathematician Arthur Cayley. We may skip the dot to simplify the composition of τ and μ as $\tau\mu$. We call the permutation set $\{s, e\}$ and the compositions among them as S_2. It describes the reflexive symmetry in a plane. For any reflexive symmetric thing a, $s(a) = a$ holds. For example, consider three points with coordinate $A = (-1, 0), B = (1, 0), C = (2, 0)$. Because $s(AB) = AB$, segment AB is reflexive symmetric, but $s(AC) \neq AC$; hence, AC is not.

What symmetry can we obtain when extend from a segment to a triangle? Let the three end points be 1, 2, and 3. There are six permutations: $(2, 1, 3)$, $(3, 2, 1)$, $(1, 3, 2)$, $(2, 3, 1)$, $(3, 1, 2)$, $(1, 2, 3)$, where the last one is the identity e, keeping points unchanged. As shown in Fig. 3.2, the first permutation swaps 1 and 2, while keeping 3 fixed. It flips the triangle against the middle axis, denoted as s. The fourth permutation rotates the triangle clockwise, denoted as r. The fifth permutation rotates counter clockwise, it's r^{-1}. We call these six permutations together with their compositions as S_3. The three triangles in Fig. 3.2 have different symmetries. The left triangle is unchanged only under the permutation of identity e. It has the least symmetry. The middle triangle is unchanged under s and e. It has more symmetries than the left. The right equilateral triangle is unchanged under all six permutations. It has the most symmetries (reflection against three axes, two rotations of $\pm\frac{2\pi}{3}$, and identity). We can measure the symmetry of triangle with S_3.

The six permutations are composable. Rotate counter clockwise r^{-1} first, then mirror s; we get the second permutation, i.e., $(2, 1, 3)(3, 1, 2) = (3, 2, 1)$, while the clockwise rotation is equivalent to rotating counter clockwise twice, i.e., $r^{-1} = rr$. The six permutations can be expressed with s, r, and e as below Cayley table of S_3:

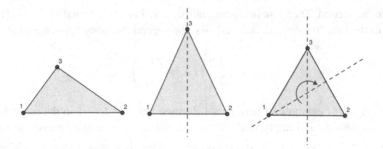

Fig. 3.2 The symmetry of triangle

	s	sr^{-1}	sr	r	r^{-1}	e
s	e	r^{-1}	r	sr	sr^{-1}	s
sr^{-1}	r	e	r^{-1}	s	sr	sr^{-1}
sr	r^{-1}	r	e	sr^{-1}	s	sr
r	sr	s	sr^{-1}	r^{-1}	e	r
r^{-1}	sr^{-1}	sr	s	e	r	r^{-1}
e	s	sr^{-1}	sr	r	r^{-1}	e

S_2 describes the symmetry of segment; S_3 describes the symmetry of triangle. We can also use them to study the symmetry of abstract things. Consider equation $x^2 + px + q = 0$ with its two roots x_1 and x_2. According to Vieta's theorem in school math:

$$\begin{cases} x_1 + x_2 = -p \\ x_1 x_2 = q \end{cases} \tag{3.1}$$

Apply S_2; s swaps the two roots; the equation and Vieta's theorem all hold. Together with identity e, we can see S_2 describes the symmetry of quadratic equation. Vieta further expanded his theorem to cubic equation $x^3 + rx^2 + px + q = 0$.

$$\begin{cases} x_1 + x_2 + x_3 = -r \\ x_1 x_2 + x_2 x_3 + x_3 x_1 = p \\ x_1 x_2 x_3 = -q \end{cases} \tag{3.2}$$

where x_1, x_2, and x_3 are the three roots. Obviously the cubic equation and Vieta's theorem all hold under the six permutations in S_3. Similarly, S_3 describes the symmetry of cubic equation. Such finding led to a new mathematics branch in history. People developed groups to define, explain, and measure symmetry. Surprisingly, some hard problems, like the solvability of equation, are ultimately solved by revealing the symmetry of roots; so as other famous problems, like the three classic geometric problems in ancient Greece.[1] This chapter provides a simple introduction to the basic abstract algebraic structures, like groups. They are not only abstraction to numbers but also applicable to concepts, properties, and relationships. They were valuable things that arose from many great minds. Some content challenges our abstract thinking. I insert stories about those great mathematicians and how they made breakthroughs with unbelievable difficulties. I believe they are encouraging.

Exercise 3.1

3.1.1 Write a program to test whether a binary tree (Sect. 2.6) is reflexive symmetric.

Answer: page 324

3.2 Group

The group theory has a history back to equations. From Rhind Mathematical Papyrus and Babylonian clay tablets, ancient Egyptians and Babylonians developed method to solve linear and quadratic equations with one unknown. However, people hadn't found the way to solve the generic cubic equations until the sixteenth century. Several Italian mathematicians made great progress. Gerolamo Cardano finally published the radical solutions to generic cubic equation and quartic in his 1545 book *Ars Magna*. It was not only about to pursue higher orders, but also refreshed our understanding to numbers. People in that time discarded negative roots because they believed negative numbers were meaningless; they also thought the coefficients must be positive. The equation $x^2 - 7x + 8 = 0$ is quite common to us today, but it had to be written in the form of $x^2 + 8 = 7x$ to ensure the coefficients are positive. After Cardano listed 20 different types of quartic equation in *Ars Magna*, he said there were another 67 types of quartic equation that could not be given because the coefficient is either negative or zero [17]. It was French mathematician François Viète who unified different forms of equations. Although most people thought the negative square root made no sense, Cardano noticed an interesting thing when solving the cubic equation $x^3 = 15x + 4$. His formula gave the intermediate result of $\sqrt[3]{2 + \sqrt{-121}} + \sqrt[3]{2 - \sqrt{-121}}$, then generating next three real roots of 4,

---

[1] They are (1) square the circle, (2) trisect arbitrary angle, and (3) double a cube, all limit to compass and edge construction.

$-2 \pm \sqrt{3}$. Such finding expanded our view to the imaginary number, and finally, Gauss developed the fundamental theorem of algebra.[2]

Gerolamo Cardano, 1501–1576

Carl Friedrich Gauss (1777–1855) in 10 Mark

People encountered surprisingly difficulty when seeking for radical solutions for generic quintic and higher-order equations in the next 300 years. The breakthrough happened in the nineteenth century with unexpected result. Around 1770, Joseph-Louis Lagrange began the groundwork that unified the method to solve equations; he introduced the new idea of permuting roots in the form of Lagrange's resolvents. But he didn't consider the combination among the permutations. Paolo Ruffini, an Italian mathematician, marked a major improvement in 1799, developing Lagrange's work on permutation theory. However, in general, Ruffini's proof was not considered

---

[2] Gauss proved the fundamental theorem of algebra several times along his life. In 1799 at the age of 22, he proved in his doctor thesis that every single-variable polynomial of degree n with real coefficient has at least one complex root and thus deduced the single-variable equation of degree n has and only has n complex roots (counted with multiplicity for the same ones). Gauss gave another two different proofs in 1815 and 1816. In 1849, to celebrate the 50th anniversary that Gauss received his doctor degree, he published the fourth proof and extended the coefficient to complex number.

convincing nor widely accepted in that time. Niels Abel, a young Norwegian mathematician, first completed the proof in 1824, demonstrating the impossibility of solving the general quintic in radicals. It is called "Abel-Ruffini theorem" nowadays. We know there are radical solutions to the special quintic equation $x^5 - 1 = 0$; in what condition is a polynomial solvable in radicals? This problem was completely solved by French young genius Évariste Galois. He developed an innovative idea, being able to determine a necessary and sufficient condition for a polynomial to be solvable by radicals while still in his teens [18].

Évariste Galois, 1811–1832

It was a tragedy in Galois's short 20 years life, but his work laid the foundations of abstract algebra. Galois was born on October 25, 1811, in Paris. His mother, the daughter of a jurist, was a fluent reader of Latin and classical literature and was responsible for her son's education for his first 12 years. In 1823, he entered a prestigious school in Paris. At the age of 14, he began to take a serious interest in mathematics. He found a copy of Euclid's *Elements* adapted by Legendre which, it is said, he read "like a novel" and mastered at the first reading [37]. At 15, he was reading the original papers of Joseph-Louis Lagrange, which likely motivated his later work on equation theory, yet his classwork remained uninspired, and his teachers accused him of affecting ambition and originality in a negative way.

In 1828, he attempted the entrance examination for the École Polytechnique, the most prestigious institution in France at the time, without the usual preparation, and failed for lack of explanations on the oral examination. In April 1829, Galois's first paper, on continued fractions, was published. It was around the same time that he began making fundamental discoveries in the theory of polynomial equations. He submitted two papers on this topic to the Academy of Sciences. But both were rejected due to some reasons.[3]

---

[3] Some said the paper was lost by Cauchy. Actually, Cauchy refereed these papers, but refused to accept them for publication for reasons that still remain unclear. However, in spite of many claims to the contrary, it is widely held that Cauchy recognized the importance of Galois's work, and that he merely suggested combining the two papers into one in order to enter it in the competition for the Academy's Grand Prize in mathematics. Cauchy, an eminent mathematician of the time, though

Fig. 3.3 Eugène Delacroix,
Liberty Leading the People,
1830 (Louvre)

On July 28, 1829, Galois's father, a mayor of the village, committed suicide
after a bitter political dispute with the village priest [19]. On August 3, Galois
made his second and last attempt to enter the Polytechnique and failed again. It
was undisputed that Galois was more than qualified; however, accounts differ on
why he failed. More plausible accounts state that Galois made too many logical
leaps and baffled the incompetent examiner, which enraged Galois. Having been
denied, Galois took the baccalaureate examinations and entered the École Normale,
a far inferior institution at that time. His examiner in mathematics reported, "This
pupil is sometimes obscure in expressing his ideas, but he is intelligent and shows a
remarkable spirit of research."

In February 1830, following Cauchy's suggestion, Galois submitted his memoir
on equation theory to the Academy's secretary Joseph Fourier, to be considered
for the Grand Prize of the Academy. Unfortunately, Fourier died soon after, and
the memoir was lost. The prize was awarded that year in June to Niels Abel
(posthumously[4]) and Carl Jacobi [11].

Galois lived in a time of political turmoil in France. The July Revolution broke
out in 1830[5] (Fig. 3.3). There was rioting in the streets of Paris and the director of
École Normale locked the students in to avoid them from taking part. Galois was

---

with political views that were at the opposite end from Galois's, considered Galois's work to be a
likely winner [19].

[4] Abel died on April 6, 1829, before the prize. He made pioneering contributions in a variety of
fields, like elliptic functions. He made his discoveries while living in poverty and died at the age
of 26 from tuberculosis.

[5] The July Revolution brought Louis-Philippe to the throne of France. The revolution was
precipitated by King Charles X's publication (July 26, 1830) of restrictive ordinances contrary to
the spirit of the Charter of 1814. Protests and demonstrations were followed by 3 days of fighting,
the abdication of Charles X, and the proclamation of Louis-Philippe as "king of the French." In the
July Revolution the upper middle class, or bourgeoisie, secured a political and social ascendancy
that was to characterize the period known as the July Monarchy.

incensed and wrote a blistering letter criticizing the director, which he submitted to the Gazette des Écoles, signing the letter with his full name. Although the Gazette's editor omitted the signature for publication, Galois was expelled.

Galois quit school and joined the staunchly Republican artillery unit of the National Guard. Due to controversy surrounding the unit, soon after Galois became a member, on December 31, 1830, the artillery of the National Guard was disbanded out of fear that they might destabilize the government. He was arrested on May 10, 1831, but was acquitted on June 15. On the Bastille Day (July 14), Galois was at the head of a protest, wearing the uniform of the disbanded artillery, and came heavily armed with several pistols, a rifle, and a dagger. He was again arrested. On October 23, he was sentenced to 6 months in prison for illegally wearing a uniform. Early in 1831, Poisson invited him to submit his work on equations, which he did on January 17, 1831. Around July 4, 1831, Poisson declared Galois's work "The argument is neither sufficiently clear nor sufficiently developed to allow us to judge its rigor"[6] While Poisson's report was made before Galois's July 14 arrest, it took until October to reach Galois in prison. It was unsurprising, in the light of his character and situation at the time, that Galois reacted violently and decided to abandon publishing his papers through the Academy and instead publish them privately through his friend Auguste Chevalier. Apparently, however, Galois did not ignore Poisson's advice, as he began collecting all his mathematical manuscripts while still in prison, and continued polishing his ideas until his release on April 29, 1832.

Shortly after released, Galois was involved in an obscure duel because of love. On May 29, Galois was so convinced of his impending death that he stayed up all night writing letters to his friends and composing what would become his mathematical testament, the famous letter to Auguste Chevalier outlining his ideas, and three attached manuscripts. Hermann Weyl commented of this testament, "This letter, if judged by the novelty and profundity of ideas it contains, is perhaps the most substantial piece of writing in the whole literature of mankind." When reading Galois's 7-page testament, there are some words that are really sad: "these subjects are not the only ones that I have explored... But I don't have time, and my ideas are not yet well developed in this area, which is immense...it would not be too much in my interest to make mistakes so that one suspects me of having announced theorems of which I would not have a complete proof." The most impressed and saddest words are: "I don't have time" At the end of the letter, he asked his friend to "Ask Jacobi or Gauss publicly to give their opinion, not as to the truth, but as to the importance of these theorems. Later there will be, I hope, some people who will find it to their advantage to decipher all this mess." [20]

---

[6] However, the rejection report ends on an encouraging note: "We would then suggest that the author should publish the whole of his work in order to form a definitive opinion." [19]

"E. Galois, le 29 mai 1832" at the bottom of the last page in his testament

Early in the morning of May 30, 1832, Galois was injured badly in the duel. He was shot in the abdomen. A passing peasant found him and sent Galois to the hospital. He died the following morning at ten o'clock, after refusing the offices of a priest. His younger brother was notified and came to the hospital. His last words to his brother Alfred were: "Don't cry, Alfred! I need all my courage to die at twenty!" We don't know the exact reason behind the duel, whether it was a love tragedy or a political murder. Whatever the reason, a great talent mathematician was killed at the age of 20. He had only studied mathematics for 5 years. Within the only 67 pages of Galois' collected works are many important ideas that have had far-reaching consequences for nearly all branches of mathematics.

Chevalier and Galois's brother published the testament in *Revue encyclopedique*, but it was not noticed. It might be too brief and hard; there was almost no any impact to the mathematics in that year. Decades passed, in 1843 Liouville reviewed Galois manuscript and declared it sound. It was finally published in the October–November 1846 issue of the *Journal de Mathématiques Pures et Appliquées*. The most famous contribution of this manuscript was a novel proof that there is no quintic formula— that is, that the fifth- and higher-degree equations are not generally solvable by radicals. Although Abel had already proved the impossibility of a quintic formula by radicals in 1824, Galois's methods led to deeper research in what is now called Galois theory. For example, one can use it to determine, for any polynomial equation, whether it has a solution by radicals. Liouville thought about this tragedy and commented in the introduction to Galois's testament: "Perhaps, his exaggerated desire for conciseness was the cause of this defect, and is something which one must endeavor to refrain from when dealing with the abstract and mysterious matters of pure Algebra. Clarity is, indeed, all the more necessary when one has intention of leading the reader away from the beaten roads into the desert... But at present all that has changed. Alas, Galois is no more! Let us cease carrying on with useless criticisms; let us leave its defects, and instead see its qualities... My zeal was soon rewarded. I experienced great pleasure the moment when, after having filled in the minor gaps, I recognized both the scope and precision of the method that Galois proved." [21]

Dedekind lectured on Galois theory at Göttingen in 1858. In 1870, Jordan wrote the book *Traité des substitutions et des équations algébriques* based on Galois theory. Galois's most significant contribution to mathematics is his development of Galois theory. He realized that the algebraic solution to a polynomial equation is related to the structure of a group of permutations associated with the roots of the polynomial, the Galois group of the polynomial. It laid the foundation of group theory and led to the development of abstract algebra and modern mathematics. Ironically "Instead of the political revolution, what Galois actually triggered was the mathematical revolution [9]."

3.2.1 Group

S_2 and S_3, which describe the symmetry of segment and triangle, are actually two examples of group. Here is the definition of group.

Definition 3.2.1 A group is a set G equipped with a binary operation "$\cdot$," which satisfies four axioms:

1. **Closure**: For all a, b in G, the result of $a \cdot b$ is in G.
2. **Associativity**: For all a, b, c in G, $(a \cdot b) \cdot c = a \cdot (b \cdot c)$.
3. **Identity element**: There exists an element e in G such that, for every a in G, $a \cdot e = e \cdot a = a$.
4. **Inverse element**: For every a in G, there exists an element a^{-1}, such that $a \cdot a^{-1} = a^{-1} \cdot a = e$.

The binary operation is often called "multiplication," and the "product" $a \cdot b$ is usually written as ab. e is the identity element. The elements in a group can be finitely or infinitely many; thus, the group is called finite group or infinite group. The *order* of a finite group is the number of the elements in that group. A group that contains infinitely many elements is said to have infinite order.

The "multiplication" may not be commutative like the normal multiplication for numbers. For example, all $n \times n$ invertible matrices with real entities, together with the matrix multiplication, form a group. However, the matrix multiplication order matters; it is not commutative. Groups for which $ab = ba$ always holds are called *abelian* groups (in honor of Abel).

Here are some examples of group.

1. Integers with addition ($\mathbb{Z}$). Elements are all integers, the binary operation is $+$, and the identity element is 0. For any integer n, its inverse is $-n$. This is one of the most familiar groups.
2. The set of remainders of all integers divided by 5, i.e., $\{0, 1, 2, 3, 4\}$. The binary operation is $+$ then modulo 5, for example, $3 + 4 = 7$ mod $5 = 2$. They form a group called addition group of integers modulo 5, denoted as $\mathbb{Z}_5$. We can consider it as partitioning of the integers by taking the remainder, called *residue class*.[7]

---

[7] Denoted as $\mathbb{Z}/5\mathbb{Z}$, where $5\mathbb{Z}$ multiplies 5 to every integer: $\{\cdots, -5, 0, 5, 10, \cdots\}$. The quotient $\{\cdots, -1, 0, 1, 2, \cdots\}/\{\cdots, -5, 0, 5, 10, \cdots\}$ contains 5 residue classes.

Fig. 3.4 There are a total of 18 rotations of Rubik cube for the 6 faces, plus the identity; these moves together with the composition operation form a group

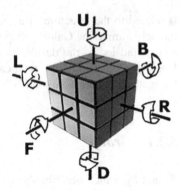

3. S_2 and S_3 are groups. Elements are permutations; they can be expressed in transformations of s, f, e, where e is the identity. The binary operation is composition. To find the inverse of an element a, we look up the Cayley table and find the column and row that correspond to a and some element b, such that they cross at e, then $b = a^{-1}$.

4. The rotations of Rubik cube form a group. The elements are all cube rotations;[8] the binary operation is the composition of cube rotations, corresponding to the result of performing one cube rotation after another (Fig. 3.4).

    In original state of the Rubik cube, all six sides have the same color for each. Let a kid play it randomly for a while. The player needs to figure out a way and restore the Rubik cube to its original state. If the kid's rotations is recorded with a camera as $\{t_0, t_1, ..., t_m\}$ and the restore process as $\{r_0, r_1, ..., r_n\}$, then the whole process to solve the Rubik cube puzzle means:

$$(r_n \cdot r_{n-1}... \cdot r_0) \cdot (t_m \cdot t_{m-1}... \cdot t_0) = e$$

    Obviously, one solution is to reverse every rotation made by the kid, that is, $r_i = t_{m-i}^{-1}$, or $r_i \cdot t_{m-i} = e$. However, a seasoned player often restores the Rubik cube through a set of "formulas." Although the above equation holds, not every r_i is the reversed rotation of some t_{m-i}; even the number of steps n may not equal to m.

5. In a plane, all the rotations around a fixed point form a group. The elements are degrees of the rotation from $-180°$ to $180°$. The binary operation is the composition, corresponding to a degree of rotation after another. The identity rotation has 0 degree; for rotation of degree θ, its reverse rotates $-\theta$ (alternatively, one may define the rotations from 0 to $360°$, then the reverse is $360° - \theta$).

    We say a shape has rotational symmetry if it looks same when it rotates for some degree (about a fixed point). A snowflake, for example, keeps same by

---

[8] There are 18 rotations of Rubik cube. A cube move rotates one of the six faces, front, back, top, bottom, left, right, 90°, 180°, or -90°. For example, rotating the left side 90°, 180°, -90° can be denoted as L, L^2, L' [22]. Plus the identity, there are a total of 19 elements.

rotating 60°, 120°, 180°, 240°, 360°, and their multiples. However, its symmetry cannot be described by the rotational group in this example. Actually, if two rotations satisfy $r_\alpha \cdot r_\beta = e$, then the two degrees either negate to each other $\alpha = -\beta$ or their sum are multiples of 360°. All symmetric shapes under this rotational group are circles at the same center. We'll give the group that describes the symmetry of snowflake in Sect. 3.2.6.

The following examples are not groups:

1. All the integers except 0, together with multiplication, do not form a group. We can't use 1 as the identity; otherwise, there will not be inverse of 3, for example (1/3 is not an integer).
2. For the same reason, remainders of modulo 5: $\{0, 1, 2, 3, 4\}$, together with modulo multiplication, do not form a group. However, when excluding all multiples of 5, then the set $\{1, 2, 3, 4\}$ and modulo multiplication form a group, as shown in below Cayley "multiplication" table:

	1	2	3	4
1	1	2	3	4
2	2	4	1	3
3	3	1	4	2
4	4	3	2	1

1 is the identity; it is also the inverse of itself; 2 and 3 inverse each other; the inverse of 4 is 4 again.
3. Although all the nonzero residues of 5 form a group under modulo multiplication, the nonzero residues of 4 do not. Observe below Cayley table:

	1	2	3
1	1	2	3
2	2	0	2
3	3	2	1

$(2 \times 2) \bmod 4 = 0$, which is not in set $\{1, 2, 3\}$. Only the remainders that are coprime to n form a group under the multiplication modulo n. This kind of groups is called a multiplicative group of integers modulo n. For any prime number p, set $\{1, 2, ..., p - 1\}$ forms a multiplicative group modulo p.
4. All the rational numbers together with multiplication do not form a group. Although every p/q has an inverse of q/p when $p, q \neq 0$, there is no inverse for 0. All the rational numbers excluding 0 form a group under multiplication.

Exercise 3.2

3.2.1 Do all the even numbers form a group under addition?

3.2.2 Can we find a subset of integers that forms a group under multiplication?

3.2.3 Do all the positive real numbers form a group under multiplication?

3.2.4 Do integers form a group under subtraction?

3.2.5 Find a group with only two elements.

3.2.6 What is the identity element for Rubik cube group? What is the inverse of F?

Answer: page 324

3.2.2 Monoid and Semigroup

The criteria to be a group are strict. The above counterexamples show some common algebraic structures don't satisfy one or more group axioms. If needn't inverse, then we get *monoid*.

Definition 3.2.2 A *monoid* is a set S together with a binary operation "$\cdot$," which satisfies two axioms:

1. **Associativity**: For any a, b, and c in S, $(a \cdot b) \cdot c = a \cdot (b \cdot c)$.
2. **Identity element**: There exists an element e in S, such that for every a in S, $a \cdot e = e \cdot a = a$.

Monoid is quite similar to group except there is no axiom about inverse. Many counterexamples of group are monoids. For example, integers under multiplication form a monoid, with 1 as the identity. Monoid often appears in computer programming; we'll revisit it as a category in Sect. 4.1.1. Here are more examples of monoid:

1. Given a character set, all finite strings under the concatenation operation form a monoid. The elements are strings; the identity element is the empty string ("").
2. Extend from string to list of type A (List A). All lists form a monoid under concatenation ($+\!\!+$); the identity element is the empty list (nil).

   With monoid as the algebraic structure, we abstract both string and list as below example program.

```
instance Monoid (List A) where
    e = nil
    (*) = (++)
```

We can abstract folding for string and list to the level of monoid (in Sect. 4.6), or define concatenation for any monoid:

$$concat = foldr\ e\ (*)$$

For example: *concat* [[1], [2], [3], [1, 2], [1, 3], [2, 3], [1, 2, 3], [1, 3, 2]]
gives: [1, 2, 3, 1, 2, 1, 3, 2, 3, 1, 2, 3, 1, 3, 2].

3. Heap is a common data structure in programming. The heap is called min-help if
   the top element is always the minimum; it is called max-heap if the top element
   is always the maximum. The skew heap is a type of heap realized in binary tree
   ([12], section 8.3).

   data SHeap A = nil | node (SHeap A, A, SHeap A)

   The definition is as the same as the binary tree except for its name. Use
   min-heap, for example, the top element is always at the root. We define merge
   operation for two heaps as below:

   $$merge(nil, h) = h$$

   $$merge(h, nil) = h$$

   $$merge(h_1, h_2) = \begin{cases} k_1 < k_2 : & node(merge(r_1, h_2), k_1, l_1) \\ \text{otherwise} : & node(merge(h_1, r_2), k_2, l_2) \end{cases}$$

   When merging two heaps, if either one is empty, the result is the other;
   otherwise, denote $h_1 = node(l_1, k_1, r_1)$ and $h_2 = node(l_2, k_2, r_2)$. To merge
   them, we firstly compare their root, select the smaller one as the new root, then
   merge the other heap with the bigger element to one of its sub-trees. Finally, we
   exchange the left and right sub-trees. For example, if $k_1 < k_2$, we select k_1 as the
   new root. Then we can either merge h_2 to l_1 or merge h_2 to r_1. Without loss of
   generality, we merge to r_1. Then, exchange the left and right sub-trees to get the
   final result $(merge(r_1, h_2), k_1, l_1)$. Note the merge operation is recursive. The
   set of all the skew heaps form a monoid under the merge operation. The identity
   element is the empty heap (nil).

4. Heap can also be realized with multi-trees (Sect. 2.6); pairing heap is such type
   of heap defined as below ([12] Section 10.3):

   data PHeap A = nil | node (A, **List** (PHeap A))

   The definition is as the same as multi-tree except for its name. A pairing heap
   is either empty or has a root and a list of sub-trees. The top element is at the root.
   Define merge operation for two pairing heaps as below:

   $$merge(nil, h) = h$$

   $$merge(h, nil) = h$$

   $$merge(h_1, h_2) = \begin{cases} k_1 < k_2 : & node(k_1, h_2 : ts_1)) \\ \text{otherwise} : & node(k_2, h_1 : ts_2)) \end{cases}$$

   If either heap is empty, the merged result is the other; otherwise, denote $h_1 = node(k_1, ts_1)$ and $h_2 = node(k_2, ts_2)$; compare their roots; select the less as the new root; let the greater be another sub-tree. The set of all the pairing heaps form a monoid under the merge operation. The identity element is the empty heap (nil).

If relaxing one more step, removing the identity requirement, then we get semigroup.

Definition 3.2.3 A *semigroup* is a set with an associative binary operation.

Associativity means for any three elements a, b, and c of the semigroup, $(ab)c = a(bc)$. Here are two examples of semigroup:

1. All positive integers form a semigroup under +, so as under ×.
2. All even numbers form a semigroup under +, so as under ×.

People often call the binary operation as "multiplication" for groups, monoids, and semigroups; therefore, it's convenient to use "power" when applying the operation multiple times, for example: $x \cdot x \cdot x = x^3$. Generally, the power of group and monoid is defined recursively as below:

$$x^n = \begin{cases} n = 0 : & e \\ \text{otherwise} : & x \cdot x^{n-1} \end{cases}$$

For semigroup, since the identity is undefined, n must not be zero:

$$x^n = \begin{cases} n = 1 : & x \\ \text{otherwise} : & x \cdot x^{n-1} \end{cases}$$

Exercise 3.3

3.3.1 The set of Boolean values {True, False} forms a monoid under the logic or ($\vee$). It is called the "Any" logic monoid. What is the identity element for this monoid?

3.3.2 The set of Boolean values {True, False} forms a monoid under the logic and ($\wedge$). It is called the "All" logic monoid. What is the identity element for this monoid?

3.3.3 For the comparable type, when comparing two elements, there can be four different results. We abstract them as {<, =, >, ?}. The first three are deterministic;[9] the last '?' means undetermined. For this set, we can define a binary operation to make it a monoid. What is the identity element for this monoid?

---

[9] Some programming languages, such as C, C++, and Java use negative number, zero, and positive number to represent them. In Haskell, they are GT, EQ, and LE, respectively.

3.3.4 Show that the power operation for group, monoid, and semigroup is commutative: $x^m x^n = x^n x^m$

Answer: page 325

3.2.3   Properties of Group

Abstract algebra allows us to focus on the inner pattern of the abstract structures
and their relations without caring about the concrete objects and their meanings.
Patterns and insights are applicable to all objects. When we know the generic group
properties, if the elements represent points, lines, and surfaces, then we obtain the
properties of geometry; if the elements represent Rubik cube rotations, then we
obtain the properties of Rubik cube transformation; if the elements represent some
data structures in programming, we obtain the properties of the algorithm. Below
are some basic group properties:

Proposition 3.2.1 *There is one and only one identity element in a group.*

Proof Assume there is another identity element e', such that $a = e'a = ae'$ holds
for every a. Substitute a with e, then $e = e'e = e'$, proving the uniqueness of the
identity element.   □

Because of this, we say *the* identity element of a group. Besides, the inverse of
every element is also unique.

Proposition 3.2.2 (Uniqueness of Inverse Element) *For every a in group, there is
one and only one inverse element a^{-1} satisfying $aa^{-1} = a^{-1}a = e$.*

Proof For every a, the inverse axiom assumes its inverse exists; we only need to
show its uniqueness. Assume there is another element b that also satisfies $ab = ba = e$, then:

$$b = be = b(aa^{-1}) = (ba)a^{-1} = ea^{-1} = a^{-1}$$

□

Remind the order of group; we can define the order of a group element a as the
minimum positive integer m, such that $a^m = e$. If such m does not exist, we say the
order is infinite. Using the Rubik cube group, for example, repeating F rotation four
times, the cube returns to its original state; the order of F is 4 (Fig. 3.5). Another
example is the integer multiplicative group modulo 5. For every element except 1,

Fig. 3.5 Restore after
rotating four times of F

the 4th power modulo 5 is 1; all their orders are 4. Actually, we have the following interesting theorem:

Proposition 3.2.3 *All elements in a finite group have finite order.*

Proof Let n denote the order of a finite group G. For any element a, construct a set $\{a, a^2, ..., a^{n+1}\}$. There are $n + 1$ elements in this set; however, the order of the group is n. According to the principle of pigeon hole, there are at least two equal elements. Denote such two as a^i and a^j, where $0 < i < j \leq n + 1$.

$$a^j = a^i$$

$$a^j(a^{-1})^i = a^i(a^{-1})^i \qquad \text{right multiply } (a^{-1})^i \text{ on both sides}$$

$$a^{j-i} = e \qquad \text{the order of } a \text{ is } j - i, \text{ finite}$$

$\square$

We say lists in programming are "isomorphic" to natural numbers to show their common structure in Chapter 1. Let's give a formal definition. For a mapping (or called morphism) f from set A to set B, denote it as $f : A \rightarrow B$, where A and B are equipped with their closed binary operation, for example, groups, monoids, or semigroups. Let a, b be two elements in A; their images in B are $f(a)$, $f(b)$, respectively. Consider $a \cdot b$; it is contained in A because the operation "$\cdot$" is closed; its image under f is $f(a \cdot b)$ in B. For the group operation "$*$" defined in B, if

$$f(a) * f(b) = f(a \cdot b)$$

always holds, we say f is a *homomorphism* from A to B. If f is *surjective* ("onto"), which means every b in B has the corresponding a in A, such that $f(a) = b$, then f is called surjective homomorphism. For example, function *odd* : $\mathbb{Z} \rightarrow$ *Bool* tests whether an integer is odd (True) or even (False). All integers form a group under addition, while the Boolean values $\{True, False\}$ also form a group under logic exclusive or ($\oplus$). We can verify that:

1. Both a and b are odd. $odd(a)$ and $odd(b)$ are both True; their sum is even, $odd(a + b)$ is False. $odd(a) \oplus odd(b) = odd(a + b)$ holds.
2. Both a and b are even. $odd(a)$ and $odd(b)$ are both False; their sum is also even, $odd(a + b)$ is False. $odd(a) \oplus odd(b) = odd(a + b)$ holds.
3. a and b are odd and even. One of $odd(a)$ and $odd(b)$ is True, the other is False; their sum is odd, $odd(a + b)$ is True. $odd(a) \oplus odd(b) = odd(a + b)$ holds.

If f is surjective and also *injective* (no two different elements map to the same image), then it's one-to-one (*bijective*) mapping, i.e., called *isomorphism*. We say A and B are *isomorphic* under f, denoted as $A \cong B$. Isomorphic is a very strong relationship. As shown in Fig. 3.6, if A is isomorphic to B, from the abstract view, they are essentially the same thing only with different names [23]. If there is a algebraic property in A, then there is exactly a same property in B. The special

Fig. 3.6 Isomorphism between semigroups

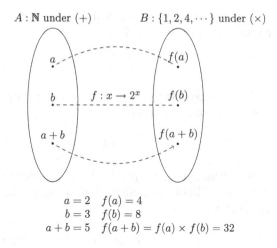

$A : \mathbb{N}$ under $(+)$ $B : \{1, 2, 4, \cdots\}$ under $(\times)$

$$a = 2 \quad f(a) = 4$$
$$b = 3 \quad f(b) = 8$$
$$a + b = 5 \quad f(a + b) = f(a) \times f(b) = 32$$

isomorphism from A to A itself is called *automorphism* of A. For example, the group of integers with addition has a automorphism under the negate operation.

One tends to use familiar concrete examples to help understand the abstract thing, like to think about numbers with "+" for groups. However, it might lead to the illusion that the elements are kind of entities analogous to numbers; the binary operation is something like arithmetic operations, usually commutative. Transformation group, on the contrary, is not abelian (not commutative), nor has entities but transformations as elements.

Transformation is a map from set A to A itself, denoted as $\tau : a \to \tau(a)$. There are variations of transformations; for example, the below lists all transformations of the Boolean set (T for true, F for false):

$$\begin{aligned}
\tau_1 &: T \to T, F \to T \\
\tau_2 &: T \to F, F \to F \\
\tau_3 &: T \to T, F \to F \\
\tau_4 &: T \to F, F \to T
\end{aligned} \qquad (3.3)$$

Among them, τ_3 and τ_4 are one to one. For a set A, all its transformations form a new set:

$$S = \{\tau, \lambda, \mu, ...\}$$

Let's define a binary operation for S and call it as multiplication. For convenience, we express $\tau(a)$ as a^τ:

$$\tau : a \to a^\tau \quad \text{where } a^\tau = \tau(a)$$

Notation a^τ does not mean the τ power of a, but transformation. For two transformations τ and λ in S:

$$\tau : a \to a^\tau \quad \lambda : a \to a^\lambda$$

Their composition $a \to (a^\tau)^\lambda = \lambda(\tau(a))$ is a transformation of A too. We define it as the "product" of τ and λ.

$$\tau\lambda : a \to (a^\tau)^\lambda = a^{\tau\lambda}$$

Such "multiplication" is actually composition of two transformations. One can choose some Boolean transformations to verify their products. The composition is associative:

$$\tau(\lambda\mu) : a \to (a^\tau)^{\lambda\mu} = ((a^\tau)^\lambda)^\mu$$

$$(\tau\lambda)\mu : a \to (a^{\tau\lambda})^\mu = ((a^\tau)^\lambda)^\mu$$

Benefitting from the exponential notation, both are $a^{\tau\lambda\mu}$. We can't say too much about a powerful notation system in history of mathematics, Leibniz and Euler set great examples of using proper symbol and notation. Among S, the identity transformation $\epsilon : a \to a$ maps every a of A to itself; it serves as the identity element under composition:

$$\epsilon\tau : a \to (a^\epsilon)^\tau = a^\tau$$

$$\tau\epsilon : a \to (a^\tau)^\epsilon = a^\tau$$

Hence, $\epsilon\tau = \tau\epsilon = \tau$. Set S almost forms a group under composition; unfortunately, it does not. This is because the inverse of a transformation τ may not necessarily exist. For example, $\tau_1 : T \to T, F \to T$ maps any Boolean value to true. None of the four transformations in Eq. (3.3) can change τ_1 back; hence, τ_1 has no inverse. Although S does not form a group, the subset G containing only one-to-one transformations does, namely, a *transformation group* of A.

Proposition 3.2.4 *All the one-to-one transformations of set A form a transformation group G.*

The bijective transformations assert inverse; it's easy to verify group axioms. Transformation groups are not necessarily abelian. For counterexample, let transformation τ_1 move (shift) a point right by 1; τ_2 rotates counter clockwise by $\pi/2$; when applying to point $(0, 0)$ in different orders:

$$\tau_1\tau_2 : (0, 0) \to (0, 1)$$
$$\tau_2\tau_1 : (0, 0) \to (1, 0)$$

Fig. 3.7 The transform order
matters

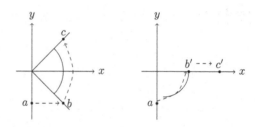

It leads to different results as in Fig. 3.7; hence, it is not abelian. Transform groups have wide applications. There is a strong fact:

Theorem 3.2.5 *Every group is isomorphic to a transformation group.*

For any abstract group, this theorem asserts an equivalent transformation group.

Exercise 3.4

3.4.1 Is the odd-even test function a homomorphism from the additive group of integers $(\mathbb{Z}, +)$ to the logic-and group of Boolean $(Bool, \wedge)$? What's about the multiplicative group of integers without zero?

3.4.2 There is a homomorphism f from groups G to G', mapping $a \rightarrow a'$; does the order of a equal to the order of a'?

3.4.3 Show that the identity element of transformation group is the identity transformation.

Answer: page 326

3.2.4 Permutation Group

S_2 and S_3 belong to a family of groups, called permutation groups. Galois developed them to study the symmetry of equation roots and was the first to name them "group." Permutation group is a special kind of transformation group. A *permutation* is a one-to-one transformation of a finite set; literally, it permutes elements of the set. The permutations form a group under the composite operation. Particularly, for a set of n elements, the group of *all* permutations ($n!$ in total from school math) is called the *symmetric group* of degree n, denoted as S_n. The order $|S_n| = n!$. A permutation maps element $a_i \rightarrow a_j$ in the set. We list n pairs of label (i, j) as $(1, k_1), (2, k_2), ..., (n, k_n)$, or in two rows as:

$$\begin{pmatrix} 1 & 2 & ... & n \\ k_1 & k_2 & ... & k_n \end{pmatrix}$$

Fig. 3.8 3-cycle permutation
(1 2 3)

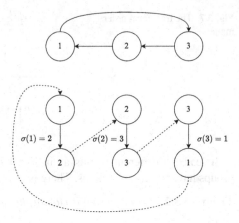

for example:

$$\begin{pmatrix} 1\ 2\ 3\ 4\ 5 \\ 2\ 5\ 4\ 3\ 1 \end{pmatrix}$$

It maps $1 \to 2, 2 \to 5, 5 \to 1$, and exchanges $3 \leftrightarrow 4$. Since the first row is always $1\ 2 \cdots n$, we further simplify the notation to $(2, 5, 4, 3, 1)$. The below lists all permutations of 3 elements, namely, S_3, with this notation:

$$(1, 2, 3), (1, 3, 2), (2, 1, 3), (2, 3, 1), (3, 1, 2), (3, 2, 1)$$

When we "multiply" (composite essentially) two permutations α and β, we determine every label in this way: for label i, first check where i is sent to in β, say j; next check where j is sent to in α, say k; the final label in the "product" is k. Note the order is to apply β first, then followed with α. For example, pick two permutations from S_3 and reverse their multiplication orders:

$$(1, 3, 2)(2, 1, 3) = (2, 3, 1)$$

$$(2, 1, 3)(1, 3, 2) = (3, 1, 2)$$

They are different; hence, S_3 is not abelian. Actually, S_3 is the smallest finite non-abelian group. In other words, any finite non-abelian group contains at least six elements.

Observe permutation $(2, 3, 1, 4, 5)$; it only rearranges the first three elements and fixes the rest two. The rearrangement as in Fig. 3.8 is cyclic.

We simplify the notation of this permutation to (1 2 3); a cycle of the first three elements leaves the remaining unchanged. Note there is no comma between labels, and we treat (1 2 3), (2 3 1), and (3 1 2) the same three cycles.[10] It's called k-cyclic permutation, for a cycle of k elements: $(i_1 i_2 ... i_k)$. The cycle maps every element to its next and maps the last element to the first:

$$i_1 \rightarrow i_2 \rightarrow i_3 \rightarrow \cdots \rightarrow i_{k-1} \rightarrow i_k \rightarrow i_1$$

The lower of Fig. 3.8 explains the k-cycle mapping, namely, σ. Follow the arrows: $1 \rightarrow \sigma(1) = 2$, $2 \rightarrow \sigma(2) = 3$, and $3 \rightarrow \sigma(3) = 1$. We can also explain k-cycle as movements (the upper of Fig. 3.8): shift every element to its predecessor, and move the first to tail to close the loop. The k elements may not be necessarily adjacent, for example, (3 9 4), nor in a fixed order, for example, (2 4 1 3). We call a cycle of two elements $(i \ j)$, a transposition (swap). The singleton cycle, for example, (1), fixes all elements and hence is the identity permutation. As the identity is unique, all singleton cycles are identical:

$$e = (1) = (2) = \cdots = (n) \tag{3.4}$$

Observe permutation (2, 1, 4, 5, 3); it contains two cycles: a transposition of (1 2) and a cycle of three (3 4 5). We can "decompose" it as below:

$$(2, 1, 4, 5, 3) = (1\ 2)(3\ 4\ 5)$$

In fact, every permutation can be expressed as the product of disjoint (without overlapped labels) cyclic permutations. As an advantage, although permutations are not necessarily commutative, disjoint k-cycles are; for example, $(1\ 2)(3\ 4\ 5) = (3\ 4\ 5)(1\ 2)$. The below lists S_3 as k-cyclic permutations:

$$(1), (1\ 2), (1\ 3), (2\ 3), (1\ 2\ 3), (1\ 3\ 2)$$

How to decompose a permutation $(k_1, k_2, ..., k_n)$ to a product of disjoint cycles? Here is the algorithm: compare every position i with its label k_i from left to right; if $i = k_i$, the element has already been in the right position; otherwise open a pair of parentheses; write down the label k_i in the parentheses, let's say $k_i = j$; next check position j against its label k_j; if they are not equal, write down k_j in the parentheses. Repeat this step until we meet some label that forms a cycle (points to the starting label k_i). At this point, we close the parentheses and continue checking from left to right. The algorithm terminates after processing all labels ([24], pp27). If $i = k_i$ for all positions, then it is the identity permutation (1). Let the input permutation π be an array of labels; below implementation decomposes it to a list of k-cycles:

---

[10] Some use (231) notation and only add spaces when exceeds 10 elements, e.g., (12 3).

```
1: function K-CYCLES(π)
2:     r ← [ ], n ← |π|
3:     for i ← 1 to n do
4:         p ← [ ], j ← i
5:         while j ≠ π[j] do
6:             APPEND(p, π[j])
7:             j ← π[j]
8:             if j = i then
9:                 break                          ▷ close the cycle
10:        if p ≠ [ ] then
11:            APPEND(r, p)
12:    if r = [ ] then
13:        r ← [[1]]                              ▷ Identity permutation
14:    return r
```

For any finite group $G = \{g_1, g_2, \cdots, g_n\}$, if apply G to itself, then every element g (just another name of some g_i) acts like a permutation:

$$\begin{pmatrix} g_1 & g_2 & \cdots & g_n \\ gg_1 & gg_2 & \cdots & gg_n \end{pmatrix}$$

Indeed, we can define the permutation as $f(g) : x \to gx$ for every x in G. f is bijective, because left multiply g^{-1} restores the permutation back. Further, it's easy to verify f is a isomorphism: $f(g)f(h) = f(gh)$; thus, G is isomorphic to the permutation group $\{f(g)|g \in G\}$.

Theorem 3.2.6 (Cayley's Theorem) *Every finite group is isomorphic to a permutation group.*

We'll later show that there is a link from the roots of an equation to permutation groups. This was exactly how Galois solved the problem to determine if an equation is radical solvable.

Exercise 3.5

3.5.1 Show that if the transform σ sends the i-th element to the j-th, i.e., $\sigma(i) = j$, then the k-cycle permutation is $(\sigma(i_1) \sigma(\sigma(i_1)) \dots \sigma^k(i_1))$.
3.5.2 List all the elements in S_4.
3.5.3 Write a program to convert the product of k-cycles back to permutation.

Answer: page 327

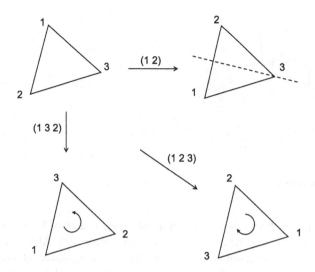

Fig. 3.9 Transform an equilateral triangle

3.2.5 Group and Symmetry

Why is the group of all permutations of a set called a symmetric group? Because it exactly defines symmetry. Let 1, 2, and 3 denote three vertexes of an equilateral triangle. We choose and apply three permutations: (1 2), (1 2 3), and (1 3 2) from S_3 to the vertexes as in Fig. 3.9.

1. The triangle transforms to the top right when applying (1 2). The vertexes change from 123 to 213, equivalent to the mirroring: flip the triangle against the axis at vertex 3. Because they are congruent before and after, the equilateral triangle is bilateral symmetric (reflexive). Besides (1 2), there are other two bilateral symmetric transformations: (1 3) and (2 3) with axes at vertexes 2 and 1.
2. The triangle transforms to the bottom right when applying (1 2 3). The vertexes change from 123 to 231, equivalent to a rotation of 120° clockwise. Because they are congruent, the equilateral triangle is rotational symmetric of 120°.
3. The triangle transforms to the bottom when applying (1 3 2). The vertexes change from 123 to 312, equivalent to a rotation of 120° counter clockwise or 240° clockwise. Because they are congruent, the equilateral triangle is symmetric of −120° or 240°.

Every symmetry of the equilateral triangle (three bilateral symmetries, two rotational symmetries, and the identity) exactly corresponds an element of S_3, a beautiful relationship between group and symmetry.

The transformations that keep the size and shape unchanged are called *congruence*. A congruence is either proper or improper. Figure 3.10a shows two sea snails: the sinistral (left-handed) shell of *Neptunea angulata* (now extinct), found mainly in

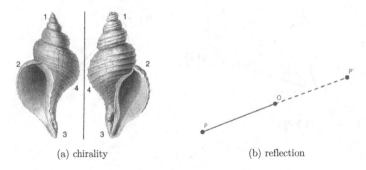

(a) chirality   (b) reflection

Fig. 3.10 Congruence

the northern hemisphere, and the dextral (right-handed) shell of *Neptunea despecta*, a similar species found mainly in the southern hemisphere. Although they look similar side by side, we cannot make them congruent for whatever rotations, flips, or motions. They "mirrored" each other. There is no way to transform a sinistral object to its mirrored dextral image in physical world. This is the difference between proper and improper congruence. The proper carries a left screw into a left screw, while the improper carries into a right screw and vice versa [36]. The reflection sends any point P to its antipode point P' with respect to O found by joining P with O and prolonging the straight line PO by its own length: $|PO| = |OP'|$ (Fig. 3.10b). Applying S_4 to the four points in Fig. 3.10a, (2 4) is an improper congruence. It carries the left-handed snail into the right-handed. In some sense, the symmetric group depicts both real and mirrored worlds.

Exercise 3.6

3.6.1   Analogous to $S3$ and equilateral triangle, what symmetry does S_4 describe?

Answer: page 328

3.2.6   Rotational Symmetry and Cyclic Group

Figure 3.11 shows a symmetric propeller. It is congruent when it rotates 60°, 120°, 180°, 240°, 300°, 360°, and their multiples. We call them rotational symmetries. Although the propeller has six blades, S_6 is not the group describing its symmetry. There are $6! = 720$ permutations in S_6, while only six rotational symmetries. Let r denote the rotation of 60°; the rotation of 120° is equivalent to rotating r twice ($r \cdot r = r^2$); similarly, rotation of 180° equals to r^3, ..., and rotation of 360° restores the propeller, hence $r^6 = e$. The six rotations are $C_6 = \{e, r, r^2, r^3, r^4, r^5\}$. They

Fig. 3.11 A propeller with
six blades

form a group isomorphic to the subset of S_6: $\{(1), (1\ 2\ 3\ 4\ 5\ 6), ..., (5\ 6\ 1\ 2\ 3\ 4)\}$.
Compare C_6 an S_3; they have different structures although both have six elements.
S_3 does not describe the symmetry of the propeller.

C_6 is an instance of cyclic group. Generalize six to n; group C_n describes the
symmetry of a propeller with n blades. The family of cyclic groups was one of the
first group families people completely understood. Their elements are all powers
of some element. For example, the integer multiplicative group modulo 5 has four
elements $\{1, 2, 3, 4\}$ (without 0, the identity is 1); they are all powers of 2 modulo
5:

$$2^1 \bmod 5 = 2 \qquad\qquad 2^2 \bmod 5 = 4$$

$$2^3 \bmod 5 = 3 \qquad\qquad 2^4 \bmod 5 = 1$$

Rearrange $\{1, 2, 3, 4\}$ to $\{1, 2, 2^2 \bmod 5, 2^3 \bmod 5\}$; this group is equivalent to
C_4. We say 2 generates this group.

Definition 3.2.4 G is a *cyclic group* if all elements are the powers of some element
a. In other words G is generated by a, denoted as: $G = (a)$; a is a generator of G.

Here are two examples:

Example 3.2.1 The additive group of integers ($\mathbb{Z}$). All integers are "powers" of 1.
The binary operation is $+$; power means repeatedly adding. For any positive integer
m:

$$m = 1 + 1 + \cdots + 1 \qquad\qquad \text{total } m$$

$$= 1 * 1 * \cdots * 1 \qquad\qquad + \text{ is the binary operation } *$$

$$= 1^m \qquad\qquad m\text{-th power}$$

Since $-1 + 1 = 0$, the inverse of 1 is -1, where 0 is the identity element.

$$
\begin{aligned}
-m &= (-1) + (-1) + \cdots + (-1) & & \text{total } m \\
&= 1^{-1} + 1^{-1} + \cdots + 1^{-1} & & -1 \text{ is the inverse of } 1 \\
&= 1^{-1} * 1^{-1} * \cdots * 1^{-1} & & + \text{ is the binary operation } * \\
&= (1^{-1})^m & & m \text{ times of inverse} \\
&= 1^{-m}
\end{aligned}
$$

It bijectively maps every integer m to 1^m and includes 0 as well because $0 = -m + m = 1^{-m} * 1^m = 1^{-m+m} = 1^0$. In summary, $\mathbb{Z} = (1)$.

The additive group of integers is an example of infinite cyclic group. Here is an example of finite cyclic group.

Example 3.2.2 The additive group of residues modulo n ($\mathbb{Z}_n$). For integer a, let $[a]$ represent the residue class that a belongs to, i.e., $[a]$ contains every m such that $m \equiv a \pmod{n}$. Denote addition modulo n as:

$$
[a] + [b] = [a + b]
$$

For example, $[3] + [4] = [2]$ modulo 5, because $3 + 4 = 7 \equiv 2 \pmod{5}$. It's easy to verify modulo add is associative, $[0]$ is the identity element, and every $[a]$ has inverse. There are n elements: $[0], [1], \cdots, [n-1]$. $\mathbb{Z}_n$ is cyclic because every $[i]$ is the i-th power of $[1]$:

$$
\begin{aligned}
[i] &= [1] + [1] + \ldots + [1] & & \text{total } i \\
&= [1]^i & & \text{repeat } i \text{ times}
\end{aligned}
$$

$[0]$ is also covered, because $0 \equiv n \pmod{n}$ implies $[0] = [n] = [1]^n$, a power of $[1]$.

We intend to choose these two examples because they actually cover all the cyclic groups.

Proposition 3.2.7 *If G is a cyclic group generated by (a), then the algebraic structure of G is completely determined by the order of a:*

1. *If the order of a is infinite, then G is isomorphic to $\mathbb{Z}$.*
2. *If the order of a is n, then G is isomorphic to $\mathbb{Z}_n$.*

Proof

1. If the order of a is infinite, then $a^h = a^k$ if and only if $h = k$. Otherwise assume $h > k$ (or we swap them); right multiply a^{-k} to both sides to give $a^{h-k} = e$,

Fig. 3.12 Snowflake crystal photos by Wilson Bentley, 1902

implying a has order of $h - k$ and conflicts with the infinite order. Therefore, we can define a bijective (one-to-one) map from $G = (a)$ to $\mathbb{Z}$:

$$f : a^k \to k$$

It's an isomorphism because $a^h a^k \to h + k$, i.e., G is isomorphic to $\mathbb{Z}$.

2. If the order of a is n, i.e., $a^n = e$, then $a^h = a^k$ if and only if $h \equiv k \pmod{n}$. We can define another bijective map from $G = (a)$ to $\mathbb{Z}_n$:

$$f : a^k \to [k]$$

It's an isomorphism because $a^h a^k = a^{h+k} \to [h + k] = [h] + [k]$, i.e., $G = (a)$ is isomorphic to $\mathbb{Z}_n$. □

From the abstract view, there is only one group that the order of generator is infinite, and there is only one group that the order of generator is some positive integer. We clearly know the algebraic structure of these cyclic groups:

1. The order of a is infinite:

   - Group elements: $\cdots, a^{-2}, a^{-1}, a^0, a^1, a^2, \cdots$
   - Binary operation: $a^h a^k = a^{h+k}$

2. The order of a is n:

   - Group elements: $a^0, a^1, a^2, ..., a^{n-1}$
   - Binary operation: $a^h a^k = a^{(h+k) \bmod n}$

We are almost ready to explain the symmetry of snowflake. However, C_6 is insufficient; it only describes the rotational symmetries: a snowflake is congruent after rotating 60°, 120°, 180°, 240°, 300°, and 360° around its center. Different from the propeller, a snowflake is also congruent when it flips the side (Fig. 3.12). For every transformation in C_6, we can follow a flip around the axis (a diameter)

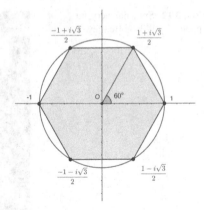

Fig. 3.13 Roots of $x^6 - 1 = 0$: $\zeta_0 = 1$, $\zeta_1 = e^{i\frac{\pi}{3}} \cdots$, $\zeta_5 = e^{i\frac{5\pi}{3}}$

by $180°$. Denote the flip as f; besides $C_6 = \{e, r, r^2, \cdots, r^5\}$, there is another six transformations of $\{f, fr, fr^2, \cdots, fr^5\}$. These 12 elements form a group called *dihedral group* D_6. Because $f^2 = e$, flipping twice restores; we can also treat $\{e, f\}$ a group, namely, C_2. Then $D_6 = C_2 \times C_6$. We can pick every element from C_2, combined with another element from C_6 to compose an element of D_6, a total of 12 combinations. In general, the dihedral group D_n describes the symmetry of regular n-polygon. Wilson Bentley (1865–1931), a US photographer, took over 5000 photographs of snowflake crystal. He donated his collection, the crystal of science and art, to the Buffalo Museum of Science. From these photos, Bentley came to the conclusion that every child knows: each snowflake is unique.

3.2.6.1 Cyclotomic Equation

We lay regular hexagon, an abstracted snowflake, to the complex plane; scale it to inscribe the unit circle. The six vertexes represent the roots of equation $x^6 - 1 = 0$ as in Fig. 3.13. Obviously $x = \pm 1$ are two roots; the other four can be given by de Moivre's formula learned in high school:

$$e^{in\theta} = \cos n\theta + i \sin n\theta = (\cos \theta + i \sin \theta)^n \tag{3.5}$$

The roots equally divide the unit circle into n arcs. For this reason, $x^n - 1 = 0$ is called *cyclotomic equation* (cycle = circle, tom = division or part). Although each root can be written as $\zeta_k = e^{i\frac{2k\pi}{n}}$, they are not necessarily algebraic solution; we neither know whether a regular n-polygon with compass and edge can be drawn. Let's start with radical solutions to the cyclotomic when n is 2, 3, 4, and 6.

(a) $n = 2$, factor $x^2 - 1 = (x + 1)(x - 1)$. The two roots are ± 1.
(b) $n = 3$, because $x = 1$ is a root, $x - 1$ divides $x^3 - 1$. Using polynomial long division, we get $x^3 - 1 = (x - 1)(x^2 + x + 1)$. Next, solve the quadratic equation

$x^2 + x + 1 = 0$, giving all three roots: $1, \frac{-1 \pm i\sqrt{3}}{2}$. We often name $\omega = \frac{-1+i\sqrt{3}}{2}$; the three roots can be written as: $1, \omega, \omega^2$.

$$
\begin{array}{r}
x^2 + x + 1 \\
\hline
x - 1 \overline{)\; x^3 \qquad\qquad - 1} \\
-x^3 + x^2 \\
\hline
x^2 \\
-x^2 + x \\
\hline
x - 1 \\
-x + 1 \\
\hline
0
\end{array}
$$

(c) $n = 4$, factor $x^4 - 1 = (x^2 - 1)(x^2 + 1) = (x - 1)(x + 1)(x - i)(x + i)$. The four roots are $\pm 1, \pm i$.

(d) $n = 6$, factor $x^6 - 1 = (x^3 - 1)(x^3 + 1)$. Therefore all the roots of cubic cyclotomic equation are also roots of $n = 6$. Obviously, $x = -1$ is another root; hence, $(x + 1)$ divides $x^3 + 1$. Using polynomial long division gives $x^3 + 1 = (x + 1)(x^2 - x + 1)$. Next, solve the quadratic equation $x^2 - x + 1 = 0$; we get all six roots: $\pm 1, \frac{\pm 1 \pm i\sqrt{3}}{2}$ in Fig. 3.13.

It's not by chance that the roots of six-degree cyclotomic equation contain all roots of cubic; the roots of quartic cyclotomic equation contain all roots of quadratic; and the roots of six degrees also contain the roots of quadratic. For this reason, we need to differentiate two things:

1. n-th root of unity: the number that n-th power is 1, i.e., $x^n = 1$.
2. n-th *primitive* root of unity: an n-th root of unity is said to be primitive if it is not any m-th root of unity for some smaller m, that is, if $x^n = 1$, and for every $m = 1, 2, ..., n - 1$, $x^m \neq 1$.

-1, for example, is the primitive 2nd root of unity; although it's also the 4-th root of unity, it is not the 4-th primitive. Taking $x^3 - 1 = (x - 1)(x^2 + x + 1)$, for example, both factors are irreducible (with rational coefficients). The root of $x - 1$, which is 1, is the first primitive root of unity, and the two roots of $x^2 + x + 1$ are the 3rd primitive root of unity. We call these irreducible factors cyclotomic polynomials. Denote the cyclotomic polynomial for k-th primitive root of unity as $\Phi_k(x)$. There is a relation between cyclotomic equation and cyclotomic polynomial:

$$
x^n - 1 = \prod_{d \mid n} \Phi_d(x) \tag{3.6}
$$

For every factor d of n, the cyclotomic equation $x^n - 1$ has a cyclotomic polynomial of degree d as a factor. There is another link from cyclotomic equation to cyclic group: all the n-th roots of unity form a cyclic group C_n under multiplication, and the n-th primitive root ζ_k is a group generator, i.e., $C_n = \langle \zeta_k \rangle$.

n	Equation	Roots	Primitives	Polynomial $\Phi_n(x)$	Cyclic group
1	$x - 1 = 0$	1	1	$x - 1$	C_1
2	$x^2 - 1 = 0$	± 1	-1	$x + 1$	C_2
3	$x^3 - 1 = 0$	$1, \omega, \omega^2$	ω, ω^2	$x^2 + x + 1$	C_3
4	$x^4 - 1 = 0$	$\pm 1, \pm i$	$\pm i$	$x^2 + 1$	C_4
6	$x^6 - 1 = 0$	$\pm 1, \frac{\pm 1 \pm i\sqrt{3}}{2}$	$\frac{1 \pm i\sqrt{3}}{2}$	$x^2 - x + 1$	C_6

Exercise 3.7

3.7.1 Proof that cyclic groups are abelian.
3.7.2 Write a program for polynomial long division.
3.7.3 Factor $x^{12} - 1$ to cyclotomic polynomials.
3.7.4 Solve the quintic cyclotomic equation in radicals. Hint: the equation $x^4 + x^3 + x^2 + x + 1 = 0$ changes to symmetric after being divided by x^2 as $x^2 + x + 1 + x^{-1} + x^{-2} = 0$.

Answer: page 330

3.2.7 Subgroup

A group may contain sub-structure of some smaller group. For example, the symmetry group of snowflakes D_6 can be decomposed into two: rotational symmetries of C_6 and reflexive symmetries of C_2; for another example, integers form a group under $+$; even numbers, a subset of integers, also form a group under $+$.

Definition 3.2.5 A subset H of group G is a *subgroup* if H forms a group under the binary operation of G.

Any group G has at least two subgroups: G itself and $\{e\}$, the singleton set of the identity. They are called trivial subgroups. Although even numbers form a subgroup under $+$, odd numbers do not(do you know why?). For another example, S_3 has a subgroup of permutations: $H = \{(1), (1\ 2)\}$; let's verify the group axioms:

1. Closure: $(1)(1) = (1)$, $(1)(1\ 2) = (1\ 2)$, $(1\ 2)(1) = (1\ 2)$, $(1\ 2)(1\ 2) = (1)$. The last one swaps twice and restores to identity.
2. Associativity applies to S_3, so as to H.
3. The identity $(1) \in H$.
4. The inverse of every element in H is also in H: $(1)(1) = (1)$, $(1\ 2)(1\ 2) = (1)$.

Here is another example; an *alternating group* A_n is a subgroup of the symmetry group S_n. For a set of n objects $x_1, x_2, ..., x_n$, this subgroup contains all the permutations that keep below product unchanged.

$$\Delta = \prod_{i<k}(x_i - x_k)$$

We call them even permutations because they flip the sign of Δ even times, hence unchanged; the other permutations (in S_n, but not in A_n) are odd. Because even · even and odd · odd are even; even · odd is odd; there are same numbers of even and odd permutations, each $n!/2$, hence $|A_n| = n!/2$. It's tedious to verify all axioms for big subset; there is alternative:

Proposition 3.2.8 *A nonempty subset H of group G forms a subgroup if and only if:*

(a) For any a, b in H, the product $ab \in H$.
(b) For every a in H, its inverse $a^{-1} \in H$.

We leave the proof to Exercise 3.8.1. It says H as a subgroup of G and shares the identity element of G; for any of its element a, the inverse a^{-1} is also in H. We combine (a) and (b) into one:

Proposition 3.2.9 *A nonempty subset H of group G forms a subgroup if and only if $ab^{-1} \in H$ for any a, b in H.*

Proof For any a in H, let $b = a$, then $e = aa^{-1} = ab^{-1} \in H$. Since $e, a \in H$, $ea^{-1} = a^{-1} \in H$, satisfying Proposition 3.2.8 (b). For any a, b in H, we just showed $b^{-1} \in H$; then $a(b^{-1})^{-1} = ab \in H$, satisfying Proposition 3.2.8 (a).

Conversely, let a, b in H, Proposition 3.2.8 (b), assert $b^{-1} \in H$, then Proposition 3.2.8 (a) asserts $ab^{-1} \in H$.   □

It simplifies subgroup verification (see Exercise 3.8.3).

Integers form a group $\mathbb{Z}$ under $+$; all multiples of n form a subgroup $H = \{kn\}$ where $k \in \mathbb{Z}$ are $0, \pm 1, \pm 2 \ldots$. We may view H is generated by multiplying n to every integer in $\mathbb{Z}$ as $H = \mathbb{Z}n$. Let's verify H is a subgroup: for any two elements kn and hn, inverse the latter to $-hn$; because $kn + (-hn) = (k - h)n \in H$, Proposition 3.2.9 asserts H is a subgroup. Actually, n partitions all integers into its residues. We treat two integers a and b equivalent if n divides $a - b$ as:

$$a \equiv b \pmod{n} \quad \text{if and only if } n|(a - b)$$

n divides $a - b$ implies $a - b \in H$, hence:

$$a \equiv b \pmod{n} \quad \text{if and only if } (a - b) \in H$$

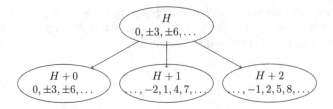

Fig. 3.14 Partition of $\mathbb{Z}$

Fig. 3.15 G is partitioned by
its subgroup H

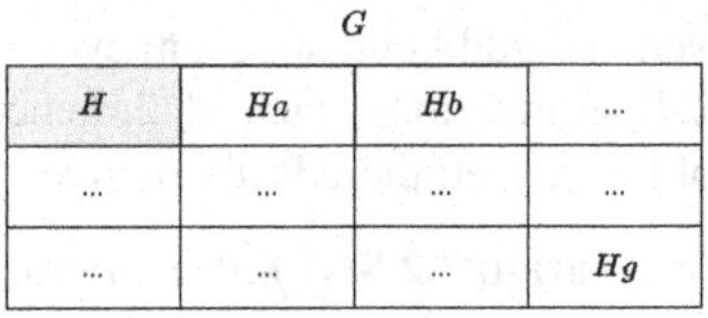

H	Ha	Hb	...
...	...	...	...
...	...	...	Hg

We can translate the statement "n partitions all integers into its residues" into: the subgroup H partitions $\mathbb{Z}$ into n residues of: $H, H+1, H+2, \ldots, H+n-1$, where $H+i = \{h+i \,|\, h \in H\}$. Figure 3.14 shows when $n = 3$. This is an instance of generic viewpoint: we can partition a group G with its subgroup H, as the subgroup defines equivalence relationship for elements as:

$$a \sim b \quad \text{if and only if } ab^{-1} \in H$$

We say a is equivalent to b, because $\sim$ satisfies all three equivalence properties:

1. **Reflexive**: $aa^{-1} = e \in H$ implies $a \sim a$; any element is equivalent to itself.
2. **Symmetric**: if $a \sim b$, then $ab^{-1} \in H$; its inverse $(ab^{-1})^{-1} = ba^{-1} \in H$ implies $b \sim a$.
3. **Transitive**: if $a \sim b, b \sim c$, then $ab^{-1} \in H, bc^{-1} \in H$; closure asserts H contains their product, i.e., $(ab^{-1})(bc^{-1}) = a(bb^{-1})c^{-1} = ac^{-1} \in H$, hence $a \sim c$.

The subgroup H partitions the group G into several subsets (Fig. 3.15). For an element a in G, let Ha be the set generated by right multiplying a to every element in H:

$$Ha = \{ha \,|\, h \in H\}$$

The set is named *right coset* because the multiplication is from right. As a subgroup, H contains e, hence $a = ea \in Ha$. If $b \sim a$ (equivalent), then $ba^{-1} = h \in H$ for some h; right multiply a to both sides: $b = ha \in Ha$. Therefore, the right coset Ha contains all elements equivalent to a; we can use a as a representative of Ha, denoted as $[a]$. For example, subgroup $H = \{3k\}$ partitions all integers into $\mathbb{Z} = [0] + [1] + [2]$.

For another example of finite non-abelian group, S_3 has six elements: $\{(1), (1\ 2), (1\ 3), (2\ 3), (1\ 2\ 3), (1\ 3\ 2)\}$. It has a subgroup $C_{2a} = \{(1), (1\ 2)\}$, which is isomorphic to the cyclic group C_2. We generate three right cosets by multiplying the identity (1), and two permutations of $(1\ 3), (2\ 3)$:

$$C_{2a}(1) = \{(1), (1\ 2)\}$$
$$C_{2a}(1\ 3) = \{(1\ 3), (1\ 2\ 3)\}$$
$$C_{2a}(2\ 3) = \{(2\ 3), (1\ 3\ 2)\}$$

We may use another three permutations to generate right cosets: $C_{2a}(1\ 2)$, $C_{2a}(1\ 2\ 3), C_{2a}(1\ 3\ 2)$; since $(1\ 2) \in C_{2a}(1), (1\ 2\ 3) \in C_{2a}(1\ 3), (1\ 3\ 2) \in C_{2a}(2\ 3)$, the cosets are the same:

$$C_{2a}(1) = C_{2a}(1\ 2)$$
$$C_{2a}(1\ 3) = C_{2a}(1\ 2\ 3)$$
$$C_{2a}(2\ 3) = C_{2a}(1\ 3\ 2)$$

Subgroup C_{2a} partitions S_3 into three disjoint right cosets: $S_3 = C_{2a}(1) \cup C_{2a}(1\ 3) \cup C_{2a}(2\ 3)$. As shown in Fig. 3.15, the term "partition" does not only mean separation but also implies the same size of every right coset: $|H| = |Ha| = |Hb| = \cdots = |Hg|$. This is because we can establish a bijective (one-to-one) map $f : h \rightarrow ha$, sending every subgroup element h to the right coset as ha; its inverse is right multiplication by a^{-1}.

Symmetric to right, there is left coset $aH = \{ah | h \in H\}$ and the corresponding equivalent relationship $\sim'$ is defined as below:

$$a \sim' b \quad \text{if and only if } b^{-1}a \in H$$

As the group "multiplication" may not be necessarily commutative (abelian or not), $\sim$ and $\sim'$ differ typically, so as are the left and right cosets. Nevertheless, there are the same number of left and right cosets (may both be infinitely many) because we can make a bijective (one-to-one) map $f : Ha \rightarrow a^{-1}H$ between them. As a subgroup evenly partitions the group into m cosets (left or right), we define m the *index* of H in G.

In above example, we partition S_3 by the subgroup $C_{2a} = \{(1), (1\ 2)\}$. S_3 has another subgroup $C_3 = \{(1), (1\ 2\ 3), (1\ 3\ 2)\}$, which is cyclic of order 3. It partitions S_3 into two right cosets:

$$C_3(1) = C_3$$
$$C_3(1\ 2) = \{(1\ 2), (1\ 3), (2\ 3)\}$$

Figures 3.16 and 3.17 show the two partitions. They have major difference: C_3 generates the same left and right cosets, while C_{2a} doesn't.

subgroup	C_3	C_{2a}
left cosets	$C_3, \{(1\ 2),(1\ 3),(2\ 3)\}$	$C_{2a}, \{(1\ 3),(1\ 3\ 2)\}, \{(2\ 3),(1\ 2\ 3)\}$
right cosets	$C_3, \{(1\ 2),(1\ 3),(2\ 3)\}$	$C_{2a}, \{(1\ 3),(1\ 2\ 3)\}, \{(2\ 3),(1\ 3\ 2)\}$

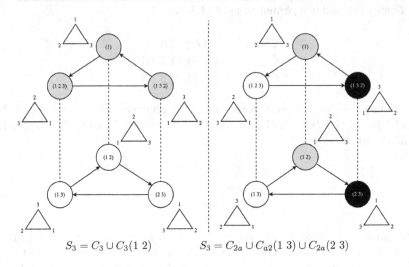

$$S_3 = C_3 \cup C_3(1\ 2) \qquad S_3 = C_{2a} \cup C_{a2}(1\ 3) \cup C_{2a}(2\ 3)$$

Fig. 3.16 Right cosets

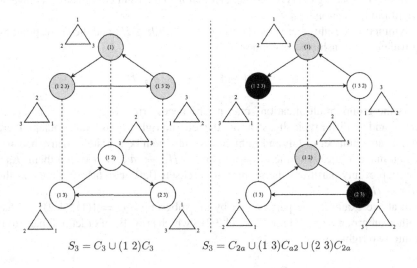

$$S_3 = C_3 \cup (1\ 2)C_3 \qquad S_3 = C_{2a} \cup (1\ 3)C_{a2} \cup (2\ 3)C_{2a}$$

Fig. 3.17 Left cosets

If a subgroup generates the same left and right cosets, it's called *normal* subgroup. It was Galois who first introduced normal subgroups to analyze if equations are solvable in radicals. Galois called them "invariant subgroups" because they give two invariant partitions of G:

$$G = Ha_1 + Ha_2 + \dots = a_1H + a_2H + \dots$$

Definition 3.2.6 A subgroup N is called normal subgroup (or invariant subgroup) of group G if for every element a in G, $Na = aN$ holds, denoted as: $N \lhd G$ or $G \rhd N$.

The left (or right) coset of a normal subgroup is called coset. As it's left/right symmetric, a normal subgroup is called the center of a group. Right multiply a^{-1} to both sides of $Na = aN$, it changes to $aNa^{-1} = N$. We can use it to test if a subgroup is normal.

Proposition 3.2.10 *A subgroup N of group G is normal if and only if $ana^{-1} \in N$ for every a in G, n in N.*

Define "multiplication" for the set of all cosets $\{aN, bN, cN, \dots\}$ as:

$$(aN)(bN) = (ab)N$$

It's easy to verify that cosets form a group under this multiplication. This group is called *quotient* group, denoted as G/N. There is a surjective map from G to the quotient group G/N, defined as $a \to aN$. Because for every a, b in G, their product $ab \to abN = (aN)(bN)$, it is a surjective homomorphism.

Definition 3.2.7 If f is a surjective homomorphism from group G to another group G', let the identity element of G' be e'; the subset N is called the *kernel* of the homomorphism if e' is the image of N. In other words, for every n in N, $f(n) = e'$.

There is an important theorem (we'll not prove it): the kernel N of the homomorphism from G to G' is a normal subgroup of G and the quotient group $G/N \cong G'$. For example, the cyclic group C_n is isomorphic to the quotient group $\mathbb{Z}/n\mathbb{Z}$. This is a powerful tool. For example, for any unfamiliar group G', if there is a homomorphism from some known group G to G', we can find the normal subgroup N of G, then G' is essentially equivalent (isomorphic) to the quotient group G/N. We can see the importance of the normal subgroup and the quotient group. Galois did use this idea to figure out whether the group of an equation is solvable.

Exercise 3.8

3.8.1 Prove Proposition 3.2.8, which tests if a subset is a subgroup.

3.8.2 Show that a nonempty subset H of a finite group G is a subgroup if and only
        if it is closed: i.e., for all $a, b \in H$, $ab \in H$.

3.8.3 Verify A_n is subgroup of S_n.

Answer: page 331

3.2.8 Lagrange's Theorem

Lagrange's theorem greatly demonstrates the power of abstract algebra. We can
gain insight into abstract structure without knowing any concrete meanings of group
elements or operations.

Joseph-Louise Lagrange (1736–1813)

Joseph-Louise Lagrange was a mathematician, physicist, and astronomer. He
was born in 1736 in Turin, Italy. Lagrange was of Italian and French descent. His
father was in charge of the king's military chest and was Treasurer of the Office
of Public Works and Fortifications in Turin. A career as a lawyer was planned
out for Lagrange by his father, but before Lagrange grew up his father had lost
most of his property in speculations. Lagrange later claimed "If I had been rich, I
probably would not have devoted myself to mathematics." Lagrange studied at the
University of Turin. At first he had no great enthusiasm for mathematics, finding
Greek geometry rather dull.

It was not until he was 17 that he showed any taste for mathematics—his interest
in mathematics being first excited by a paper by Edmond Halley which he came
across by accident. That paper introduced about the new calculus invented by
Newton. Alone and unaided he threw himself into mathematical studies.

Starting from 1754, he worked on the problem of tautochrone, discovering a
method of min-max functionals. Lagrange wrote several letters to Euler between

1754 and 1756 describing his results. His work made him one of the founders of the calculus of variations. Euler was very impressed with Lagrange's results. He tried to persuade Lagrange to come to Berlin, but Lagrange had no such intention and shyly refused the offer.

In 1766, king Frederick of Prussia wrote to Lagrange expressing the wish of "the greatest king in Europe" to have "the greatest mathematician in Europe" resident at his court. Lagrange was finally persuaded and he spent the next 20 years in Prussia, where he produced not only the long series of papers published in the Berlin and Turin transactions but also his monumental work, the *Mécanique analytique*.

Lagrange was a favorite of the king, who frequently used to discourse with him on the advantages of perfect regularity of life. The lesson went home, and thenceforth Lagrange studied his mind and body as though they were machines, and found by experiment the exact amount of work which he was able to do without breaking down. Every night he set himself a definite task for the next day, and on completing any branch of a subject he wrote a short analysis to see what points in the demonstrations or in the subject matter were capable of improvement. He always thought out the subject of his papers before he began to compose them, and usually wrote them straight off without a single erasure or correction.

Nonetheless, during his years in Berlin, Lagrange's health was rather poor on many occasions, and that of his wife Vittoria was even worse. She died in 1783 after years of illness and Lagrange was very depressed. In 1786, following King Frederick's death, Lagrange accepted the offer of Louis XVI to move to Paris. In France he was received with every mark of distinction and special apartments in the Louvre were prepared for his reception, and he became a member of the French Academy of Sciences. Lagrange married for a second time in 1792, his wife being Renée-Françoise-Adélaide Le Monnier, the daughter of one of his astronomer colleagues.

In September 1793, the French revolution broke out. Under the intervention of Antoine Lavoisier (known as the father of modern chemistry, who recognized and named oxygen and hydrogen), who himself was by then already thrown out of the Academy along with many other scholars, Lagrange was specifically exempted by name in the decree of October 1793 that ordered all foreigners to leave France. On May 4, 1794, Lavoisier, who had saved Lagrange from arrest, and 27 other tax farmers were arrested and sentenced to death and guillotined on the afternoon after the trial. Lagrange said on the death of Lavoisier on May 8 "It took only a moment to cause this head to fall and a hundred years will not suffice to produce its like." [25]

After Napoleon attained the power of France in 1799, he warmly encouraged science and mathematics studies and was a liberal benefactor of them. He loaded Lagrange with honors and distinctions. Lagrange died at the age of 77 in 1813. The funeral operation was pronounced by Laplace, represented the House of Lords, and Dean Lacépède represented the Academy of Sciences. The commemorative events were held in Italian universities.

Lagrange was the great mathematician and scientist in the eighteenth to nineteenth century. He made significant contributions to the fields of analysis, number theory, and both classical and celestial mechanics. Napoleon said "Lagrange is the lofty pyramid of the mathematical science". Lagrange made important progress in solving algebraic equations of any degree in the first decades in Berlin. He introduced a concept called Lagrange's resolvents. The significance of this method is that it exhibits the already known formulas for solving quadratic, cubic, and quartic equations as manifestations of a single principle. He failed to give a general formula for solutions to quintic, because the auxiliary equation involved had higher degree than the original one. Nevertheless, Lagrange's idea already implied the concept of permutation group. The permutations keep the resolvent invariant form a subgroup, and the order of the subgroup is a factor of the original permutation group. This is exactly the famous Lagrange's theorem in group theory. Lagrange is a pioneer of group theory. His thoughts were adopted and developed later by Galois, and it was foundational in Galois theory.

Theorem 3.2.11 (Lagrange's Theorem) *For any finite group G, the order of every subgroup H divides the order of G.*

Proof H fully partitions G into cosets. From the equivalence relation $\sim$ (or $\sim'$), there is no overlap. If c belongs to both Ha and Hb, then $c \sim a$ and $c \sim b$, hence $a \sim b$ and $Ha = Hb$. For the partition nature, every element in the group must be in some coset. All cosets have the same size of $n = |H|$ since there is bijective map $f : h \mapsto ha$ from H to any Ha (or ah for left coset). Let the number of cosets be m (the index of H), then:

$$|G| = mn$$

□

Note the converse of Lagrange's theorem is not true in general. For any divisor d of $|G|$, the subgroup of order d may not exist. For a counterexample: $|A_4| = 12$, but A_4 does not have a subgroup of order 6.

Corollary 3.2.12 *If G is a finite group, the order of every element a divides $|G|$.*

Every element of a finite group has finite order. a generates a (cyclic) subgroup of order n; applying Lagrange's theorem, n divides $|G|$.   □

Corollary 3.2.13 *If G has prime order, then G is cyclic.*

Let $p = |G|$ be a prime. Every element $a \neq e$ generates a subgroup of finite order n. Lagrange's theorem says n divides prime p; because $n \neq 1$ (a is not the identity), n must be p, the same order as G. Therefore, a is the generator of G, i.e., $G = (a)$.   □

Corollary 3.2.14 *For every element a of a finite group G, $a^{|G|} = e$.*

Proof This is because the order n of a divides G, let $|G| = nk$, then:

$$a^{|G|} = a^{nk} = (a^n)^k = e^k = e$$

□

Lagrange's theorem in group theory can be applied to prove Fermat's little theorem in number theory. The theorem is named after Fermat, who found it in 1636. In a letter he wrote to a friend[11] in 1640, Fermat stated this theorem first. It's called the "little theorem" to distinguish from Fermat's last theorem. Euler provided the first published proof in 1736.[12]

Theorem 3.2.15 (Fermat's Little Theorem) *A prime number p divides $a^{p-1} - 1$ for any integer a that $0 < a < p$.*[13]

Proof The multiplicative group of integers modulo p contains elements of 1, 2, ..., $p - 1$; hence, the order of the group is $p - 1$. Corollary 3.2.14 of Lagrange's theorem gives:

$$a^{p-1} = e$$

Since 1 is the identity element modulo p:

$$a^{p-1} \equiv 1 \ (\text{mod } p)$$

shows p divides $a^{p-1} - 1$.

□

Other methods are much more complex (e.g., Sections 5.2–5.4 in [9]). Below is a combinatorial method called proof by counting necklace [26].

Proof We are making strings of length p from pearls in a different colors. There are a total of a^p different strings since we pick p pearls each from a colors, for example, pearls in two colors: A red, B green. To make a string of 5 pearls, there are a total of $2^5 = 32$ different strings:

AAAAA, AAAAB, AAABA, ..., BBBBA, BBBBB

We are going to show that, among these a^p strings, if we remove a same colored strings (in the above example, two strings of AAAAA and BBBBB), the remaining $a^p - a$ strings can be partitioned into several sets, and each has exactly p strings; thus, p divides $a^p - a$.

---

[11] Friend and confidant, French mathematician Frénicle de Bessy.

[12] Leibniz had given virtually the same proof in an unpublished manuscript from some time before 1683.

[13] Alternative saying: p divides $a^p - a$, or $a^p \equiv a \ (\text{mod } p)$.

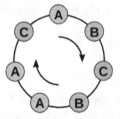
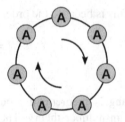

(a) A necklace in 3 colors represents 7 strings: (b) The necklace in same color only
ABCBAAC, BCBAACA, CBAACAB, BAACABC,   represent one string: AAAAAAA
AACABCB, ACABCBA, CABCBAA

Fig. 3.18 Partition strings as necklaces

We link the head and tail of a string to make it a necklace. Some different strings
become the same necklace. For example, the following five strings form the same
necklace (by rotation) (Fig. 3.18):

AAAAB, AAABA, AABAA, ABAAA, BAAAA

The 32 strings are partitioned into 2 types: (a) 5 necklaces in different colors and
(b) 2 necklaces in the same color:

[AAABB, AABBA, ABBAA, BBAAA, BAAAB]
[AABAB, ABABA, BABAA, ABAAB, BAABA]
[AABBB, ABBBA, BBBAA, BBAAB, BAABB]
[ABABB, BABBA, ABBAB, BBABA, BABAB]
[ABBBB, BBBBA, BBBAB, BBABB, BABBB]
[AAAAA]
[BBBBB]

How many strings does a necklace represent? If a string S can be split into
several same sub-strings of T, while T can't be split into sub-strings anymore,
then S represents $|T|$ different strings, where $|T|$ is the length of the sub-string.
For example, string S =ABBABBABBABB can be split into sub-string T =ABB, while
ABB can't be split anymore. If we rotate a pearl per time, there are a total of three
different strings:

ABBABBABBABB, BBABBABBABBA, BABBABBABBAB

There are not any other different strings besides these three. Since the length
of ABB is three, further rotation repeats. Basically, there are two types of the a^p
strings: (1) a strings in the same color and (2) strings in different colors. But as
the length p is prime, it cannot be generated by duplicating sub-strings. Therefore,
every necklace in different colors represents p different strings. There are a total of
$a^p - a$ strings in different colors. They can be partitioned into sets of necklaces, each
containing p strings of the same necklace. Hence, p divides $a^p - a = a(a^{p-1} - 1)$.
Since p doesn't divide a, p divides $a^{p-1} - 1$. $\square$

The proof by counting necklaces needs little prerequisite knowledge, while the
proof by Lagrange's theorem is direct. It took 100 years before Euler's proof. It is

not rare; Fermat's last theorem stimulated many talent mathematicians and took 358 years till Andrew Wiles proved it in 1995. The main tools Wiles used include elliptic curves, modularity theory, and Galois representations [11]. These conjectures left by Fermat are a rich treasure in mathematics.

Pierre de Fermat (1601–1665)

Pierre de Fermat was a French mathematician, born about August 1601. His father was a wealthy leather merchant. He became a lawyer at the Parlement of Toulouse French when he grew up. In 1630, he bought the office of a councillor at the Parlement of Toulouse, one of the High Courts of Judicature in France, and was sworn in by the Grand Chambre in May 1631. He held this office for the rest of his life. Fermat thereby became entitled to change his name from Pierre Fermat to Pierre de Fermat. He was fluent in six languages. Fermat studied mathematics in his spare time. But the mathematical achievements made by Fermat were the peak of his time. Fermat's pioneering work in analytic geometry was circulated in a manuscript form in 1636.[14] Together with René Descartes, he was one of the two leading mathematicians of the first half of the seventeenth century developing analytic geometry. Fermat and Blaise Pascal laid the foundation for the theory of probability through their correspondence in 1654. They are now regarded as joint founders of probability theory. In physics, Fermat generalized the principle of least time as "light travels between two given points along the path of shortest time." The term Fermat's principle was named in recognition of him. Fermat also contributed to the early development of calculus.

But Fermat's crowning achievement was in the theory of numbers. Fermat was inspired by ancient Greek mathematician, Diophantus's work *Arithmetica*. Claude Bachet translated it to Latin and published in France in 1621. Fermat bought this book in Paris and was deeply attracted by the puzzles of numbers. There was a wide

---

[14] The 8-page paper, *Introduction to Plane and Solid Loci*, was completed in 1630, but it was published posthumously in 1679, 14 years after Fermat's death.

margin left in the pages of this edition; it turned out to be Fermat's "note book". When he studied Diophantus's problems and answers, Fermat often got inspired by wider and deeper problems, then wrote his thoughts and comments in the margin.

Fermat published nearly nothing in his lifetime, although it is unbelievable to the people nowadays. It was common in Fermat's time that sometimes he wrote mails to his scholar friends about his findings. Some of the most striking of his results were found after his death on loose sheets of paper or written in the margins of works which he had read and annotated, and are unaccompanied by any proof. He was constitutionally modest and retiring and didn't intend to publish his papers. Fermat was totally driven by strong curiosity to explore the mathematical mysteries. It's pure because he treated mathematics study as a hobby. When he found the result never been touched, Fermat was truly exciting and self-satisfied. It was not significant to him to publish the result and get recognition [11]. Interestingly, this silent genius sometimes liked to tease people. He often challenged other mathematicians in mails by asking them to prove his discoveries.

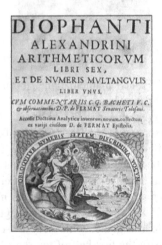

The 1670 edition with Fermat's annotation

When Fermat died in 1665, his research results were scattered here and there. His son Clément-Samuel Fermat spent 5 years to collect Fermat's mails and notes, then produced a special edition of *Arithmetica* containing his father's achievement. On the cover page, it printed "augmented with Fermat's commentary". This edition includes 48 comments by Fermat. In 1679, Samuel collected and published the second volume of Fermat's work. His research results were finally circulated, which greatly enriched the mathematics in the seventeenth century and impacted the development of mathematics later.

Before Fermat, the theory of numbers was basically a collection of problems. It was Fermat who first systematically studied the theory of numbers. He proposed a large number of theorems and introduced generalized methods and principles, thus bringing the theory of numbers to the modern development. It was Fermat's work that the theory of numbers really became a branch of mathematics. Before Gauss's *Disquisitiones Arithmeticae*, the development of number theory was originally driven by Fermat. He was called the "father of modern number theory".

However, for many propositions conjectured, Fermat only provided some key part or even without any proof. Some of them were found wrong;[15] many were later proved by Euler.

Euler's theorem is more generic than Fermat's little theorem. Euler was not satisfied to merely prove Fermat's little theorem; what if p is not a prime? He extended the theorem to cover composite numbers.

Theorem 3.2.16 (Euler's Theorem) *If $0 < a < n$; a and n are coprime, then n divides $a^{\phi(n)} - 1$,*

where $\phi(n)$ is Euler's totient function, defined as the number of positive integers that are less than n and coprime to n.

$$\phi(n) = |\{i \mid 0 < i < n, \gcd(i, n) = 1\}|$$

As the application of Lagrange's theorem, there is an elegant proof.

Proof Among the nonzero residues modulo n, pick those which are mutually inverses (any $0 < a, b < n$, satisfying $ab \equiv 1 \pmod{n}$). They form a multiplicative group modulo n. Moreover, they are all coprime to n, because $ab \equiv 1 \pmod{n}$ implies n divides $ab - 1$; as $\gcd(n, a)$ divides both $ab - 1$ and ab, $\gcd(n, a) = 1$. From the definition of Euler's totient function, the order of this group is $\phi(n)$. Apply Corollary 3.2.14 of Lagrange's theorem:

$$a^{\phi(n)} = e$$

Therefore, $a^{\phi(n)} \equiv 1 \pmod{n}$ implies n divides $a^{\phi(n)} - 1$. □

For example, below is the multiplication table for all the nonzero residues modulo 10. Locate identity 1 and mark the cell underline. From that cell, along with the row and column, find two numbers and mark both bold. Their modulo product is 1 and both are coprime to 10. For any residue that is not coprime to 10, the row and column where it is in also contain 0. But 0 is not a group element. All residues coprime to 10 are exactly the group elements of $1, 3, 7, 9$.

---

[15] For example, Fermat number, named after Fermat who first studied them in 1640, is a positive number of the form $2^{2^n} + 1$. Fermat claimed all such numbers are primes. It is true when n is $0, 1, 2, 3, 4$; the corresponding numbers are $3, 5, 17, 257, 65537$. However, Euler calculated in 1732 that $2^{2^5} + 1 = 641 \times 6700417$ is not a prime number. As of 2017, people have found 243 counterexamples without finding the 6th Fermat prime number. It is still an unsolved conjecture whether there are any other Fermat primes.

	1	2	3	4	5	6	7	8	9
1	1	2	3	4	5	6	7	8	9
2	2	4	6	8	0	2	4	6	8
3	3	6	9	2	5	8	1	4	7
4	4	8	2	6	0	4	8	2	6
5	5	0	5	0	5	0	5	0	5
6	6	2	8	4	0	6	2	8	4
7	7	4	1	8	5	2	9	6	3
8	8	6	4	2	0	8	6	4	2
9	9	8	7	6	5	4	3	2	1

How to evaluate Euler's $\phi(n)$ function except exhaustive counting? As n can be factored as powers of prime number, we start from evaluating $\phi(p^m)$ for prime p. To count how many numbers from 1 to $p^m - 1$ are coprime to p^m, we remove all multiples of p; they are $p, 2p, 3p, ..., p^m - p$. Divide each by p changes to $1, 2, 3, ..., p^{m-1} - 1$, hence total $p^{m-1} - 1$ numbers being removed. Count the remaining for Euler's totient function:

$$\phi(p^m) = (p^m - 1) - (p^{m-1} - 1)$$

$$= p^m - p^{m-1}$$

$$= p^m(1 - \frac{1}{p})$$

For $n = p^u q^v$, where p and q are distinct primes, among $1, 2, ..., n - 1$, there are $n/p - 1$ multiples of p, $n/q - 1$ multiples of q, and $n/(pq) - 1$ multiples of pq. We remove all multiples of p and q and add multiples of pq back to avoid removing twice (principle of inclusion and exclusion).

$$\phi(p^u q^v) = (n - 1) - (\frac{n}{p} - 1) - (\frac{n}{q} - 1) + (\frac{n}{pq} - 1)$$

$$= n(1 - \frac{1}{p})(1 - \frac{1}{q})$$

$$= p^u(1 - \frac{1}{p})q^v(1 - \frac{1}{q}) \qquad \text{substitute } n = p^u q^v$$

$$= \phi(p^u)\phi(q^v)$$

Particularly when both u and v are 1, $\phi(pq) = \phi(p)\phi(q)$. Extend to multiple powers of prime numbers $n = p_1^{k_1}, p_2^{k_2}...p_m^{k_m}$:

$$\phi(n) = n(1 - \frac{1}{p_1})(1 - \frac{1}{p_2})...(1 - \frac{1}{p_m})$$
$$= \phi(p_1^{k_1})\phi(p_2^{k_2})...\phi(p_m^{k_m})$$

We can develop an algorithm to compute Euler's totient function based on this (see Exercise 3.9.5).

Leonhard Euler was a great mathematician and scientist. He is held to be one of the greatest mathematicians in history. Euler was born in 1707 in Basel, Switzerland. As a pastor, his father wanted him to study theology and sent him to the University of Basel at the age of 14. During that time, he was receiving private lessons every Saturday afternoon from Johann Bernoulli, the foremost mathematician in Europe. He was also a good friend of Leonhard's father. Bernoulli quickly discovered Euler's incredible talent in mathematics and convinced his father that Leonhard was destined to become a great mathematician.

Leonhard Euler (1707–1783)

In 1727, Euler became a member of Imperial Russian Academy of Sciences in St. Petersburg. He devoted himself to research work and later became the head of mathematics department after his friend, Daniel Bernoulli, left for Basel. During the 14 years in Russia, Euler studied analytics, number theory, and mechanics. In 1747, Euler took up a position at the Berlin Academy, which he had been offered by King Frederick the Great of Prussia. He lived for 25 years in Berlin, where he wrote over 380 papers. During this time, his research was more extensive, involving planetary motion, rigid body movement, thermodynamics, ballistics, and demography. Euler made groundbreakings in differential equations and surface differential geometry. After the political situation in Russia stabilized after Catherine the Great's accession to the throne, in 1766 Euler accepted an invitation to return to the St. Petersburg Academy. He spent his rest life in Russia.

Euler worked in almost all areas of mathematics, such as geometry, calculus, trigonometry, algebra, and number theory, as well as continuum physics, lunar theory, and other areas of physics. He is a seminal figure in the history of mathematics. He published 856 papers and 31 books[16] in half a century from age of 19 to 76 and listed as the top productive mathematician.[17] Euler's name is associated with almost every area in mathematics [28].

Euler had an unbelievable willpower. His right eye began to have problem from a fever when he was 31. He lost his right eye in 1740. But even worse, his left eye lost sight too in 1771. Just as deafness did not stop Beethoven in music, blindness did not stop Euler in mathematics [11]. Euler remarked on his loss of vision "Now I will have fewer distractions." As he compensated for it with his mental calculation skills and exceptional memory. He could remember the first 100 prime numbers, their squares, cubes, and even higher powers and perform complex mental arithmetic. With the aid of his scribes, Euler's productivity on many areas actually increased. He published, on average, a paper every week in 1775. Half of Euler's work was dictated after his eyes were completely blind. François Arago said "Euler calculated without any apparent effort, just as men breathe, as eagles sustain themselves in the air." Euler could work in any bad environments. He often held the child on the lap to write papers, regardless of any noise around.

Among Euler's work, there are difficult monographs as well as readings for the general public. Euler wrote over 200 letters to a German Princess in the early 1760s, which were later compiled into a best-selling volume entitled *Letters of Euler on different Subjects in Natural Philosophy Addressed to a German Princess*. This book was widely read and published across Europe and United States. Euler also wrote a course on elementary algebra for readers of non-mathematics background, which is still in print today. Many popular notations were carefully introduced by Euler, for example, π (1736); the imaginary unit i (1777); the base of the natural logarithm e, also known as Euler's number (1748); function sin and cos (1748); tg (1753); Δx (1755); summation $\sum$ (1755); function $f(x)$ (1734); and so on [11].

On September 18, 1783, after lunch with his family, Euler was discussing the newly discovered planet Uranus and its orbit with a fellow academician. As usual, he played with one of his granddaughters while having tea. Suddenly, the pipe drop from his hand. He said "My pipe", then bent over to pick it, but was not able to stand. "He ceased to calculate and to live."[18]

Fermat little theorem is widely used in our everyday life, from Internet shopping to electronic payment. In 1977, Ron Rivest, Adi Shamir, and Leonard Adleman in Massachusetts Institute of Technology developed a one-way function that was hard to invert based on the theory of numbers. The algorithm is now known as RSA—the initials of their surnames.

---

[16] These did not count the loss of the fire in St. Petersburg in 1771.

[17] This record was refreshed by Paul Erdős with 1525 papers and 32 books in the twentieth century.

[18] In the eulogy for the French Academy, Marquis de Condorcet wrote this.

RSA is based on the idea that one can easily create a composite number from two large primes, while there is practical difficulty to factor it. This is known as the *factoring problem*. For a large number over 200 digits, even the most powerful super computer would spend time longer than the age of universe. In order to generate an encrypt key hard to break, we need a method to find large prime numbers fast. However, people do not know the exact pattern of prime number distribution; there is no formula to enumerate prime numbers. The brute force method is to pick a number n randomly; check any one from 1 to $\sqrt{n}$ divides n. But this primality test is very inefficient; the time would exceed the age of universe too. A better way is Eratosthenes's sieve algorithm: enumerate all numbers from 2 to n, starting from 2, the first prime, and remove all multiples of 2; the next number, 3, is prime, then remove all multiples of 3. Repeat the removals from the next number that hasn't been filtered out, to $\sqrt{n}$; the remaining are prime numbers to n. However, it is still only applicable for small n, but can't serve for large primality test.

Fermat's little theorem provides a way for fast primality test. For n, randomly select a positive integer $a < n$ as a "witness". Then check whether $a^{n-1} \equiv 1$ (mod n). If not, n must not be prime by Fermat's little theorem; otherwise, n *may* be prime. *Fermat primality test* algorithm (or Fermat test for short) is based on this idea:

1: **function** PRIMALITY(n)
2:     Random select a positive integer $a < n$
3:     **if** $a^{n-1} \equiv 1$ (mod n) **then**
4:         **return** prime
5:     **else**
6:         **return** composite

We needn't compute exactly $(n - 1)$-th power of a, then divide n to get the remainder; instead, use modular arithmetic to speed up. The intermediate result can be largely reused. After calculating $b = a^2$ mod n, we next get b^2 mod n, which equals to a^4 mod n. For example, evaluate a^{11} mod n, since:

$$a^{11} = a^{8+2+1} = ((a^2)^2)^2 a^2 a \pmod{n}$$

We only need compute a^2 mod n, $(a^2)^2$ mod n, and $((a^2)^2)^2$ mod n. Basically, express n in binary format and only iterate calculating the modulo product for the digits of 1. The complexity of this algorithm is $O(\lg n)$ (proportion to the logarithm of n). Fermat test is very fast because of this.

However, a number is not necessarily prime even if it passes Fermat test. For example $341 = 11 \times 31$ is composite, but $2^{340} \equiv 1$ (mod 341) passes the test. To reduce the probability of such failure, people developed several improvements, such as using more witnesses. Below theorem shows that, if Fermat test fails for some witness, it fails too for at least half of n witnesses ([30], pp26).

Theorem 3.2.17 *If $a^{n-1} \not\equiv 1$ (mod n) for some positive integer a less than n and coprime to n, then it is also the case for at least half selected $a < n$.*

Fig. 3.19 Map from the set
that passes Fermat test to the
set that does not

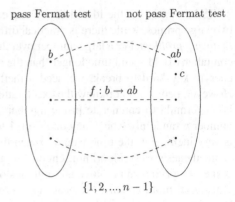

$\{1, 2, ..., n - 1\}$

Proof If $a^{n-1} \not\equiv 1 \pmod{n}$ for some a, then for any witness b that passes Fermat test (i.e. $b^{n-1} \equiv 1 \pmod{n}$), there is a negative case of ab.

$$(ab)^{n-1} \equiv a^{n-1}b^{n-1} \equiv a^{n-1}1 \equiv a^{n-1} \not\equiv 1 \pmod{n}$$

And if $i \neq j$ then $a \cdot i \neq a \cdot j$; hence, all these negative cases are distinct. As shown in Fig. 3.19, if it can't pass Fermat test for some witness, then there are as many negative cases as positive cases. □

If we select k distinct witnesses for Fermat test, we can reduce the probability of false positive to $\frac{1}{2^k}$. However, there exists special composite number n, such that $a^{n-1} \equiv 1 \pmod{n}$ for any a less than n and coprime to n, passing Fermat test for whatever witness! Carmichael found first such number in 1910 as $561 = 3 \times 11 \times 17$. They are called Carmichael numbers or Fermat pseudoprimes.[19] Erdős conjectured there were infinitely many Carmichael numbers. In 1994, people proved that for sufficient large n, there were at least $n^{2/7}$ Carmichael numbers from 1 to n, thus explaining Erdős' conjecture [29].

The primality test in RSA is Miller-Rabin algorithm. It is also a probabilistic algorithm.[20] By Theorem 3.2.17, with over 100 distinct witnesses, the probability of failure is expected less than $\frac{1}{2^{100}}$. Donald Knuth commented "far less, for instance, than the probability that a random cosmic ray will sabotage the computer during the computation!"

Diagram in Fig. 3.20 highlights monoid, semigroup, group, and their relations.

---

[19] Czech mathematician Václav Šimerka found the first 7 Fermat pseudoprimes: $561 = 3 \times 7 \times 11$, $1105 = 5 \times 13 \times 7$, $1729 = 7 \times 13 \times 19$, $2465 = 5 \times 17 \times 29$, $2821 = 7 \times 13 \times 31$, $6601 = 7 \times 23 \times 41$, $8911 = 7 \times 19 \times 67$. However, his work was not well known.

[20] There is a deterministic version of Miller-Rabin algorithm, but the correctness depends on the generalized Riemann hypothesis (GRH) [31].

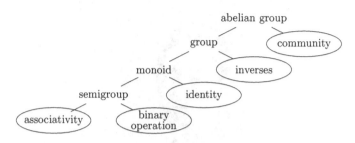

Fig. 3.20 Monoid, semigroup, and group

Exercise 3.9

3.9.1 Today is Sunday; what day will it be after 2^{100} days?

3.9.2 Given two strings (or list), write a program to test if they form the same necklace.

3.9.3 Weil thought Fermat might use the binomial theorem to derive the little theorem [27]. Prove the Fermat's little theorem in this way.

3.9.4 Write a program to realize Eratosthenes sieve.

3.9.5 Extend the idea of Eratosthenes sieve algorithm; write a program to generate Euler's ϕ function values for all numbers from 2 to 100.

3.9.6 From Euler's theorem, how many n-th primitive roots of unity are there?

3.9.7 Write a program to realize fast modulo multiplication and Fermat's primality test.

Answer: page 332

3.3   Ring and Field

The modern theory of abstract algebra, particularly about rings and ideals, was greatly developed by Emmy Noether. Noether was born to a Jewish family in Erlangen Germany on March 23, 1882. Her father, Max Noether, was a mathematician in the University of Erlangen. She originally planned to teach French and English after passing the required examinations in 1900, but she chose instead to continue her studies at the University of Erlangen.

Emmy Noether (1882–1935)

This was an unconventional decision. The Academic Senate of the university declared that allowing mixed-sex education would "overthrow all academic order". As one of two women out of 986 students, Noether was only allowed to audit classes rather than participate, and required the permission from individual professors. Despite these obstacles, she passed the graduation exam in 1903. After that, Noether studied at the University of Göttingen, attending lectures given by Hermann Minkowski, Felix Klein, and David Hilbert. She returned to Erlangen in October 1904. Under the supervision of Paul Gordan, she received a Ph.D. degree from Erlangen in 1907 with a dissertation on algebraic invariants. For the next 7 years (1908–1915) she taught at the university without pay only because she was a woman.

Hilbert and Klein invited Noether to Göttingen in 1915. She soon used her knowledge of invariants to explore the mathematics behind Albert Einstein's recently published theory of general relativity. The effort to recruit her, however, was blocked by some faculty. They insisted women should not be teaching at the university. Hilbert responded with indignation "I do not see that the sex of the candidate is an argument. After all, we are a university, not a bath house."

During her first years at Göttingen, she did not have an official position and was not paid; her family paid for her room and board. Her lectures were often advertised under Hilbert's name. Despite this, she soon proved the theorem now known as Noether's theorem, which shows that a conservation law is associated with any differentiable symmetry of a physical system. Today, physicists are still formulating new theories that rely on Noether's work. She won formal admission as an academic lecturer in 1919, but still unpaid.

Noether's work in algebra began in 1920. She published the important paper *Theory of Ideals in Ring Domains* in 1921. This revolutionary work gave rise to the term "Noetherian ring" and the naming of several other mathematical objects as Noetherian. In 1931, Noether's student, Van der Waerden, published *Moderne Algebra*, a central text in the field developed by Noether. Although Noether did not seek recognition, Van der Waerden included a note "based in part on lectures by E. Artin and E. Noether". This classic book influenced a generation of young mathematicians in that time. In November 1932, Noether delivered an hour speech at the 9th International Congress of Mathematicians (ICM) in Zürich.

However, the reputation and recognition did not improve her difficult situation as a woman. Her colleagues were frustrated for the fact that she was never promoted to the position of full professor and was unpaid for her work.

When the Nazi came to power in Germany in 1933, Noether and many other Jewish professors at Göttingen were dismissed. Noether provided support to others during this difficult time. Hermann Weyl later wrote "Emmy Noether—her courage, her frankness, her unconcern about her own fate, her conciliatory spirit—was in the midst of all the hatred and meanness, despair and sorrow surrounding us, a moral solace."

As a woman, it was very hard to get a position outside of Germany. She left for the United States to become a visiting professor of mathematics at Bryn Mawr College, a girl's university. Although she was invited to Princeton University to give lectures, she remarked that she was not welcome at "the men's university, where nothing female is admitted."

In April 1935, doctors discovered a tumor in Noether's pelvis. She died after the surgery on April 14 at the age of 53. Noether is best remembered for her contributions to abstract algebra. Einstein wrote shortly after her death that "Noether was the most significant creative mathematical genius thus far produced since the higher education of women began." [32].

3.3.1 Ring

Groups are sets with an operation; rings are enriched with two operations: addition ($+$) and multiplication ($\times$). Both are closed and satisfy below rules:

1. Rules for $+$:

    (a) Associative: $a + (b + c) = (a + b) + c$.
    (b) Commutative: $a + b = b + a$.
    (c) There are unique identity element and inverse of every element.

2. Rules for $\times$:

    (a) Associative: $a(bc) = (ab)c$

3. Distributive:

    (a) $a(b + c) = ab + ac$
    (b) $(b + c)a = ba + ca$

A ring R essentially consists of an additive abelian group and a multiplicative semigroup. For example, integers form a ring $\mathbb{Z}$ under normal addition and multiplication; so as polynomials and matrices form ring. The definition only requires + is commutative (a ring is an abelian group under +), but × needn't be commutative. When the multiplication is also commutative, it's called *commutative ring*, and the following equation holds for any positive integer n and two elements:

$$a^n b^n = (ab)^n$$

The definition does not require an identity element for × too, but if there exists some e satisfying $ea = ae = a$ for all a in R, we call it a unit ring or "ring with identity."[21] Traditionally, we use 1 to represent the identity for × and use 0 as the identity for +. They are just symbols, but not the actual number 1 or 0. With 1, we can also define multiplicative inverse: if there are $ab = ba = 1$, we say b is the multiplicative inverse of a, denoted as $b = a^{-1}$. We use negate notation $-a$ for the additive inverse. 1 is unique if exists, so as the multiplicative inverse is unique. However, it is not necessary that all elements are invertible. For example, $\mathbb{Z}$, no integer except ± 1, is invertible. We call any invertible element a *unit*.

From the distributive axiom:

$$(a - a)a = a(a - a) = aa - aa = 0$$

in short:

$$0a = a0 = 0$$

For any two elements a, b in the ring, their product $ab = 0$ if either one is 0. But the converse is not true. We cannot deduce from $ab = 0$ to $a = 0$ or $b = 0$. For a counterexample, the residue class modulo n, where + is modulo addition: $[a] + [b] = [a + b]$, which takes modulo n on normal sum, it forms an abelian additive group. × is defined as: $[a][b] = [ab]$, which takes modulo n on product. It's easy to verify this is a ring, called residue ring modulo n. If n is not prime, for example, 10, the product of two nonzero elements $[5][2] = [5 \times 2] = [0]$. In fact, the modulo product of any two factors of n must be zero.

In a commutative ring, if there exists $a \neq 0, b \neq 0$, but $ab = 0$, we call a and b as *zero divisors* of the ring. Of course, it's possible there is no zero divisor, for example, the integer ring $\mathbb{Z}$. We can deduce from $ab = 0$ that $a = 0$ or $b = 0$ only if there is no zero divisor in the ring.

Proposition 3.3.1 *The multiplicative cancellation law holds in a ring if and only if there is no zero divisor.*

(a) *If $a \neq 0$, and $ab = ac$, then $b = c$.*
(b) *If $a \neq 0$, and $ba = ca$, then $b = c$.*

---

[21] More people define a ring to have a multiplicative identity nowadays and use symbol *rng* (without 'i') for a ring without multiplicative identity.

Conversely, if one cancellation rule holds in a ring, then the other cancellation rule also holds, and there is no zero divisor in this ring.

We've seen three additional conditions: (1) $\times$ is commutative, (2) 1 exists, and (3) no zero divisor. A ring that satisfies these three is called *integral domain* (nonzero commutative ring with identity). For example, integer ring $\mathbb{Z}$ is an integral domain.

Analogous to semigroup, by removing the additive inverse, we obtain *semiring*.

Definition 3.3.1 A set R under the addition and multiplication forms a semiring if it satisfies the following axioms:

1. R forms a commutative monoid under addition with the identity element 0.
2. R forms a monoid under multiplication with the identity element 1.
3. Distributive:

$$a(b + c) = ab + ac$$

$$(b + c)a = ba + ca$$

4. Any element multiplies 0 gives 0, $a0 = 0a = 0$.

Natural number $\mathbb{N}$ is an example of semiring. The two Boolean values form a semiring under logic and and or.

Exercise 3.10

3.10.1 Prove Proposition 3.3.1 that the cancellation law holds in a nonzero ring (ring without zero divisor).
3.10.2 Prove that all real numbers in the form of $a + b\sqrt{2}$, where a, b are integers, form an integral domain under the normal addition and multiplication.

Answer: page 335

3.3.2   Division Ring and Field

Not all elements are invertible in a ring. If every nonzero element has inverse, then it forms a special ring called *division ring*. For example, in $\mathbb{Q}$, the ring of all rational numbers under normal addition and multiplication, every nonzero a has inverse of $\frac{1}{a}$.

Definition 3.3.2 A ring R is a division ring if it satisfies the following:

1. R has at least one nonzero element.
2. R has the multiplicative identity.
3. Every nonzero element is invertible (unit).

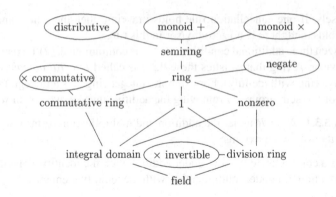

Fig. 3.21 Semiring, ring, integral domain, and field

Definition 3.3.3 A commutative division ring is a **field**.

By this definition, all rational numbers form a field $\mathbb{Q}$. Similarly, all real numbers $\mathbb{R}$ and all complex numbers $\mathbb{C}$ are fields. There are some interesting properties of division ring and field. For example, there is no zero divisor in a division ring or field. Because if $a \neq 0$, and $ab = 0$, then multiply the inverse of a from left to both sides:

$$a^{-1}ab = b = 0$$

b must be zero; hence, it's a nonzero ring. Further, all nonzero elements form a group under multiplication, denoted as $R*$, namely, the multiplicative group of division ring R. A division ring or a field consists of two groups: addition group and multiplication group. The distributive law bridges the two groups together. Figure 3.21 summarizes ring, semiring, integral domain, division ring, field, and their relationship.

I can only scratch the surface of rings and fields in this book and had to skip some important topics like subring, ideal, etc. As rings and fields have rich structures, they have plenty of interesting properties and in-depth results.

3.4 Galois Theory

Galois theory provides a connection between group and field: certain problems in field can be reduced with groups, in some sense simpler and easier to understand. A field can contain infinitely many elements with $+, -, \times, /$. It's a complex mathematical object. Galois theory simplifies the problems in field to a corresponding problem in finite group with only one operation. This is the key idea of Galois theory.

Galois theory is famous for its difficulty and hard to understand. Compared to its original form, the modern Galois theory is no longer ambiguous and much clearer and elegant. This is because more than two dozens of mathematicians greatly developed and improved it. The most important contributors are Jordan, Dedekind, and Emil Artin. Jordan and Dedekind first systematically studied and lectured Galois theory in France and Germany. The modern form of Galois theory is transformed by Artin [33]. We adopted a gentle explanation that is friendly to beginners [38] in this short section. It's quite common that one can't understand it in the first time reading. It's always helpful to "read the masters". Our life is not linear. It's always helpful to revisit what we learned before and find more great books to read. Our understanding changes along the time, and there will be some "aha!" moments in the future.

3.4.1 Field Extension

By definition, all rational numbers form a field $\mathbb{Q}$. Consider the set of all the numbers in the form of $a + b\sqrt{2}$, where a and b are rational numbers [35]. Obviously, $\mathbb{Q}$ is a subset of it (just let $b = 0$). It's easy to verify, for any two such numbers, the results of $+, -, \times, /$ are still some $a + b\sqrt{2}$. For example, to verify $x = \frac{1}{a+b\sqrt{2}}$, multiply both nominator and denominator by $a - b\sqrt{2}$:

$$x = \frac{a - b\sqrt{2}}{(a + b\sqrt{2})(a - b\sqrt{2})}$$
$$= \frac{a - b\sqrt{2}}{a^2 - 2b^2}$$

Let $p = a^2 - 2b^2$, then x equals $(a/p) - (b/p)\sqrt{2}$. It shows this set is really a field as every nonzero element has multiplicative inverse. Denote it as $\mathbb{Q}(\sqrt{2})$. Similarly, $\mathbb{Q}(\sqrt{3})$ is also a field. They are both examples of field extension.

Definition 3.4.1 If field E contains field F, then E is *field extension* of F, denoted as $F \subseteq E$ or E/F.

For example, the field of real numbers is the field extension of rational numbers; $\mathbb{Q}(\sqrt{2})$ is the field extension of $\mathbb{Q}$. Basically, for field F, if $\alpha \in F$, but $\sqrt{\alpha} \notin F$, then $F_1 = F(\sqrt{\alpha})$ is also a field, and a field extension of F. We can keep extending with this method. If $\beta \in F_1$, but $\sqrt{\beta} \notin F_1$, then:

$$F_2 = F_1(\sqrt{\beta})$$
$$= F(\sqrt{\alpha})(\sqrt{\beta})$$
$$= F(\sqrt{\alpha}, \sqrt{\beta})$$

All numbers of $a + b\sqrt{\beta}$ where $a, b \in F_1$ form a higher-level field extension. $F(\sqrt{\alpha}, \sqrt{\beta})$ contains linear combinations of $\sqrt{\alpha}$, $\sqrt{\beta}$, and $\sqrt{\alpha\beta}$ (numbers in $a + b\sqrt{\alpha} + c\sqrt{\beta} + d\sqrt{\alpha\beta}$, where $a, b, c, d \in F$). Starting from rational numbers, we obtain a series of field extensions $\mathbb{Q} \subset F_1 \subset F_2 \cdots \subset F_n$.

Why does field extension matter? For example, equation $x^2 - 2 = 0$ cannot be solved in rational numbers (there are no rational numbers satisfying this equation); however, if it extends to $\mathbb{Q}(\sqrt{2})$, it has a pair of solutions of $x = \pm\sqrt{2}$. For another example, equation $x^4 - 5x^2 + 6 = 0$ has no rational solution, but it has two solutions in extension $\mathbb{Q}(\sqrt{2})$, which are $\pm\sqrt{2}$; if further extended to $\mathbb{Q}(\sqrt{2}, \sqrt{3})$, then one gets all the four roots of $\pm\sqrt{2}$, and $\pm\sqrt{3}$. This field is called the splitting field.

Definition 3.4.2 For equation $p(x) = 0$, the smallest field extension that contains all the roots is called the *splitting* field (or root field) of $p(x)$.

Thus, the splitting field of $x^2 - 2 = 0$ is $\mathbb{Q}(\sqrt{2})$. The term "splitting" describes the fact that the polynomial $x^2 - 2$ cannot be decomposed in the field of rational numbers $\mathbb{Q}$; however, in field extension $\mathbb{Q}(\sqrt{2})$ it *splits* to:

$$(x + \sqrt{2})(x - \sqrt{2})$$

For a given polynomial equation, starting from the basic field of rational numbers, with a series of field extensions, if it reaches to the splitting field, then the equation is radical solvable. We've seen the example of square root. There are more complex cases. For example, the cyclotomic equation $x^p - 1 = 0$ has p roots of $1, \zeta, \zeta^2, ..., \zeta^{p-1}$ in the unit circle of the complex plane. We need to define the field extension to cover such cases.

Let element b be in F and $\alpha^m = b$ for some integer m, or in radical as $\alpha = \sqrt[m]{b}$. If m is composite, factor it as a product of primes and break into field extension for each prime. As such, we simplify α_i in every extension $F(\alpha_i)$ as p-th root for some prime p. For example, instead of adding $\sqrt[6]{\alpha}$, break it into two elements of $\sqrt{\alpha} = \beta$ followed by $\sqrt[3]{\beta}$. For equation $x^3 - 2 = 0$, it's insufficient to only add $\sqrt[3]{2}$ because the other two roots $\omega\sqrt[3]{2}$, $\omega^2\sqrt[3]{2}$ are not in the extension $\mathbb{Q}(\sqrt[3]{2})$ (obviously, the extension only contains real numbers, while $\omega = \frac{-1+i\sqrt{3}}{2}$ is a complex number). We need to add $\sqrt[3]{2}$ and ω to extend the field to $\mathbb{Q}(\sqrt[3]{2}, \omega)$. We needn't add $\omega^2 = \frac{-1-i\sqrt{3}}{2}$ as $\omega^2 = -1 - \omega$ is included already. The field extension that contains *all* roots of the polynomial (the splitting field) is called *normal* extension.

Extend the field as a series to $F(\alpha_1)(\alpha_2) \ldots (\alpha_k)$. If each α_i is radical, we call the field extension $F(\alpha_1, \alpha_2, \ldots, \alpha_k)$ as *radical* extension of F. Let the equation roots be $x_1, x_2, \ldots, x_n$ (note x_i may not be radical form); if the splitting field $E = \mathbb{Q}(x_1, x_2, \ldots, x_n)$ is radical extension, then the equation is radical solvable.

Exercise 3.11

3.11.1 Prove that $\mathbb{Q}(a, b) = \mathbb{Q}(a)(b)$, where $\mathbb{Q}(a, b)$ contains all the expressions combined with a and b, such as $2ab, a + a^2b$, etc.

Answer: page 336

3.4.2 From Newton, Lagrange to Galois

Great ideas were often advanced by joint efforts of many people. Unlike what Eric Temple Bell wrote [37], Galois theory did not suddenly appear from scratch in one night. Tracing backwards, there were a series of great steps taken by Viète, Newton, Vandermonde, Lagrange, and Gauss. Galois carried over the torch.

3.4.2.1 Symmetric Polynomials and Newton's Theorem

Newton was famous for his great discoveries of universal law of gravitation, calculus, physics, planetary motion, etc. He studied symmetric polynomials after Viète. A portion of his work was published in Newton's *Arithmetica Universals* in 1707. Gauss referenced it as "Newton's theorem." It's hard for people nowadays to understand why Newton did not announce his findings or publish the *Mathematical Principles of Natural Philosophy* unless Halley kept persuading and pushing him. Same thing happened to Newton's theorem; people found below result in his notebook when he was 23 years old (in 1665 or 1666) [34].

Isaac Newton (1642–1726)

Let the three roots for cubic equation $x^3 + bx^2 + cx + d = 0$ be r, s, t, then

$$x^3 + bx^2 + cx + d = (x - r)(x - s)(x - t)$$

and

$$r + s + t = -b$$

$$r^2 + s^2 + t^2 = b^2 - 2c$$

$$r^3 + s^3 + t^3 = -b^3 + 3bc - 3d$$

$$rs + st + rt = c$$

$$r^2s + s^2t + t^2r + r^2t + t^2s + s^2r = -bc + 3d$$

$$\ldots \ldots$$

$$r^3s^3t^3 = -d^3$$

Newton list 19 symmetric polynomials. He even extended the degree to 8 and calculated result for equation $x^8 + px^7 + qx^6 + rx^5 + sx^4 + tx^3 + vx^2 + yx + z = 0$. Among these symmetric polynomials, Newton found some were "elementary." Generalize to the equation of degree n, there are n roots.

$$x^n + b_1 x^{n-1} + b_2 x^{n-2} + \ldots + b_n = (x - r_1)(x - r_2)\ldots(x - r_n)$$

The elementary symmetric polynomials are:

$$\sigma_1 = \qquad\qquad r_1 + r_2 + \cdots + r_n = -b_1$$

$$\sigma_2 = \qquad r_1 r_2 + r_2 r_3 + \cdots + r_{n-1} r_n = b_2$$

$$\sigma_3 = \quad r_1 r_2 r_3 + r_1 r_2 r_4 + \cdots + r_{n-2} r_{n-1} r_n = -b_3 \qquad\qquad (3.7)$$

$$\ldots$$

$$\sigma_n = \qquad\qquad r_1 r_2 \cdots r_n = (-1)^n b_n$$

Why are they elementary symmetric? Newton found them acted as the basic building blocks of other symmetric polynomials, i.e., any symmetric polynomial can be expressed as a combination of the elementary ones. For example, $(a - b)$ is not symmetric (the $\pm$ sign changes when swap a and b), but its square $(a - b)^2$ can be expressed as:

$$(a - b)^2 = (a + b)^2 - 4ab \qquad \text{in two symmetric polynomials}$$

$$= \sigma_1^2 - 4\sigma_2 \qquad \text{quadratic elementary symmetric}$$

Theorem 3.4.1 (Newton's Theorem) *Any polynomial of $r_1, r_2, \ldots, r_n$ can be expressed with the elementary symmetric polynomials of $\sigma_1, \sigma_2, \ldots, \sigma_n$.*

3.4.2.2 Lagrange's Resolvent

We skip the proof of Newton's theorem. Vandermonde and Lagrange studied following this direction independently. Their methods to solve cubic equation were essentially the same. Lagrange's influential paper has 220 pages; we can only brief it shortly (Galois read this paper[22] at his age of 15). For generic cubic equation $ay^3 + by^2 + cy + d = 0$, substitute $y = x - b/3a$; convert to $x^3 + px + q = 0$; denote the three roots by r_1, r_2, r_3. Lagrange also considered the 3rd primitive root of unity $\omega = \frac{-1+i\sqrt{3}}{2}$. By cyclotomic equation: $\omega^3 = 1$ and $1 + \omega + \omega^2 = 0$. Lagrange introduced a key concept, *resolvent*:

$$t = r_1 + \omega r_2 + \omega^2 r_3 \tag{3.8}$$

When permuting the three roots, Lagrange found, although there were six permutations, the cubic of the resolvent (t^3) had only two distinct values. From the view point of groups, though it was not known in 1770, t^3 has two values under the six permutations of group S_3.

$$\begin{cases} t_1^3 = (r_1 + \omega r_2 + \omega^2 r_3)^3 = L^3 & (1) \\ t_2^3 = (\omega r_1 + \omega^2 r_2 + r_3)^3 = (\omega L)^3 = L^3 & (132) \\ t_3^3 = (\omega^2 r_1 + r_2 + \omega r_3)^3 = (\omega^2 L)^3 = L^3 & (123) \end{cases}$$

The left side list the cubic of the resolvent; the right side are the permutations of S_3. For example, in the second row, since $\omega^3 = 1$, $r_3 \omega^3 = r_3$. Multiply by 1, ω and ω^2 correspond to the permutations of $\{(1), (132), (123)\}$, which is the invariant (normal) subgroup C_3.

$$\begin{cases} t_4^3 = (r_1 + \omega^2 r_2 + \omega r_3)^3 = R^3 & (23) \\ t_5^3 = (\omega r_1 + r_2 + \omega^2 r_3)^3 = (\omega R)^3 = R^3 & (12) \\ t_6^3 = (\omega^2 r_1 + \omega r_2 + r_3)^3 = (\omega^2 R)^3 = R^3 & (13) \end{cases}$$

These three permutations fix a root and transpose the other two, corresponding to the permutations of $\{(23), (12), (13)\}$ in the coset $C_3(23)$. Lagrange next introduced the resolvent equation:

$$(X - t_1)(X - t_2)(X - t_3)(X - t_4)(X - t_5)(X - t_6) = 0 \tag{3.9}$$

The six-degree equation looks complex; actually it can be converted to a quadratic equation. Collect the first three factors in one set and the rest three in another set, where $(X - t_1)(X - t_2)(X - t_3) = (X - L)(X - \omega L)(X - \omega^2 L) =$

---

[22] *Réflexions sur la résolution algébrique des équations* was published in 1770.

$X^3 - L^3$; similarly, the rest three factors simplify to $X^3 - R^3$. As such, the resolvent equation changes to $(X^3 - L)(X^3 - R) = 0$. Expand it to:

$$X^6 - (L^3 + R^3)X^3 + L^3 R^3 = 0 \qquad (3.10)$$

Substituting X^3 as Y, it changes to a quadratic equation $Y^2 - (L^3 + R^3)Y + L^3 R^3 = 0$. Express the cubic of resolvent in the coefficients of the original equation:

$$\begin{cases} L^3 + R^3 = -27q \\ L^3 R^3 = -27p^3 \end{cases} \qquad (3.11)$$

Substituting p, q in, one can solve the L, R from the quadratic equation of Y. Revisit the equation $x^3 + px + q = 0$; the coefficient of the quadratic term is 0. By Viète's formula: $r_1 + r_2 + r_3 = 0$; list it as the third equation:

$$\begin{cases} r_1 + \omega r_2 + \omega r_3 = L \\ r_1 + \omega^2 r_2 + \omega r_3 = R \\ r_1 + r_2 + r_3 = 0 \end{cases} \qquad (3.12)$$

Solving this system gives the three roots expressed in Lagrange's resolvent:

$$\begin{cases} r_1 = \dfrac{\omega^2 L + \omega R}{3} \\[3mm] r_2 = \dfrac{\omega L + \omega^2 R}{3} \\[3mm] r_3 = \dfrac{L + R}{3} \end{cases} \qquad (3.13)$$

Substituting L, R solves the original cubic equation finally. Lagrange's method is significant. Using the quartic equation, for example, define the resolvent with the 4-th primitive roots of unity $\pm 1, \pm i$:

$$t = r_1 + i r_2 - r_3 - i r_4 \qquad (3.14)$$

Below resolvent equation is invariant under the $4! = 24$ permutations of the symmetric group S_4:

$$\Phi(X) = (X - t_1)(X - t_2)...(X - t_{24}) \qquad (3.15)$$

By Newton's theorem, all its coefficients are known. Lagrange successfully simplified this equation of degree 24 to a cubic equation and solved the original quartic equation. However, Lagrange was stuck when applying his method to the quintic equation. The resolvent equation $\Phi(X) = (X - r_1)(X - r_2)...(X - r_{120})$ has the degree of $5! = 120$. Lagrange simplified it to a lower degree, but got another quintic equation. It looped back to the same problem of quintic equation. The next critical step was taken by Galois.

3.4.2.3 Galois' Memoir

Going back to the midnight of May 29, 1832, Galois wrote down a special, depressed, unbelievable memoir. We skip the proof, but only illustrate his thought with some verification examples. Galois firstly defined what irreducible polynomial was: a polynomial $f(x)$ with its coefficients in field[23] F is reducible if $f(x)$ can be factored in F, else it's irreducible. For example, $x^2 + 1$ is irreducible in field $\mathbb{Q}$ of rationals, but it can be factored as $(x + i)(x - i)$ in field $\mathbb{C}$ of complex numbers. For another example, $x^2 - 2$ is irreducible in $\mathbb{Q}$, but it can be factored as $(x + \sqrt{2})(x - \sqrt{2})$ in field extension $\mathbb{Q}(\sqrt{2})$. For some element α in field F, the monic (the coefficient of the highest degree is 1) irreducible polynomial of the lowest degree, such that α is a root, is called the minimal polynomial. For example, $x^2 - 2$ is the minimal polynomial in $\mathbb{Q}$ with $\sqrt{2}$ as a root. Galois next defined what a known value was: the added specific values from rational field and their rational expressions (arithmetic $+, -, \times, /$) are known. This is essentially the concept of field extension nowadays.

Galois next introduced the concept of group; particularly, he referred to the permutation group. He was the first one to use the French word "groupe" and pointed out the composite of two permutations was also a permutation. Essentially, Galois was using S_n and the subgroups to permute the roots in their rational expressions. For example, applying $\sigma = (123)$ to rational expression $x_1 x_2 + x_3$ gives:

$$\sigma(x_1 x_2 + x_3) = x_2 x_3 + x_1$$

Galois next gave an important fact of irreducible polynomial:

Lemma 3.4.2 (Property of the Irreducible Polynomial) *Let $f(x)$ be a polynomial and $p(x)$ be an irreducible polynomial in field F. If $f(x)$ and $p(x)$ have a common root, then $p(x)$ divides $f(x)$.*

Analogous to integer division: polynomials correspond to integers; irreducible polynomials correspond to prime numbers; a common root is like a common prime factor. For example, $f(x) = x^3 - 1$ and $p(x) = x^2 + x + 1$, where $p(x)$ is irreducible in $\mathbb{Q}$. They have two common roots of $\omega = \frac{-1+i\sqrt{3}}{2}$ because:

$$\omega^3 - 1 = \omega^2 + \omega + 1 = 0$$

$$p(x) \text{ divides } f(x) \text{ since } f(x) = x^3 - 1 = (x - 1)(x^2 + x + 1) = (x - 1)p(x).$$

---

[23] There was no concept of field by that time until Dedekind defined it later.

What Galois next did looked like extending from Lagrange's resolvent; he defined "Galois's resolvent."

Lemma 3.4.3 (Galois Resolvent t) *Let $f(x)$ be a polynomial in field F with roots of $x_1, x_2, ..., x_n$. There exists some rational expression of the roots $t = \varphi(x_1, x_2, ..., x_n)$; its value changes under different root permutations.*

The permutations of n roots lead to different values of t. For example, $f(x) = x^2 + 1$ with coefficients in rational field has two roots of $x_1 = i, x_2 = -i$; we can find a rational expression: $t = x_1 - x_2$, which takes different values under the two permutations of (1) and (12) in S_2:

$$\begin{cases} t_1 = x_1 - x_2 = 2i & \text{permutation (1)} \\ t_2 = x_2 - x_1 = -2i & \text{permutation (12)} \end{cases}$$

The two values are different. Galois only said there existed t, and gave a method to construct it through linear combination $t = k_1 x_1 + k_2 x_2 + ... + k_n x_n$, but he did not prove this lemma. He excluded the case of duplicated roots, as the value wouldn't change under any permutation. Analogous to what Lagrange did, Galois next expressed the roots in resolvent.

Lemma 3.4.4 (Express Root in Galois's Resolvent t) *There exists rational expression $\varphi_1(x), \varphi_2(x), ..., \varphi_n(x)$ in field F, where the roots $x_1, x_2, ..., x_n$ of polynomial $f(x)$ can be expressed in Galois's resolvent t:*

$$x_1 = \varphi_1(t), x_2 = \varphi_2(t), ..., x_n = \varphi_n(t)$$

For the same example, the two roots of $f(x) = x^2 + 1$ are $x_1 = i, x_2 = -i$; the resolvent $t = x_1 - x_2 = 2i$; one can find $\varphi_1(x) = \frac{x}{2}, \varphi_2(x) = -\frac{x}{2}$, such that the two roots can be expressed in t as:

$$x_1 = \frac{t}{2}, \ x_2 = -\frac{t}{2}$$

As such, Galois linked the splitting field (root field), $F(x_1, x_2, ..., x_n)$ of the equation to the field extension $F(t)$. When permuting n roots, Galois's resolvent t gets different values. Following Lagrange's idea, t is some root of another resolvent equation $f_t(X) = 0$ (which has many roots).

Lemma 3.4.5 *In field F, construct a minimal polynomial $f_t(X)$ with t as a root, let the roots of $f_t(X)$ be $t_1, t_2, ..., t_m$, then*

$$\varphi_1(t_k), \varphi_2(t_k), ..., \varphi_n(t_k)$$

is some permutation of the roots of the original equation $f(x) = 0$, where $k = 1, 2, ..., m$.

Note the difference of the two equations: $f(x) = 0$ is the original equation, while $f_t(X) = 0$ is the constructed equation, where $f_t(X)$ is the minimal polynomial with t as a root. The original equation has n roots of $x_1, \ldots, x_n$, while the resolvent equation has m roots of $t_1, t_2, \ldots, t_m$

$$(X - t_1)(X - t_2)\ldots(X - t_m) = 0$$

The relationship between them is: $t = \varphi(x_1, \ldots, x_n)$, and $x_i = \varphi_i(t)$. For example, equation $x^3 - x^2 - 2x + 2 = 0$ has three roots of $1, \pm\sqrt{2}$. *Step 1*: construct Galois's resolvent $t = \varphi(x_1, x_2, x_3) = x_1 + 2x_2 + 4x_3$. It has six different values when applying S_3:

Permutation	$\varphi(x_1, x_2, x_3)$	t
$(1, 2, 3)$	$x_1 + 2x_2 + 4x_3$	$1 - 2\sqrt{2}$
$(1, 3, 2)$	$x_1 + 2x_4 + 4x_2$	$1 + 2\sqrt{2}$
$(2, 1, 3)$	$x_2 + 2x_1 + 4x_3$	$2 - 3\sqrt{2}$
$(2, 3, 1)$	$x_2 + 2x_3 + 4x_1$	$4 - \sqrt{2}$
$(3, 1, 2)$	$x_3 + 2x_1 + 4x_2$	$2 + 3\sqrt{2}$
$(3, 2, 1)$	$x_3 + 2x_2 + 4x_1$	$4 + \sqrt{2}$

Step two: select $t = 1 - 2\sqrt{2}$ and the first three rows to form equations:

$$\begin{cases} x_1 + 2x_2 + 4x_3 & = t \\ x_1 + 2x_3 + 4x_2 & = 2 - t \\ x_2 + 2x_1 + 4x_3 & = \dfrac{3t + 1}{2} \end{cases}$$

Solve the them and express the three roots in t:

$$\begin{cases} x_1 = \varphi_1(t) = 1 \\ x_2 = \varphi_2(t) = \dfrac{1 - t}{2} \\ x_3 = \varphi_3(t) = \dfrac{t - 1}{2} \end{cases}$$

Step three: use $t = 1 - 2\sqrt{2}$ as a root; construct a minimal polynomial $f_t(X) = X^2 - 2X - 7$. It has two roots: $t_{1,2} = 1 \pm 2\sqrt{2}$. Substitute $\varphi_1(t), \varphi_2(t), \varphi_3(t)$ to generate two permutations:

t_i	$\varphi_1(t_i), \varphi_2(t_i), \varphi_3(t_i)$	Permutation in S_3
t_1	$1, \sqrt{2}, -\sqrt{2}$	(1)
t_2	$1, -\sqrt{2}, \sqrt{2}$	(23)

Although influenced by Lagrange, Galois's resolvent is not given in a fixed way. Galois only said it existed. The resolvent equation was also defined with its existence by some minimal polynomial. Another difference from Lagrange's method, m does not necessarily equal to $n! = |S_n|$. Galois realized not all permutations in S_n act equally; some special permutations reflected the symmetry of equation, as such he introduced Galois group.

Definition 3.4.3 (Galois Group of a Equation) Use the roots of polynomial $f(x)$ in field F; construct rational expression r. If the value of r is in F, then it is *known* value ($r \in F$). When permuting the roots with some σ, and the value of r is unchanged, then r is *invariant* ($\sigma(r) = r$). For all such r, there exists group G satisfying the following property:

r is unchanged under all permutations in G $\iff$ r is known value.

G is Galois group of the equation $f(x) = 0$ in field F.

It is summarized as "known if and only if invariant." [39] Note r is a rational expression, but not a rational number; it consists of $+, -, \times, /$ to variables with rational coefficients. For example, $r = \frac{3a^2+2b+1}{a^2-2b}$ with all coefficients in $\mathbb{Q}$. When assigning values to these variables, if the value of r is in F, then it's known; otherwise it (the value is not in F) is unknown. For example, let $a = 1, b = \sqrt{2}$, then $r = -\frac{12}{7} - \frac{10\sqrt{2}}{7}$; it's unknown in field $\mathbb{Q}$, but it's known in field extension $\mathbb{Q}(\sqrt{2})$.

For example, equation $x^2 - 2x - 1 = 0$ has two roots of $1 \pm \sqrt{2}$. Its Galois group in rational field $\mathbb{Q}$ is $S_2 = \{(1), (12)\}$, including the identity $e = (1)$ and transpose (12). As any rational expression is unchanged under identity permutation; let us focus on the transpose (12). If a rational expression r is unchanged when swapping the two variables, it must be a symmetric polynomial. By Newton's theorem, r can be expressed with the elementary symmetric polynomials of $x_1 x_2 = -1$ and $x_1 + x_2 = 2$; hence, it is a rational number and known value. Revisiting the previous example, the Galois group of equation $x^3 - x^2 - 2x + 2 = 0$ is $\{(1), (23)\}$, including the identity permutation, and another one that fixes the first root and swaps the other two roots: $\pm\sqrt{2}$. It's a subgroup of S_3.

Exercise 3.12

3.12.1 Prove Eq. (3.11); give the relation between Lagrange's resolvent and the coefficients of the original equation. Hint: compute $(L + R)(L + \omega R)(L + \omega^2 R)$.

3.12.2 Verify the Galois group of equation $x^3 - x^2 - 2x + 2 = 0$ is $\{(1), (23)\}$.
Hint: verify "known if and only if invariant."

Answer: page 337

3.4.3 Automorphism and Galois Group

In 1938, Emil Artin (Fig. 3.22) reformulated Galois's idea on top of field and gave the modern form of Galois theory. Comparing Artin's version and Galois's original one, we see how Galois theory gradually changed to clear, concise, and elegant.

Equation $x^2 - 2 = 0$ has two roots of $\pm\sqrt{2}$. The equation holds when substituting $\sqrt{2}$ with $-\sqrt{2}$. What's more, $\sqrt{2}^2 + \sqrt{2} + 1 = 3 + \sqrt{2}$ also holds with this substitution. The expression $\alpha^2 + \alpha + 1 = 3 + \alpha$ is symmetric under $\pm\sqrt{2}$. Any rational expression of $\sqrt{2}$ is unchanged when transposing $\pm\sqrt{2}$. Group, as the best tool for symmetry, was introduced by Galois to study the symmetry of equations. Artin found the relationship that linked Galois group to field automorphism. For example, define a function from field extension $\mathbb{Q}(\sqrt{2})$ to itself: $f : \mathbb{Q}(\sqrt{2}) \to \mathbb{Q}(\sqrt{2})$. It flips the sign of $\sqrt{2}$:

$$f(a + b\sqrt{2}) = a - b\sqrt{2}$$

f is a field automorphism.

Definition 3.4.4 A field automorphism is an invertible function f that maps a field to itself, satisfying $f(x + y) = f(x) + f(y)$, $f(ax) = f(a)f(x)$, and $f(1/x) = 1/f(x)$.

It's easy to verify $f(a + b\sqrt{2}) = a - b\sqrt{2}$ satisfies all the three conditions. Flipping the sign is equivalent to permuting the roots $x_{1,2} = \pm\sqrt{2}$. The idea behind field automorphism is that one can permute some elements in the field without any impact to the field structure.

Fig. 3.22 Emil Artin
(1898–1962)

Definition 3.4.5 (F-Automorphism) Let E be the field extension of F; if an automorphism f of field E also satisfies $f(x) = x$ for any x in F, then f is F-automorphism of E.

F-automorphism precisely defines the symmetry of roots. f fixes all elements in F (keep unchanged); the new elements and only those new elements in field extension E are changed under f. It well explains "known if and only if invariant" by Galois. An element x is known ($x \in F$) if and only if it is unchanged under the automorphism ($f(x) = x$). The changes among elements are exactly permutations; the relations that keep unchanged under permutation are exactly symmetries. For example, in $\mathbb{Q}(\sqrt{2})$, there are two $\mathbb{Q}$-automorphisms: (1) the identity function $e(x) = x$ and (2) $f(a + b\sqrt{2}) = a - b\sqrt{2}$. They form a group $\{e, f\}$, which is isomorphic to cyclic group C_2 and is also isomorphic to symmetric group S_2. We can write it as $\{f, f^2\}$ because $f^2 = f \cdot f = e$; we can also write it as permutations $\{(1), (1\ 2)\}$, where (1) fixes the two roots and (12) transposes them.

For another example, the splitting field of equation $x^4 - 5x^2 + 6 = 0$ is $\mathbb{Q}(\sqrt{2}, \sqrt{3})$. Automorphism f transposes $\pm\sqrt{2}$. To transpose $\pm\sqrt{3}$, define automorphism $g(a + b\sqrt{3}) = a - b\sqrt{3}$. There does not exist automorphism transposes $\sqrt{2}$ and $\sqrt{3}$; otherwise, assume $h(\sqrt{2}) = \sqrt{3}$, then $h(\sqrt{2})^2 = h(\sqrt{2}^2) = h(2) = 2$. As h is $\mathbb{Q}$-automorphism, $h(x) = x$ for any rational x; since $h(\sqrt{2}) = \sqrt{3}$, then $h(\sqrt{2})^2 = \sqrt{3}^2 = 3$, implying $2 = 3$ which is absurd. The counterexample of h shows not all permutations in S_4 reflect the symmetry of the equation. It's Galois group that contains four automorphisms: $\{e, f, g, f \cdot g\}$, where e is the identity; f transposes $\pm\sqrt{2}$; g transposes $\pm\sqrt{3}$, and the composition of $f \cdot g$ transposes the two pairs simultaneously.

With the link from field extension to a group of automorphisms under composition, Artin reformulated the definition of Galois group.

Definition 3.4.6 (Galois Group) For field extension E from F, there is a group G of all the F-automorphisms of E under composition defined as $(f \cdot g)(x) = f(g(x))$; G is Galois group of field extension E/F, denoted as $Gal(E/F)$.

It's a critical idea to study the equation through the symmetry of its roots in Galois theory. Here is the definition of symmetry in mathematics: the symmetry of object is the automorphisms to the object itself while preserving all its structures.

Exercise 3.13

3.13.1 Prove that for any polynomial $p(x)$ with rational coefficients, $E/\mathbb{Q}$ is a field extension, f is a $\mathbb{Q}$-automorphism of E, then equation $f(p(x)) = p(f(x))$ holds.

3.13.2 What is the splitting field for polynomial $p(x) = x^4 - 1$? Is it $\mathbb{C}$? What are its $\mathbb{Q}$-automorphisms?

3.13.3 What's the Galois group for quadratic equation $x^2 - bx + c = 0$?

3.13.4 Show that, if p is a prime number, Galois group of equation $x^p - 1$ is cyclic group C_{p-1}.

3.13.5 Let α be a root of equation $x^3 + x^2 - 4x + 1$; verify that $2 - 2\alpha - \alpha^2$ is also a root. What's Galois group of this equation in rational field?

Answer: page 338

3.4.4 Fundamental Theorem of Galois Theory

We arrive at the center of Galois theory. Starting from the field F of equation coefficients, keep extending to the splitting field: $F \subset F_1 \subset F_2 \cdots \subset E$. The corresponding Galois group is $Gal(E/F)$. Galois found there was one-to-one correspondence between Galois subgroups and the intermediate fields $F_1, F_2, \ldots$ in *reversed* order.

Theorem 3.4.6 (Fundamental Theorem of Galois Theory) *If E/F is normal extension, G is the corresponding Galois group. There is one-to-one correspondence between subgroups of G and intermediate fields. If $F \subset L \subset E$, and $Gal(E/L) = H$, then H is the subgroup of $Gal(E/F)$.*

As normal extension (114), E is the splitting field for polynomials with coefficients in F. It's in reversed order for the group gets smaller when extending the field. The starting point of field extension is F, and the corresponding Galois group is $G = Gal(E/F)$; the end point is the splitting field E, while the corresponding group only contains one element, the identity $\{e\}$. For the intermediate field L, the corresponding group is $H = Gal(E/L)$. It's a subgroup, $Gal(E/L) \subset Gal(E/F)$. If H is a normal subgroup (see Definition 3.2.6), then its quotient group is $G/H = Gal(L/F)$.

For example, equation $x^4 - 8x^2 + 15 = 0$ can be factored as $(x^2 - 3)(x^2 - 5) = 0$. The coefficients are rationals in field $\mathbb{Q}$, and the splitting field is $E = \mathbb{Q}(\sqrt{3}, \sqrt{5})$; its Galois group $Gal(E/\mathbb{Q})$ is isomorphic to a Klein-four group (with order 4), often denoted as V or K_4 (Fig. 3.23). There are three intermediate field extensions: $\mathbb{Q}(\sqrt{3})$, $\mathbb{Q}(\sqrt{5})$, and $\mathbb{Q}(\sqrt{15})$. The orders of corresponding Galois subgroups to them are 2, while Galois group V has only one subgroup of order 2. Besides that, it hasn't any other nontrivial subgroups. By fundamental theorem of Galois theory, there is no other field extension besides these three.

Fig. 3.23 Klein four.
Symmetric when flip
horizontally, vertically, or
simultaneously

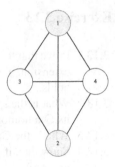

Fig. 3.24 Galois
correspondence

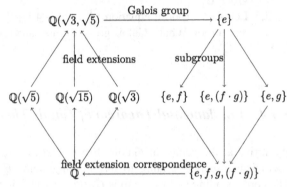

As shown in Fig. 3.24, any element in the splitting field is in the form of:

$$\alpha = (a + b\sqrt{3}) + (c + d\sqrt{3})\sqrt{5}$$
$$= a + b\sqrt{3} + c\sqrt{5} + d\sqrt{15}$$

where a, b, c, d are rational numbers. For any element in the three intermediate
field extensions, there are at least two of b, c, d that are 0. Define following
automorphisms:

(1) f negates $\sqrt{3}$:

$$f((a + b\sqrt{3}) + (c + d\sqrt{3})\sqrt{5}) = (a - b\sqrt{3}) + (c - d\sqrt{3})\sqrt{5}$$
$$= a - b\sqrt{3} + c\sqrt{5} - d\sqrt{15}$$

(2) g negates $\sqrt{5}$:

$$g((a + b\sqrt{3}) + (c + d\sqrt{3})\sqrt{5}) = (a + b\sqrt{3}) - (c + d\sqrt{3})\sqrt{5}$$
$$= a + b\sqrt{3} - c\sqrt{5} - d\sqrt{15}$$

(3) The composite morphism $f \cdot g$ negates $\sqrt{3}$ and $\sqrt{5}$ simultaneously.

$$(f \cdot g)((a + b\sqrt{3}) + (c + d\sqrt{3})\sqrt{5}) = (a - b\sqrt{3}) - (c - d\sqrt{3})\sqrt{5}$$
$$= a - b\sqrt{3} - c\sqrt{5} + d\sqrt{15}$$

Including the identity automorphism e, there are four elements in Galois group: $G = \{e, f, g, (f \cdot g)\}$ for the base field $\mathbb{Q}$. This group is isomorphic to Klein-four group V, and V is the product of two cyclic groups of C_2. It has five subgroups, each corresponding to a field extension:

1. The subgroup only contains the identity element $\{e\}$, corresponding to the splitting field $\mathbb{Q}(\sqrt{3}, \sqrt{5})$.
2. G itself, corresponding to the rational field $\mathbb{Q}$.
3. Order 2 subgroup $\{e, f\}$, corresponding to field extension $\mathbb{Q}(\sqrt{5})$. f only negates $\sqrt{3}$ and leaves $\sqrt{5}$ unchanged.
4. Order 2 subgroup $\{e, g\}$, corresponding to field extension $\mathbb{Q}(\sqrt{3})$. g only negates $\sqrt{5}$ and leaves $\sqrt{3}$ unchanged.
5. Order 2 subgroup $\{e, (f \cdot g)\}$, corresponding to field extension $\mathbb{Q}(\sqrt{15})$. $(f \cdot g)$ negates $\sqrt{5}$ and $\sqrt{3}$ simultaneously, and leaves $\sqrt{15}$ unchanged.

Up to isomorphic, we can write Galois group of this equation as a permutation group of $\{(1), (1\ 2), (3\ 4), (1\ 2)(3\ 4)\}$, where (1) keeps all the four roots unchanged; $(1\ 2)$ swaps $\pm\sqrt{3}$; $(3\ 4)$ swaps $\pm\sqrt{5}$; and $(1\ 2)(3\ 4)$ swaps the two pairs simultaneously. Through the fundamental theorem, Galois converted the equation problem in field to an equivalent problem about permutation group. The last attack was to reveal the solvability essence with groups.

3.4.5 Solvability

Definition 3.4.7 The polynomial equation is radical solvable if and only if its splitting field E is the radical extension of its base field F.

Start from base field $F = \mathbb{Q}$; repeatedly apply radical extension by adding the primitive root of unity or $\sqrt[p]{\alpha}$ where p is a prime number, and α is some element in previous field. If it finally obtains the splitting (root) field $E = F(\alpha_1, \ldots, \alpha_k)$, it is radical solvable, and they form a tower of radical extension series:

$$F = F_0 \subseteq F_1 \subseteq F_2 \subseteq \cdots \subseteq F_k = E$$

where every $F_i = F_{i-1}(\alpha_i)$ and α_i is either a primitive root of unity or p-th root (p is prime) of some element in F_{i-1}. Galois correspondence connects the radical extension with group structure. Along the tower of extension series, there is a tower of reduced groups:

$$Gal(F_k/F_0) = G_0 \rhd G_1 \rhd \cdots \rhd G_k = Gal(F_k/F_k) = \{e\}$$

where every group is a normal subgroup of the previous one. Each step from G_{i-1} to its subgroup G_i corresponds to extending the field F with some α_i. Galois had the insight between radical extension and the reduction of Galois groups. Every time, Galois added all the p roots of the minimal polynomial $f_t(X)$ to the field, where p is prime. With normal extension, the corresponding Galois group H partitions the previous group into p cosets:

$$G = \sigma_1 H + \sigma_2 H + \cdots + \sigma_p H$$

where H is invariant (normal) subgroup and the quotient group G/H consists of these cosets and has prime order p. By Corollary 3.2.13 of Lagrange's theorem, any group of prime order is cyclic, and any cyclic group is abelian (by Exercise 3.7.1). Therefore, the reduced quotient group after radical extension is abelian.

Definition 3.4.8 (Solvable Group) A finite group G is solvable, if it has a subnormal series:

$$G = G_0 \rhd G_1 \rhd \cdots \rhd G_k = \{e\}$$

such that every quotient group G_i/G_{i+1} is abelian. The group series is called solvable group series.

From this definition, any subgroup in solvable series is solvable. Because for any abelian group G, $G \rhd \{e\}$ holds, all abelian groups are solvable. Using his theory, Galois gave the final answer to radical solvability for any polynomial equation:

Theorem 3.4.7 *A polynomial equation is radical solvable if and only if its Galois group is solvable.*

For general cubic equation, all permutations of the three roots form the group S_3; it has a normal subgroup C_3 (Fig. 3.16), and C_3 is a simple group; it has the only subgroup $\{e\}$. There is a group series:

$$S_3 \rhd C_3 \rhd \{e\}$$

The order of the quotient groups: $S_3/C_3 = 2$, $C_3/\{e\} = 3$ are all primes; hence, it is solvable group series. The general cubic equation is radical solvable.

For general quartic equation, all 24 permutations of the four roots form the group S_4. It has a normal subgroup A_4, an alternating group of 12 elements (Fig. 3.25). The next subgroup is Klein-four group V with four elements. It has a normal subgroup of C_2 (isomorphic to S_2); it's a simple group, with the only subgroup of $\{e\}$. The group series are:

$$S_4 \rhd A_4 \rhd V \rhd C_2 \rhd \{e\}$$

The orders of the corresponding quotient groups are 2, 3, 2, 2, all primes. The group series is solvable; the general quartic equation is solvable.

Fig. 3.25 Cayley graph of alternating group A_4. Cut the three vertex angles of a triangle pyramid

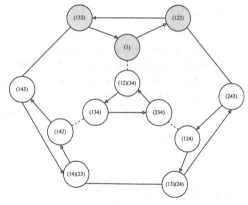

Fig. 3.26 Cayley graph of alternating group A_5 similar to a football

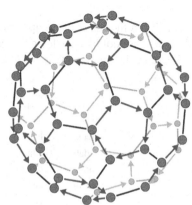

As an application of Galois theory, Abel-Ruffini theorem says there is no solution in radicals to general quintic polynomial equation. All 120 permutations of the 5 roots form the group S_5. Its only normal subgroup is the alternating group A_5 with 60 elements as shown in Fig. 3.26. A_5 is a simple group; its only normal subgroup is $\{e\}$. The group series are:

$$S_5 \triangleright A_5 \triangleright \{e\}$$

The quotient group $A_5/\{e\}$ is isomorphic to A_5; however, A_5 is not abelian. The order of the quotient group is 60, which is not a prime. It is not a solvable group. The general quintic polynomial equation is not solvable in radicals.

We can verify that S_n is not solvable when $n \geq 5$ (see Exercise 3.14.2. Use reduction of absurdity, it leads to the contradict of result that $\{e\}$ contains three cycles). Galois theory tells there is no solution in radicals to general polynomial equations of degree five or higher.

Galois studied the symmetry of equations and developed his theory. Galois theory is a powerful tool in abstract algebra. It is used to solve many problems, like to determine whether a regular n-gon can be constructed with compass and edge and prove that it's impossible to trisect an arbitrary angle.

Exercise 3.14

3.14.1 The quintic cyclotomic equation $x^5 - 1 = 0$ is radical solvable. What is its Galois group and what are the solvable group series?

3.14.2 Show that, when $n \geq 5$, assume the subgroup $G \subset S_n$ contains all three cycles like $(a\ b\ c)$; if N is a normal subgroup of G, and the quotient group G/N is abelian, then N contains all three cycles too.

Answer: page 341

Appendix

Galois listed the tower of group series for general quartic equation in his memoir. There was no tool like Cayley graph at his time. He denoted the four roots as *abcd* and permuted them to explain his idea.

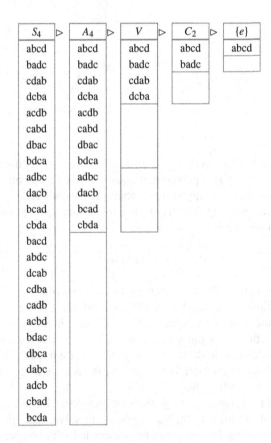

S_4 $\triangleright$	A_4 $\triangleright$	V $\triangleright$	C_2 $\triangleright$	$\{e\}$
abcd	abcd	abcd	abcd	abcd
badc	badc	badc	badc	
cdab	cdab	cdab		
dcba	dcba	dcba		
acdb	acdb			
cabd	cabd			
dbac	dbac			
bdca	bdca			
adbc	adbc			
dacb	dacb			
bcad	bcad			
cbda	cbda			
bacd				
abdc				
dcab				
cdba				
cadb				
acbd				
bdac				
dbca				
dabc				
adcb				
cbad				
bcda				

Chapter 4
Category

> *Mathematics is the art of giving the same name to different things.*
>
> *Henri Poincaré*

Welcome to the world of category! A new garden of abstraction. The path to this garden is developed by many talents. People first abstracted numbers and shapes from the concrete things and then, as the second step, removed the meanings of numbers, shapes, and arithmetic operations, abstracting them to algebraic structures (e.g., group) and relations (e.g., homomorphism); category theory is another step of abstraction.

Escher, Dewdrop, 1948

What is category? Why does it matter? Any relationship between category and programming? Category theory was a "side product" when mathematicians studied homological algebra in the 1940s. In recent decades, category theory has been widely used in a variety of areas. Because of the powerful abstraction capability, it is applicable to many problems. It may also be used as an axiomatic foundation for mathematics, as an alternative to set theory and other proposed foundations. Category theory has practical applications in programming language theory. More and more mainstream programming languages adopted category concepts. The usage of monads is realized in more than 20 languages [41]. Speaking in language of category, "a monad is a monoid in the category of endofunctors." Most basic

© China Machine Press 2024
X. Liu, *Mathematics in Programming*,
https://doi.org/10.1007/978-981-97-2432-1_4

computations have been abstracted in this way nowadays. For example, the generic list folding from right:

$$foldr\ f\ z\ [\ ] \quad = z$$
$$foldr\ f\ z\ (x{:}xs) = f\ x\ (foldr\ f\ z\ xs)$$

is written in the language of categories (Sect. 4.6) as:

$$foldr\ f\ z\ t\ = appEndo\ (foldMap\ (Endo \circ f)\ t)\ z$$

The most important thing is abstraction. Hermann Weyl said "Our mathematics of the last few decades has wallowed in generalities and formalizations." The same thing happens in programming. The problems in modern computer science are challenging us with the complexity we never seen before. Big data, distributed computation, high concurrency, as well as the requirement to consistency, security, and integrity. We can't catch them up in the traditional way, like brute-force exhaustive search, pragmatic engineering practice, and smart idea plus some luck. It forces us to adopt the latest methods and tools from mathematics and science. As Dieudonne said "This abstraction which has accrued in no way sprang from a perverse desire among mathematicians to isolate themselves from the scientific community by the use of a hermetic language. Their task was to find solutions for problems *handed down by the 'classical' age*, or arising directly from new discoveries in physics. They found that this was possible, but only through the *creation* of new objects and new methods, whose abstract character was *indispensable* to their success." [40]

Category theory was developed by Eilenberg and Mac Lane in the 1940s.

Samuel Eilenberg (1913–1998)

Samuel Eilenberg was born in Warsaw, Kingdom of Poland, to a Jewish family. His father was a brewer. Eilenberg studied at the University of Warsaw. A remarkable collection of mathematicians were on the staff there. He earned his Ph.D. in 1936. Lvov was the second mathematical center in Poland at that time. It was there that Eilenberg met Banach, who led the Lvov mathematicians. He joined the community of mathematicians working and drinking in the Scottish

Café, and he contributed problems to the Scottish Book, the famous book in which the mathematicians working in the Café entered unsolved problems. In 1939, Eilenberg's father convinced him that the right course of action was to emigrate to the United States. Once there he went to Princeton. In 1940, he was appointed as an instructor at the University of Michigan. In 1949 André Weil was working at the University of Chicago, and he contacted Eilenberg to ask him to collaborate on writing about homotopy groups and fiber spaces as part of the Bourbaki project. Eilenberg became a member of the Bourbaki team. He wrote the 1956 book *Homological Algebra* with Henri Cartan. Eilenberg mainly studied algebraic topology. He worked on the axiomatic treatment of homology theory with Norman Steenrod and on homological algebra with Saunders Mac Lane. In the process, Eilenberg and Mac Lane created category theory. Eilenberg spent much of his career as a professor at Columbia University. Later in life he worked mainly in pure category theory, being one of the founders of the field. He was awarded Wolf Prize in 1987 for his fundamental work in algebraic topology and homological algebra. Eilenberg died in New York City in January 1998.

Eilenberg was also a prominent collector of Asian art. His collection mainly consisted of small sculptures and other artifacts from India, Indonesia, Nepal, Thailand, Cambodia, Sri Lanka, and Central Asia. He donated more than 400 items to the Metropolitan Museum of Art in 1992 [42].

Saunders Mac Lane (1909–2005)

Saunders Mac Lane was born in Norwich, Connecticut, in the United States in 1909. He was christened "Leslie Saunders MacLane," but "Leslie" was later removed because his parents dislike it. He began inserting a space into his surname because his wife found it difficult to type the name without a space.

In high school, Mac Lane's favorite subject was chemistry. While in high school, his father died, and he came under his grandfather's care. His half-uncle helped to send him to Yale University and paid his way there beginning in 1926. His mathematics teacher, Lester S. Hill, coached him for a local mathematics competition which he won, setting the direction for his future work. He studied

mathematics and physics as a double major and graduated from Yale with a B.A.
in 1930. In 1929, at a party of Yale football supporters in New Jersey, Mac Lane
was awarded a prize for having the best grade point average yet recorded at Yale.
He met Robert Maynard Hutchins, the new president of the University of Chicago,
who encouraged him to go there for his graduate studies [43]. Mac Lane Joined
University of Chicago.[1] At the University of Chicago he was influenced by Eliakim
Moore, who was nearly 70 years old. He advised Mac Lane to study for a doctorate
at Göttingen in Germany certainly persuading Mac Lane to work at the foremost
mathematical research center in the world at that time. In 1931, having earned his
master's degree, Mac Lane earned a fellowship from the Institute of International
Education and became one of the last Americans to study at the University of
Göttingen prior to its decline under the Nazis.

At Göttingen, Mac Lane studied with Paul Bernays and Hermann Weyl. Before
he finished his doctorate, Bernays had been forced to leave the university because
he was Jewish, and Weyl became his main examiner. Mac Lane also studied with
Gustav Herglotz and Emmy Noether. In 1934, he finished his doctorate degree and
returned to United States.[2]

In 1944 and 1945, Mac Lane also directed Columbia University's Applied
Mathematics Group, which was involved in the war effort. Mac Lane was vice
president of the National Academy of Sciences and the American Philosophical
Society and president of the American Mathematical Society. While presiding over
the Mathematical Association of America in the 1950s, he initiated its activities
aimed at improving the teaching of modern mathematics. He was a member of the
National Science Board, 1974–1980, advising the American government. In 1976,
he led a delegation of mathematicians to China to study the conditions affecting
mathematics there. Mac Lane was elected to the National Academy of Sciences in
1949 and received the National Medal of Science in 1989.

Mac Lane's early work was in field theory and valuation theory. In 1941, while
giving a series of visiting lectures at the University of Michigan, he met Samuel
Eilenberg and began what would become a fruitful collaboration on the interplay
between algebra and topology. He and Eilenberg originated category theory in 1945.

Mac Lane died on April 14th, 2005 in San Francisco, California, USA.

4.1　Category

There are a variety of living species in the wild, like lion, jackal, wild cat, kite, goat,
owl, snake, rabbit, mouse, plants, etc. called objects. There are relations among these
living things, for example, a rabbit eats the grass; a line eats the rabbit; when the lion

---

[1] Hutchins soon offered Mac Lane a scholarship after the party. But Mac Lane neglected to actually
apply to the program, but showed up and was admitted anyway.

[2] Within days of finishing his degree, Mac Lane married Dorothy Jones, from Chicago, who had
typed his thesis. It was a small ceremony followed by a wedding dinner with a couple of friends in
the Rathaus Keller.

Fig. 4.1 An ecosystem of living objects and food chain arrows

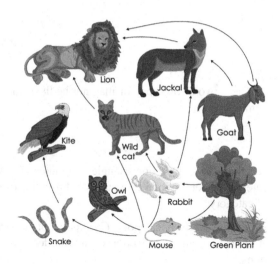

dies, bacteria break down its body, returning it to soil where it provides nutrients for plants like grass. They form an ecosystem of food chains, with arrows from grass to rabbit, from rabbit to lion, and etc. These objects together with the arrows form a structured and connected system (Fig. 4.1).

To become a category, this system needs to satisfy two axioms: (1) Every object has an arrow pointing to itself. Such self-pointed arrow is called the identity arrow. In the food web, the arrow from lion to rabbit means the rabbit is downstream to the lion along the food chain. We define every species is downstream to itself (it does not mean one eats itself, but does not eat itself), such that every object has an arrow pointing to itself. (2) Arrows are composite. For example, let the arrow from grass to rabbit be f, the arrow from rabbit to lion be g; compose them to $g \circ f$ (read as g after f), which means grass is downstream to lion along the food chain. This composed arrow can be further composed with a third one. For example, bacteria decompose the dead lion; denote the corresponding arrow from the lion to bacteria by h. Compose them together: $h \circ (g \circ f)$. It is identical to the arrow $(h \circ g) \circ f$. The upstream/downstream relations are associative.

Below diagram shows the arrows among three objects.

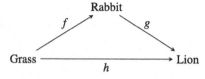

There are two paths from grass to lion. One is the composite arrow $g \circ f$, which means grass is downstream to the rabbit and rabbit is downstream to the lion; the other is arrow h, which means grass is downstream to lion, showing them as a pair of parallel arrows.

$$\text{Grass} \xrightarrow[\;h\;]{\;g \circ f\;} \text{Lion}$$

These two arrows may or may not be the same; when they are, we say they are commutative:

$$h = g \circ f$$

and the grass–rabbit–lion triangle commutes. The living things in the wild form a category under the food chain arrows. From this example, we abstract the definition of category.

Definition 4.1.1 A category C consists of a collection of objects,[3] denoted as $A, B, C, \ldots$, and a collection of arrows denoted as $f, g, h, \ldots$, with four operations:

(a) Two total operations:[4] *source* and *target*.[5] $A \xrightarrow{f} B$ means assign object A as the source of arrow f, and object B as the target of f;

(b) The third total operation is identity arrow.[6] $A \xrightarrow{id_A} A$ assigns identity arrow from A to itself for every object A (self-pointed);

(c) A partial operation: composition. Compose two arrows $B \xrightarrow{f} C$ and $A \xrightarrow{g} B$ to $A \xrightarrow{f \circ g} C$, or $f \circ g$ for short, read as "f after g."

Besides, a category satisfies the following two axioms:

(1) **Associativity**: Arrows are associative. For every three composable arrows f, g, h:

$$f \circ (g \circ h) = (f \circ g) \circ h$$

Associativity allows simplification without parentheses as $f \, g \, h$.

(2) **Identity**: The identity arrows act as the identity for composition. For every arrow $A \xrightarrow{f} B$:

$$f \circ id_A = f = id_B \circ f$$

---

[3] It has nothing related to the object oriented programming. Here the object means abstract thing.

[4] Total operation is defined for every object without exception; its counterpart is partial operation, which is not defined from some objects. For example, the negate $x \mapsto -x$ is a total operation for natural numbers, while the reciprocal $x \mapsto 1/x$ is not defined for 0; hence it is a partial operation.

[5] Should be treated as verb: assign source and assign target.

[6] Should also be treated as verb: assign identity arrow.

The identity axiom explains $A \xrightarrow{id_A} A \xrightarrow{f} B \xrightarrow{id_B} B$; f is identical to the composite of the first two and last two arrows. The subscript of id differentiates the source/target of each identity arrow. For example, the arrow $f : x \mapsto e^{ix}$ is from A of real number axis to B of the unit circle in complex plan. id_A self-points to the axis, while id_B self-points to the unit circle. They are different identity arrows.

4.1.1 Examples of Categories

In mathematics, a set is called a monoid if it is defined with a special identity element and associative binary operation (Definition 3.2.2). The integers, for example, with 0 as the identity element and + as the binary operation, form a monoid.

A monoid may contain other things besides numbers. For the set of words and sentences (called strings in programming), define the binary operation that concatenates two strings, for example, `"red"` $+\!\!+$ `"apple"` = `"redapple"`. Strings form a monoid, where the identity element is the empty string `""`. It's easy to verify the monoid axioms:

(1) The string concatenation is associative.

   `"red"` $+\!\!+$ (`"apple"` $+\!\!+$ `"tree"`) = (`"red"` $+\!\!+$ `"apple"`) $+\!\!+$ `"tree"`

(2) Concatenating any string with empty string (the identity element) gives the string itself:

   `""` $+\!\!+$ `"apple"` = `"apple"` = `"apple"` $+\!\!+$ `""`

Put the monoid of strings aside, consider another monoid. The elements are sets of characters; the binary operation is set union; the identity element is the empty set. Union of two character sets gives a bigger set, for example,

$$\{a, b, c, 1, 2\} \cup \{X, Y, Z, 0, 1\} = \{a, b, c, X, Y, Z, 0, 1, 2\}$$

Together with the first example of additive monoid of integers, we have three monoids on hand. We next setup transforms among them (called morphisms in mathematics). First is the map from monoid of strings to the monoid of character sets. From any string (word or sentence), collect the distinct characters to form a set. Call this operation *chars*. It satisfies the following:

chars(`"red"` $+\!\!+$ `"apple"`) = *chars*(`"red"`) $\cup$ *chars*(`"apple"`) and *chars*(`""`) = $\varnothing$

The distinct characters of a concatenated string are identical to the union of each character set; empty string does not contain any characters (corresponds to the empty set). *chars* with these strong properties is a *homomorphism* (Sect. 3.2.3).

Fig. 4.2 The monoid as a
category with only one object

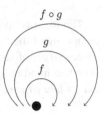

Second, define a transform from the monoid of character sets to the additive monoid of integers: count the size (how many characters) of a character set. The empty set contains 0 character. Call this operation *count*. It is not a homomorphism from set to integer because $|\{r, e, d\}| + |\{a, p, l, e\}| = 7 \neq |\{r, e, d\} \cup \{a, p, l, e\}| = |\{a, e, d, l, p, r\}| = 6$. Next, compose the two transforms:

$$\text{String} \xrightarrow{\text{chars}} \text{Character set} \xrightarrow{\text{count}} \text{Integer}$$

to

$$\text{String} \xrightarrow{\text{count} \circ \text{chars}} \text{Integer}$$

Obviously, the composite result *count* ∘ *chars* is a transform too. It first collects the distinct characters from a string and then counts the number of characters. We obtain the *monoid category **Mon***. The objects are a variety of monoids; the arrows are morphisms (transforms) among objects; the identity arrows self-point to every monoid.

This is a very big category. It contains *all* monoids.[7] On the contrary, category can also be small of only one monoid. For example, the monoid of strings; there is only one object. It doesn't matter what this object is. It can be any fixed set or even the set of all strings (although it does not look small). For any string like `"hello"`, define the prefix operation, that prepands `"hello"` to any other string. Call this operation (*prefix* `"hello"`). Apply (*prefix* `"hello"`) to word `"Alice"` gives string `"helloAlice"`; Similarly, if define (*prefix* `"hi"`), and apply it to `"Alice"`, gives `"hiAlice"`. Take them as two arrows, then their composition is:

(*prefix* `"hello"`) ∘ (*prefix* `"hi"`) = (*prefix* `"hellohi"`)

It first prepends `"hi"`, then prepends `"hello"`. It's easy to verify that, such arrows are associative. For the identity arrow, any string keeps the same when prepending with empty string; hence (*prepend* `""`) is the identity arrow. This monoid is a category with only one object as in Fig. 4.2. Every arrow in the category corresponds to an element in the monoid, the arrow composition is given by the

---

[7] One may think about Russel's paradox. Strictly speaking, **Mon** contains all "small" monoids.

monoid binary operation ($+\!\!+$ in this example), and the identity arrow corresponds to the identity element of the monoid ($"\,"$ in this example).

For monoid, we see both the category **Mon**, that contains all monoids, and the category of only one monoid. It is like a hidden world in a piece of sand. Interestingly, for any object A in a category C, define $hom(A, A)$ as the set of all arrows from A to A itself. Then this set of arrows forms a monoid under composition, where the identity element is the identity arrow. Such symmetry emerges surprisingly.

The second example is set. Treat every set as an object;[8] arrows are functions (or maps), for example, f from set A to set B. A is the domain of f, and B is the range (or codomain)[9] of f. Arrow composition is defined as function composition: $z = (g \circ f)(x) = g(f(x))$ for $y = f(x)$ and $z = g(y)$. It's associative. The identity arrows are the identity functions of each set: $id_A(x) = x$ of type $A \rightarrow A$. This is the category **Set** of all sets and functions.

The third example contains a pair: *partial order set* and *pre-order set*. An order defines the comparable relation among set elements, often denoted by $\leq$. The notation does not necessarily mean less than or equal, but is generic; for example, a set is subset of another, a string is prefix of another, a person is descendant of another, etc. A relation $\leq$ is pre-order if it satisfies the following:

(a) **reflexive**: $a \leq a$ for every a;
(b) **transitive**: If $a \leq b$ and $b \leq c$, then $a \leq c$;

Besides these two, $\leq$ is partial order if follows an additional one:

(a) **anti-symmetric**: if $a \leq b$ and $b \leq a$, then $a = b$;

A set satisfies pre-order and a set satisfies partial order are

$$\text{preset} \qquad \text{poset}$$

Two elements may not be comparable. For example, Duck Donald's family is a poset under descendant relation (Fig. 4.3). Donald $\leq$ Quackmore, but neither Huey and Donald, nor Donald and Scrooge are comparable. Although everyone has his/her ancestor (treat the roots as ancestors of themselves by the reflexive rule), those at the same level or in different branches are not comparable.

For another example, Fig. 4.4 shows a poset consists of all subsets of $\{x, y, z\}$ under inclusion (*subseteq*). Although every subset has its superset, $\{x\}$ and $\{y, z\}$ are not comparable, nor those at the same level.

In general, any poset is a preset, but the converse is not necessarily true. A monotone function never decreases: if $x \leq y$, then $f(x) \leq f(y)$. By composing

---

[8] The set of all sets may end up with Russel's paradox "the set all sets that does not contain itself"; see Sect. 7.2.

[9] Assume total function, which is applicable to every element of A.

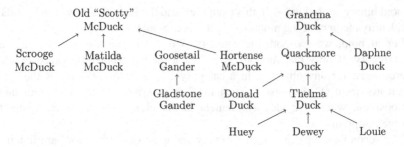

Fig. 4.3 Donald Duck's family tree

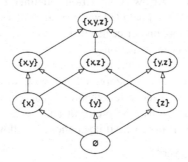

Fig. 4.4 All subsets form a partial order set under ⊆

monotone functions for poset or preset, we obtain a pair of categories:

$$\textbf{\textit{Pre}} \qquad \textbf{\textit{Pos}}$$

Objects are *presets* and *posets*; arrows are monotone functions; the identity function, as is also monotone, is the identity arrow in the category.

We intend to choose categories of monoids and preset as two examples. They are the two most simple categories. The study of monoids is the study of composition in the miniature. The study of presets is the study of comparison in the miniature. It is the core of the entire category to study the composition and comparison of mathematical structures. In a sense, every category is an amalgam of certain monoid and preset [44].

Analogue to monoid, **Pre** is a big category that contains all the presets on one hand; a solely preset as category can also be small on the other hand. Denote the elements by $i, j, k, \ldots$; if i and j are comparable, and $i \leq j$, define an arrow:

$$i \longrightarrow j$$

Treat elements as objects. For every two objects, either there is no arrow in between, which means they are not comparable; or there exists one, which means the $\leq$ relationship. In summary, there is at most one arrow between any two objects.

It's easy to verify such arrows are composable. Since $i \leq i$ for every i by the reflexive condition, self-pointed arrows exist. The preset itself is a category. We see the symmetry between monoid and preset as categories: monoid category contains only one object, but has many arrows; while preset category contains many objects, it has at most one arrow between objects.

Exercise 4.1

4.1.1 Show that the identity arrow is unique for each object (hint: refer to the uniqueness of identity element of group in Sect. 3.2.3).

4.1.2 Verify the monoid $(S, \cup, \varnothing)$ (the elements are sets; the binary operation is set union; the identity element is empty set) and $(\mathbb{N}, +, 0)$ (elements are natural numbers; the binary operation is add; the identity element is zero) are all categories with only one object.

4.1.3 Introduce Peano's axioms for natural numbers and things with the same structure of nature numbers, like linked-list. They can be described in categories. This was found by Dedekind although the category theory was not established by his time. We named this category as Peano category, denoted as **Pno** nowadays. The objects in this category is (A, f, z), where A is a set, for example, natural numbers $\mathbb{N}$; $f : A \to A$ is a successor function. It is *succ* for natural numbers; $z \in A$ is the starting element; it is zero for natural numbers. Given any two Peano objects (A, f, z) and (B, g, c), define morphism from A to B as:

$$A \xrightarrow{\phi} B$$

It satisfies:

$$\phi \circ f = g \circ \phi \quad \text{and} \quad \phi(z) = c$$

Verify that **Pno** is a category.

Answer: page 141

4.1.2   *Arrow $\neq$ Function*

In most examples so far, arrows are functions (maps or morphisms) or comparisons. It gives an illusion that arrows are function like things. The example of relation category helps to avoid such deception. In this category, objects are sets; the arrow

$A \xrightarrow{R} B$ from set A to set B is defined as:

$$R \subseteq B \times A$$

What does it mean? The product of $B \times A$ consists of all pairs of (b, a) from B and A:

$$B \times A = \{(b, a)|b \in B, a \in A\}$$

Use the Donald Duck's family, for example, set $A = \{$ Scrooge McDuck, Matilda McDuck, Hortense McDuck$\}$ contains three members of the McDucks; $B = \{$Goosetail Gander, Quackmore Duck$\}$ contains two external members to the McDucks. The product of $B \times A$ consists of 3×2 pairs:

(b, a)	Goosetail Gander	Quackmore Duck
Scrooge McDuck	(Goosetail Gander, Scrooge McDuck)	(Quackmore Duck, Scrooge McDuck)
Matilda McDuck	(Goosetail Gander, Matilda McDuck)	(Quackmore Duck, Matilda McDuck)
Hortense McDuck	(Goosetail Gander, Hortense McDuck)	(Quackmore Duck, Hortense McDuck)

R is a subset of $B \times A$; it represents some relation between A and B. If a in A and b in B satisfy the relation R, then the pair $(b, a) \in R$, denoted as aRb (treat R as infix operator). For example, $R = \{$(Goosetail Gander, Matilda McDuck), (Quackmore Duck, Hortense MuDuck)$\}$ defines the relation of couples (In the story of Donald Duck, Goosetail married Matilda, they adopted Gladstone as their child; Quackmore married Hortense, they are parents of Donald).

In summary, $B \times A$ contains all possible combinations of (b, a) and is a complete set of relations; a subset R of it as an arrow defines some relation aRb if $(b, a) \in R$; the set of all arrows $\{A \xrightarrow{R} B\}$ represents all possible relations between A and B.

There is composition of arrows:

$$A \xrightarrow{R} B \xrightarrow{S} C$$

if both relations bRa and cSb hold for some element b in set B. It implies cTa, where $T = S \circ R$. We can view it as a special intersection of R and S, that is, $T = \{(c, a)|(b, a) \in R$ and $(c, b) \in S\}$. For example in Donald Duck's family, let set $C = \{$Gladstone, Donald, Thelma$\}$; relation R is same as above; relation $S = \{$(Donald, Quackmore), (Thelma, Quackmore)$\}$ are children and father. The composite arrow $S \circ R$ is a set of $\{$(Donald, Hortense), (Thelma, Hortense)$\}$, each pair of (c, a) stands for child and mother. Both relations of couple (R) and child–father (S) are satisfied through Quackmore; Donald and Thelma are children of their mother Hortense.

For identity arrows, every set A has the identical relation to itself $id_A = \{(a,a)|a \in A\}$; it is the special subset of $A \times A$ which equals to $\{(a,a')|a = a'\}$ for all a, a' in A.

We can generate a new category from any category C by reversing its arrows, namely, the dual *opposite* category C^{op}. When understanding a category, we understand its dual category too.

4.2 Functors

Categories abstract monoids and maps; posets and monotone functions; sets and relations. The next question is how to bridge categories and compare them? Functor[10] is used to compare categories and their inner relations.

We may consider functor as the transform between categories. However, it does not only map objects from one category to another, but also maps arrows. This makes functor different from normal morphisms (e.g., between sets). We often use **F** as the notation for functor. Since functor likes the morphism for category, it must faithfully preserve the structure and relations between categories. In order to achieve this, a functor needs to satisfy two conditions:

1. Any identity arrow in one category is transformed to identity arrow in another:

$$A \xrightarrow{id} A \quad \longmapsto \quad FA \xrightarrow{id} FA$$

2. Any composition of arrows in one category is transformed to composition in another.[11]

$$\mathbf{F}(g \circ f) = \mathbf{F}(g) \circ \mathbf{F}(f)$$

In summary, functor preserves identity and composition. One may compose functors too, for example, functor $(\mathbf{FG})(f)$ means first apply **G** to arrow f, next apply **F** to **G**(f).

---

[10] Some programming languages use the term functor for function object, an entity that acts like a function. It has nothing to do with category theory.

[11] There are two types of transformation, covariance and contravariance; the latter reverts the arrows. The two terms are also used in type system of programming languages. We limit to covariance in this book.

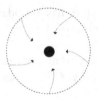

Constant functor acts like a blackhole

Let's use some examples to understand functor.

Example 4.2.1 A functor is an *endo-functor*[12] if it maps a category to itself. The simplest functor is the identity functor, which is an endo-functor, denoted as id : $C \rightarrow C$. It can be applied to any category, maps object A to A, and maps arrow f to f.

Example 4.2.2 The second simple functor is the constant functor. It acts like a blackhole. Denote it by $\mathbf{K}_B : C \rightarrow B$. When applied to a category, it maps all objects to the blackhole object B and maps all arrows to the identity arrow in the blackhole id_B. The blackhole category has only one identity arrow, satisfying composition law: $id_B \circ id_B = id_B$.

Example 4.2.3 (Maybe Functor) Tony Hoare (Sir Charles Antony Richard Hoare), computer scientist and winner of ACM Turing award in 1980, had an interesting speaking apologized for inventing the null reference.[13]

Tony Hoare

I call it my billion-dollar mistake. It was the invention of the null reference in 1965. At that time, I was designing the first comprehensive type system for references in an object oriented language (ALGOL W). My goal was to ensure that all use of references should be

---

[12] Analogue to automorphism (see Sect. 3.2.3).

[13] At a software conference QCon London in 2009.

Fig. 4.5 Maybe functor

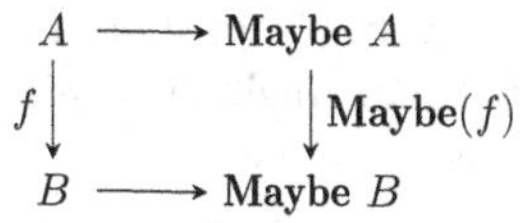

$$
\begin{array}{ccc}
A & \longrightarrow & \text{Maybe } A \\
f \downarrow & & \downarrow \text{Maybe}(f) \\
B & \longrightarrow & \text{Maybe } B
\end{array}
$$

absolutely safe, with checking performed automatically by the compiler. But I couldn't resist the temptation to put in a null reference, simply because it was so easy to implement. This has led to innumerable errors, vulnerabilities, and system crashes, which have probably caused a billion dollars of pain and damage in the last forty years [45].

From 2015, mainstream programming environments gradually adopted the maybe concept to replace the error-prone null.[14] Diagram in Fig. 4.5 shows how **Maybe** functor behaves.

Let the left side objects A and B be programming data types, like *Int*, *Bool*; the right side objects are mapped types through functor **Maybe**. If A is *Int*, then its right side corresponds to **Maybe** *Int*; if B is *Bool*, then its right side corresponds to **Maybe** *Bool*. Below is how the maybe functor maps an object:

```
data Maybe A = Nothing | Just A
```

The functor maps type A as an object to type **Maybe** A. Note the object is a type, but not a value. The value of **Maybe** A is either empty, denoted by *Nothing*, or a value of *Just a* for some a in A. For example of *Int* type, the maybe functor maps it to **Maybe** *Int*; the value could be *Nothing*, *Just* 5, or *Just* 0.

Consider a binary search tree (Sect. 2.6) that contains elements of type A. When look it up a value, as the value may not exists, the type of the result is **Maybe** A (see Sect. 4.6 for the complete example program):

$$
\begin{aligned}
lookup\ Nil\ \_ &= Nothing \\
lookup\ (Br\ l\ k\ r)\ x &= \begin{cases} x < k : & lookup\ l\ x \\ x > k : & lookup\ r\ x \\ x = k : & Just\ k \end{cases}
\end{aligned}
\tag{4.1}
$$

For **Maybe** type data, we must handle two possible cases of value. Below function, for example, tests whether a maybe value holds anything:

$$
\begin{aligned}
elem\ Nothing &= False \\
elem\ (Just\ x) &= True
\end{aligned}
$$

Functor maps objects and arrows. We see how **Maybe** maps objects; how does it map arrows? In Fig. 4.5, there is a vertical arrow $A \xrightarrow{f} B$ on the left and another vertical arrow **Maybe** $A \xrightarrow{\text{Maybe}(f)}$ **Maybe** B on the right. Denote the right side

---

[14] For example, the `Optional<T>` in Java and C++.

arrow by f' for short. If we know the behavior of arrow f, then what does arrow f' behave? Analogue to *elem*, it must handle two possible cases for any data of **Maybe** type:

$$f' \, Nothing = Nothing$$
$$f' \, (Just \, x) = Just \, (f \, x)$$

It is exactly what **Maybe** functor does for an arrow: maps f to f'. In programming environment, we often use $fmap$ to define the mapping of arrows for every functor **F** as:

$$fmap : (A \to B) \rightsquigarrow (\mathbf{F}A \to \mathbf{F}B)$$

or make the squiggle arrow straight and put the label above:

$$(A \to B) \xrightarrow{fmap} (\mathbf{F}A \to \mathbf{F}B) \tag{4.2}$$

$fmap$ maps an arrow from A to B to the arrow from $\mathbf{F}A$ to $\mathbf{F}B$ for functor **F**. For **Maybe** functor, the corresponding $fmap$ is defined as below:

$$fmap : (A \to B) \to (\mathbf{Maybe} \, A \to \mathbf{Maybe} \, B)$$
$$fmap \, f \, Nothing = Nothing \tag{4.3}$$
$$fmap \, f \, (Just \, x) = Just \, (f \, x)$$

Back to the binary search tree example, let the elements in the tree be integers. We want to search some value and convert it to binary format if find; otherwise return *Nothing*. There has already existed a function that converts a decimal number to its binary format:

$$binary(n) = \begin{cases} n < 2 : & [n] \\ \text{Otherwise} : & binary\left(\left\lfloor \dfrac{n}{2} \right\rfloor\right) + [n \bmod 2] \end{cases}$$

Here is the corresponding example program, implemented in tail recursive[15] manner.

```
binary = bin [] where
    bin xs 0 = 0 : xs
    bin xs 1 = 1 : xs
    bin xs n = bin ((n `mod` 2) : xs) (n `div` 2)
```

---

[15] A recursive function is tail recursive if the final result of the recursive call is the final result of the function itself. If the result of the recursive call must be further processed (e.g., by adding 1), it is not tail recursive.

Fig. 4.6 *fmap* as **Maybe**
functor lifts *binary* arrow up

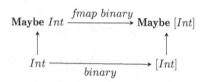

With functor, we can "lift" *binary* function as an arrow up:

Further, we can apply the lifted *binary* to the *lookup* result (Eq. (4.1)) of a binary search tree:

$$fmap\ binary\ (lookup\ t\ x)$$

There are two worlds (categories) in diagram Fig. 4.6. The lower world on the ground is threatened by the null reference; the upper world in the sky is free from null and safe. With maybe functor, all legacy programs on the ground, even they are not capable to handle null, are lifted up to the safe world dominated by maybe.

Proof *(Readers may skip this box)* To show **Maybe** is a functor, we verify the two conditions about arrow mapping:

$$\begin{aligned} fmap\ id\ \ &= id \\ fmap\ (f \circ g) &= fmap\ f \circ fmap\ g \end{aligned} \tag{4.4}$$

where $id\ x = x$ is identity arrow. For the first condition:

$$\begin{aligned} fmap\ id\ Nothing &= Nothing &&\text{by Eq. (4.3)} \\ &= id\ Nothing &&\text{reverse of } id \end{aligned}$$

and

$$\begin{aligned} fmap\ id\ (Just\ x) &= Just\ (id\ x) &&\text{by Eq. (4.3)} \\ &= Just\ x &&\text{definition of } id \\ &= id\ (Just\ x) &&\text{reverse of } id \end{aligned}$$

(continued)

For the second condition:

$$fmap\ (f \circ g)\ Nothing = Nothing \qquad\qquad \text{by Eq. (4.3)}$$
$$= fmap\ f\ Nothing \qquad\qquad \text{reverse of}$$
$$fmap$$
$$= fmap\ f\ (fmap\ g\ Nothing) \qquad \text{reverse of}$$
$$fmap$$
$$= (fmap\ f \circ fmap\ g)\ Nothing \quad \text{composition}$$

and

$$fmap\ (f \circ g)\ (Just\ x) = Just\ ((f \circ g)\ x) \qquad\qquad \text{by Eq. (4.3)}$$
$$= Just\ (f\ (g\ x)) \qquad\qquad \text{reverse of}$$
$$\text{composition}$$
$$= fmap\ f\ (Just\ (g\ x)) \qquad \text{reverse of}$$
$$fmap$$
$$= fmap\ f\ (fmap\ g\ (Just\ x)) \qquad \text{reverse of}$$
$$fmap$$
$$= (fmap\ f \circ fmap\ g)\ (Just\ x) \quad \text{composition}$$

□

Example 4.2.4 (List Functor) We reuse the definition of list in Sect. 1.5 to map objects.

```
data List A = Nil | Cons(A, List A)
```

From programming perspective, it defines the linked-list data structure. The elements stored are of type A, so it's called link-list of type A. From category perspective, a list functor maps both objects and arrows. Here objects are types; arrows are total functions (see Fig. 4.7):

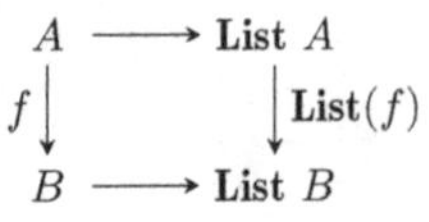

Fig. 4.7 List functor

$$
\begin{array}{ccc}
A & \longrightarrow & \textbf{List}\ A \\
f\ \downarrow & & \downarrow\ \textbf{List}(f) \\
B & \longrightarrow & \textbf{List}\ B
\end{array}
$$

Objects A, B on the left side are data types, like Int, $Bool$, or even complex types such as **Maybe** $Char$. Objects on the right side are types mapped by list functor. If A is Int, then its right side corresponds to **List** Int; if B is $Char$, then its right side corresponds to **List** $Char$, which is $String$ essentially.

We highlight again that object is type, but not value. A may be Int, but cannot be 5, a particular value. **List** Int corresponds to list of integers, but not a particular list, like [1, 1, 2, 3, 5]. How to generate a list of values? Use function Nil and $Cons$ to generate empty list or list like $List(1, List(1, List(2, Nil)))$.

We know how list functor maps objects, but how does it map arrows? Given a function $f : A \rightarrow B$, how to get another function g : **List** $A \rightarrow$ **List** B through list functor? Analogue to maybe functor, we need another instance of $fmap$, which sends arrow f to arrow g:

$$fmap : (A \rightarrow B) \rightsquigarrow (\textbf{List } A \rightarrow \textbf{List } B)$$

or make the squiggle arrow straight and put the label above:

$$(A \rightarrow B) \xrightarrow{fmap} (\textbf{List } A \rightarrow \textbf{List } B)$$

We next figure out how g behaves. For the trivial case, g sends empty (Nil) of **List** A to empty of **List** B for whatever f:

$$fmap\ f\ Nil = Nil$$

For the recursive case $Cons(x, xs)$, where x is some value of type A, and xs is a sub-list of type **List** A. Let $f(x) = y$, which sends x of type A to y of type B. We take three steps: (1) apply f to x to get y; (2) recursively apply f to every element of the sub-list xs; convert xs to a new sub-list ys of B; (3) link y to ys by $Cons$ to get the final result:

$$fmap\ f\ Cons(x, xs) = Cons(f\ x, fmap\ f\ xs)$$

Summarize the two cases together to complete the definition of $fmap$ for list:

$$
\begin{aligned}
&fmap : (A \rightarrow B) &&\rightarrow (\textbf{List } A \rightarrow \textbf{List } B) \\
&fmap\ f\ Nil &&= Nil \\
&fmap\ f\ Cons(x, xs) &&= Cons(f\ x, fmap\ f\ xs)
\end{aligned}
\tag{4.5}
$$

Or with the simplified infix notation ":" for $Cons$, and [ ] for Nil (Sect. 1.5):

$$
\begin{aligned}
&fmap : (A \rightarrow B) \rightarrow (\textbf{List } A \rightarrow \textbf{List } B) \\
&fmap\ f\ [\,] &&= [\,] \\
&fmap\ f\ (x{:}xs) &&= (f\ x){:}(fmap\ f\ xs)
\end{aligned}
\tag{4.6}
$$

Compare with the list mapping (for each map in Eq. (1.21)), they behave essentially same. It's perfectly okay to reuse list mapping as an alias of list functor, for example as below program:

```
instance Functor [] where
    fmap = map
```

(Readers may skip the proof in this box) To close this example, we verify the arrow mapping preserves identity and composition:

$$fmap\ id\ \ \ \ = id$$
$$fmap\ (f \circ g) = fmap\ f \circ fmap\ g$$

Proof Use mathematical induction on the length of list, first for 0 of empty list ([ ]):

$$fmap\ id\ [\ ] = [\ ] \qquad \text{by Eq. (4.6)}$$
$$= id\ [\ ] \qquad \text{reverse of } id$$

Assume $fmap\ id\ xs = id\ xs$ holds for xs of length n; we show it holds for $n + 1$ of list $x{:}xs$:

$$fmap\ id\ (x{:}xs) = (id\ x){:}(fmap\ id\ xs) \qquad \text{by Eq. (4.6)}$$
$$= (id\ x){:}(id\ xs) \qquad \text{induction assumption}$$
$$= x{:}xs \qquad \text{definition of } id$$
$$= id\ (x{:}xs) \qquad \text{reverse of } id$$

For composition, use mathematical induction again on the length of list. For 0 of empty list ([ ]):

$$fmap\ (f \circ g)\ [\ ] = [\ ] \qquad \text{by Eq. (4.6)}$$
$$= fmap\ f\ [\ ] \qquad \text{reverse of } fmap$$
$$= fmap\ f\ (fmap\ g\ [\ ]) \qquad \text{reverse of } fmap$$
$$= (fmap\ f \circ fmap\ g)\ [\ ] \qquad \text{reverse composition}$$

(continued)

Assume $fmap\ (f \circ g)\ xs = (fmap\ f \circ fmap\ g)\ xs$ holds for xs of length n; we show it holds for $n + 1$ of list $x:xs$:

$$fmap\ (f \circ g)\ (x:xs) = ((f \circ g)\ x):(fmap\ (f \circ g)\ xs) \qquad \text{by Eq. (4.6)}$$

$$= ((f \circ g)\ x):((fmap\ f \circ fmap\ g)\ xs) \quad \text{induction}$$

assumption

$$= (f(g\ x)):(fmap\ f\ (fmap\ g\ xs)) \qquad \text{composition}$$

$$= fmap\ f\ ((g\ x):(fmap\ g\ xs)) \qquad \text{reverse of}$$

$fmap$

$$= fmap\ f\ (fmap\ g\ (x:xs)) \qquad \text{reverse of}$$

$fmap$

$$= (fmap\ f \circ fmap\ g)\ (x:xs) \qquad \text{reverse}$$

composition

$\square$

Exercise 4.2

4.2.1 For the list functor, define the arrow map with $foldr$.
4.2.2 Verify that the functor compositions of **Maybe** ○ **List** and **List** ○ **Maybe** are all functors.
4.2.3 Show that the composition for any functors **G** ○ **F** is still a functor.
4.2.4 Give an example functor for preset.
4.2.5 Define functor for binary tree (see Sect. 2.6).

Answer: page 151

4.3 Products and Coproducts

We start from set as example. Given two set A and B, their Cartesian product $A \times B$ is the set of all ordered pairs (a, b).

$$\{(a, b) | a \in A, b \in B\}$$

For example, the product of sets $\{1, 2, 3\}$ and $\{a, b\}$ is:

$$\{(1, a), (2, a), (3, a), (1, b), (2, b), (3, b)\}$$

As mentioned in 4.1.2, the relation arrow is exactly defined as subset of product. Set A and B may have algebraic structure of the same kind, like group, ring, etc.; then their product carries the structure. The product can be decomposed through projections to each component as below:

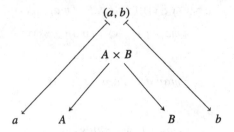

Symmetric to Cartesian product, there is a dual construction. From two sets A and B without overlapping, we generate a disjoint union set (sum) $A + B$. In order to distinguish from which set, A or B, an element in the sum comes from, we add a tag:

$$A + B = (A \times \{0\}) \cup (B \times \{1\})$$

For example, the tagged sum of $\{1, 2, 3\}$ and $\{a, b\}$ is:

$$\{(1, 0), (2, 0), (3, 0), (a, 1), (b, 1)\}$$

For every element (x, tag) in $A + B$, if the tag is 0, we know x is from A; else if the tag is 1, then x is from B. Below example program builds $A + B$ from two lists:

```
A + B = zip A [0, 0, ...] ++ zip B [1, 1, ...]
```

Similarly, if set A and B have algebraic structure of the same kind, their sum carries the structure. The two embeddings locate disjoint copies of the parent sets within the sum as below:

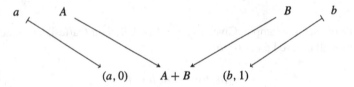

The two constructions are symmetric. If we rotate the two diagrams by 90°, laying *A* on top of *B*, then the two problems appear left and right. They are symmetric like our two hands. We can find many dual constructions in category theory. Our world is full of symmetric things. When we understand one, we understand the dual one.

René Descartes (1596–1650). Portrait after Frans Hals, Louvre Museum

Cartesian product, also known as direct product, is named after René Descartes, a philosopher, mathematician, and scientist. Descartes also had a Latinized name: Cartesius; its adjective form is Cartesian. This is the reason why things are said to be Cartesian, for example, Cartesian coordinate, Cartesian product, etc.

Descartes was born in La Haye, France, in 1596. His father was a member of the Parlement of Brittany at Rennes. His mother died a year after giving birth to him, and so he was not expected to survive. His father then remarried in 1600. Descartes lived with his grandmother at La Haye. His health was poor when he was a child. Throughout his childhood, up to his 20s, he was pale and had a persistent cough.

In 1607, he entered the Jesuit College of La Flèche, where he was introduced to mathematics and physics, including Galileo's work. While in the school his health was poor, and, instead of rising in the morning like the other boys, he was granted permission to remain in bed until 11:00 AM, a custom he maintained until the year of his death.

After graduation in 1614, he studied for 2 years at the University of Poitiers. He received a law degree in 1616 to comply with his father's wishes, but he quickly decided that this was not the path he wanted to follow. He returned to Paris and then became a volunteer in the army of Maurice of Nassau. In the army, Descartes started studying mathematics and mechanics under the Dutch scientist Isaac Beeckman and began to seek a unified science of nature.

From 1620 to 1628 Descartes traveled through Europe, spending time in Bohemia, Hungary, Germany, and Holland, through Switzerland to Italy and then Venice and Rome. He returned to Paris in 1625. His Paris home became a meeting place for philosophers and mathematicians and steadily became more and more busy. By 1628 Descartes, tired of the bustle of Paris, the house full of people, and the life of traveling he had before, decided to settle down where he could work in solitude. He gave much thought and chose Holland. What he longed for was somewhere peaceful where he could work away from the distractions of a city such as Paris yet still have access to the facilities of a city. He told his friend Marin Mersenne[16] where he was living so that he might keep in touch with the mathematical world, but otherwise he kept his place of residence a secret. Descartes wrote all his major work during his over 20 years in the Netherlands, initiating a revolution in mathematics and philosophy. In 1633, Galileo was condemned by the Italian Inquisition, and Descartes abandoned plans to publish *Treatise on the World*, his work of the previous 4 years. Nevertheless, in 1637 he published parts of this work in three essays, *The Meteors*, *Dioptrics*, and *Geometry*, preceded by an introduction, his famous *Discourse on the Method*.

In *Geometry*, Descartes exploited the discoveries he made with Pierre de Fermat. This later became known as Cartesian Geometry. Descartes continued to publish works concerning both mathematics and philosophy for the rest of his life. In 1641 he published a metaphysics treatise, *Meditations on First Philosophy*. It was followed in 1644 by *Principles of Philosophy*. He became the most influential philosophers in Europe.

In 1649 Queen Christina of Sweden persuaded Descartes to go to Stockholm. However the Queen wanted to draw tangents at 5 AM, and Descartes broke the habit of his lifetime of getting up at 11 o'clock. After only a few months in the cold northern climate, walking to the palace for 5 o'clock every morning, he died of pneumonia February 1650 at the age of 54.

Descartes left the best-known philosophical statement "I think, therefore I am." He laid the foundation for seventeenth-century continental rationalism. He was well-versed in mathematics as well as philosophy and contributed greatly to science as well. He is credited as the father of analytical geometry, the bridge between algebra and geometry—used in the discovery of infinitesimal calculus and analysis. Descartes was also one of the key figures in the Scientific Revolution.

4.3.1 Definition of Product and Coproduct

Definition 4.3.1 For a pair of objects A and B in category C, a *wedge*

to   from

---

[16] Mersenne was a French polymath, whose works touched a wide variety of fields. He is perhaps best known today among mathematicians for Mersenne prime numbers, in the form $M_n = 2^n - 1$ for some integer n. He had many contacts in the scientific world and has been called "the center of the world of science and mathematics during the first half of the 1600s."

the pair A, B is an object X together with a pair of arrows

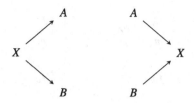

in the parent category C.

For a given pair of objects A, B in the category, there may be many wedges on one side or the other. We look for the "best possible" wedge, the one that is as "near" the pair as possible. It leads to the below pair of definitions:

Definition 4.3.2   Given a pair of objects A, B in category C, a

                     product   coproduct

of the pair is a wedge:

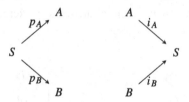

with the following universal property, for each wedge:

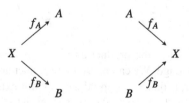

there is a unique arrow:

$$X \xrightarrow{m} S \qquad S \xrightarrow{m} X$$

such that:

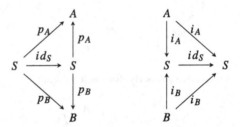

commutes. This arrow m is the *mediating arrow* (or mediator) for the wedge on X.

In this definition, the product or coproduct is not just an object S, but an object together with a pair of arrows. For any given X, the mediating arrow m is unique. When we say the diagram commutes, it means the arrows satisfying the following:

$$\begin{cases} f_A & = p_A \circ m \\ f_B & = p_B \circ m \end{cases} \qquad\qquad \begin{cases} f_A & = m \circ i_A \\ f_B & = m \circ i_B \end{cases}$$

It immediately leads to this special case: if $X = S$, then m is an endo-arrow (self-pointed): $S \xrightarrow{m} S$, hence, is the identity arrow (see Exercise 4.1.1). The diagram then simplifies to:

Among the many wedges, the product and coproduct are special; they are the "closest" or the "best" wedge. We can prove that product and coproduct are unique (Appendix C). However, product or coproduct may not exist at the same time; it's also possible that neither does exist. We next show what product and coproduct mean to set.

Proposition 4.3.1 *Let A, B be a pair of sets.*

$A \times B$ *(Cartesian product)* $\qquad\qquad$ $A + B$ *(disjoint union) furnished with canonical functions form the*

$$product \qquad coproduct$$

of the pair in the category Set.

See Appendix D for the proof. In programming environment, product is often realized with pair (a, b) and functions of *fst, snd* (see Eq. (1.10)). However, coproduct is sometimes realized with a dedicated data type.

```
data Either A B = Left A | Right B
```

The advantage is that we needn't the explicit 0, 1 tags to mark if x comes from A or B, because it is encoded in the type **Either** A B. Below example program handles the value of coproduct through pattern matching:

```
either :: (a → c) → (b → c) → Either a b → c
either f _ (Left x)    = f x
either _ g (Right y)   = g y
```

For example, the coproduct **Either String Int** is either a string, like s = **Left** "hello", or an integer, like n = **Right** 8. If it is a string, we want to count its length; if it is a number, then double it. To do this, we call the *either* function: **either length** (*2) x.

Thus, **either length** (*2) s counts the length of string "hello" which gives 5, while **either length** (*2) n doubles 8 to give 16. Some programming environments have concept of union or enum data type to realize coproduct partially. We call the two arrows of the coproduct *left* and *right*, respectively, in the rest of this chapter.

4.3.2 The Properties of Product and Coproduct

By definition of product and coproduct, for any wedge consist of X and arrows, the mediator arrow m is uniquely determined. In category *Set*, the mediator arrow can be defined as the following:

$$product \qquad\qquad coproduct$$

$$m(x) = (a, b) \qquad \begin{cases} m(a, 0) = p(a) \\ m(b, 1) = q(b) \end{cases}$$

For a generic category, how to define the mediator arrow? To do that, we introduce two dedicated operations for arrows. For any wedge:

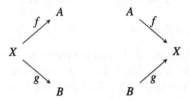

define:

$$m = \langle f, g \rangle \qquad m = [f, g]$$

which satisfy:

$$\begin{cases} fst \circ m = f \\ snd \circ m = g \end{cases} \qquad \begin{cases} m \circ left = f \\ m \circ right = g \end{cases}$$

Hence the following diagram commutes:

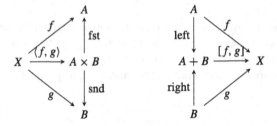

This pair of diagrams gives some important properties for product and coproduct. First is the *cancellation law*:

$$\begin{cases} fst \circ \langle f, g \rangle = f \\ snd \circ \langle f, g \rangle = g \end{cases} \qquad \begin{cases} [f, g] \circ left = f \\ [f, g] \circ right = g \end{cases} \qquad (4.7)$$

For product, if arrow $f = fst$ and arrow $g = snd$, then we obtain the special case of identity arrow mentioned in previous section. Similarly for the coproduct, if $f = left$ and $g = right$, then the mediator is also the identity arrow. We call this property the *reflection law*:

$$id = \langle fst, snd \rangle \qquad id = [left, right] \qquad (4.8)$$

If there are another wedge Y, arrows h and k, together with f and g, satisfying the following conditions:

$$
\begin{array}{cc}
\text{product} & \text{coproduct} \\
\begin{cases} h \circ \phi = f \\ k \circ \phi = g \end{cases} & \begin{cases} \phi \circ h = f \\ \phi \circ k = g \end{cases}
\end{array}
$$

for:

$$
\langle h, k \rangle \circ \phi \qquad\qquad\qquad \phi \circ [h, k]
$$

Substitute m, and apply cancellation law; then we obtain the *fusion law*:

$$
\begin{array}{cc}
\text{product} & \text{coproduct} \\
\begin{cases} h \circ \phi = f \\ k \circ \phi = g \end{cases} \Rightarrow \langle h, k \rangle \circ \phi = \langle f, g \rangle & \begin{cases} \phi \circ h = f \\ \phi \circ k = g \end{cases} \Rightarrow \phi \circ [h, k] = [f, g]
\end{array}
$$

$$(4.9)$$

It means:

$$
\langle h, k \rangle \circ \phi = \langle h \circ \phi, k \circ \phi \rangle \qquad \phi \circ [h, k] = [\phi \circ h, \phi \circ k] \qquad (4.10)
$$

We'll see later, these laws play important roles during algorithm reasoning, simplification, and optimization.

4.3.3   Functors from Products and Coproducts

Product allows us to extend from normal functor to *bifunctor* (also known as binary functor). The functors we've seen so far send objects and arrows from one category to another category. A bifunctor applies to the product of two categories C and D ($C \times D$ as the source), which sends it to the following:

$$
\begin{array}{ll}
C \times D \longrightarrow E & \text{categories} \\
A \times B \longmapsto \mathbf{F}(A \times B) & \text{objects}
\end{array}
$$

The bifunctor as a functor need transform arrows as well. For the arrow f in category C, and g in category D, what does the bifunctor behaves? Observe below diagram:

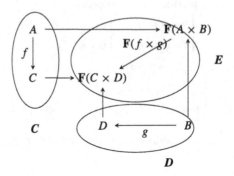

The bifunctor $\mathbf{F}$ sends arrow $A \xrightarrow{f} C$ and $B \xrightarrow{g} D$ to a new arrows in category E. For this new arrow, the source object is $\mathbf{F}(A \times B)$, and the target object is $\mathbf{F}(C \times D)$. Define the product of arrows f and g that applies to the product of A and B: it sends every pair of (a, b) in $A \times B$ to pair $(f(a), g(b))$ in $\mathbf{F}(C \times D)$:

$$(f \times g)(a, b) = (f(a), g(b))$$

The bifunctor $\mathbf{F}$ applies to $f \times g$, which sends this product of arrows to the new arrow $\mathbf{F}(f \times g)$. Below is the complete definition of bifunctor:

$$C \times D \longrightarrow E \qquad \text{category}$$
$$A \times B \longmapsto \mathbf{F}(A \times B) \qquad \text{object} \qquad (4.11)$$
$$f \times g \longmapsto \mathbf{F}(f \times g) \qquad \text{arrow}$$

(Readers may skip this box) We need to verify the bifunctor satisfies the two functor conditions: identity and composition.

$$\mathbf{F}(id \times id) = id$$
$$\mathbf{F}((f \circ f') \times (g \circ g')) = \mathbf{F}(f \times g) \circ \mathbf{F}(f' \times g')$$

Proof We treat a product as an object. If we prove the following two conditions, then the proof for the bifunctor follows through the known result of normal functor.

$$id \times id = id$$

(continued)

$$(f \circ f') \times (g \circ g') = (f \times g) \circ (f' \times g')$$

We start from proving the identity condition. For all $(a, b) \in A \times B$:

$$(id \times id)(a, b) = (id(a), id(b)) \qquad \text{product of arrows}$$
$$= (a, b) \qquad \text{definition of } id$$
$$= id(a, b) \qquad \text{reverse of } id$$

Next prove the composition condition.

$$((f \circ f') \times (g \circ g'))(a, b)$$
$$= ((f \circ f')\, a, (g \circ g')\, b) \qquad \text{product of arrows}$$
$$= (f(f'(a)), g(g'(b))) \qquad \text{composition}$$
$$= (f \times g)(f'(a), g'(b)) \qquad \text{reverse of product}$$
$$= (f \times g)((f' \times g')(a, b)) \qquad \text{reverse of product}$$
$$= ((f \times g) \circ (f' \times g'))(a, b)$$

$\square$

Many programming environments have limitation to use the same notation for both object and arrow mapping. Analogous to *fmap*, we define a dedicated *bimap* for bifunctor arrow mapping. It demands every bifunctor $\mathbf{F}$ is an instance of the following:

$$bimap : (A \to C) \to (B \to D) \to (\mathbf{F}\, A \times B \to \mathbf{F}\, C \times D)$$

It says $\mathbf{F}$ as a bifunctor sends the product of arrows $A \xrightarrow{g} C$ and $B \xrightarrow{h} D$ to an arrow which maps from $\mathbf{F}\, A \times B$ to $\mathbf{F}\, C \times D$. With bifunctor, we can define product and coproduct functors. We'll adopt the infix notation, denote:

$$\text{product functor by } \times \qquad \text{coproduct functor by } +$$

Given two objects, the product functor maps them to their product; the coproduct functor maps them to their coproduct. For arrows, define:

$$f \times g = \langle f \circ fst, g \circ snd \rangle \qquad f + g = [left \circ f, right \circ g] \qquad (4.12)$$

Proof *(Readers may skip this box)* We need verify the two functor conditions. For the identity condition, substitute f, g with id:

$$
\begin{array}{l|ll}
id \times id & id + id & \\
= \langle id \circ fst, id \circ snd \rangle & = [left \circ id, right \circ id] & \text{by Eq. (4.12)} \\
= \langle fst, snd \rangle & = [left, right] & \text{definition of } id \\
= id & = id & \text{reflection law}
\end{array}
$$

Next to verify the composition condition:

$$
\overset{\text{product}}{(f \times g) \circ (f' \times g') = f \circ f' \times g \circ g'} \quad \overset{\text{coproduct}}{(f + g) \circ (f' + g') = f \circ f' + g \circ g'}
$$

we first prove the *absorption law*:

$$
\overset{\text{product}}{(f \times g) \circ \langle p, q \rangle = \langle f \circ p, g \circ q \rangle} \quad \overset{\text{coproduct}}{[p, q] \circ (f + g) = [p \circ f, q \circ g]}
$$

$$(4.13)$$

We only prove the product law on the left; the coproduct law is similar (see Exercise 4.3.2).

$$
\begin{array}{ll}
(f \times g) \circ \langle p, q \rangle & \\
= \langle f \circ fst, g \circ snd \rangle \circ \langle p, q \rangle & \text{definition of } \times \\
= \langle f \circ fst \circ \langle p, q \rangle, g \circ snd \circ \langle p, q \rangle \rangle & \text{fusion law} \\
= \langle f \circ p, g \circ q \rangle & \text{cancellation law}
\end{array}
$$

With the absorption law, let $p = f' \circ fst, q = g' \circ snd$; we next verify the composition condition:

$$
\begin{array}{ll}
(f \times g) \circ (f' \times g') & \\
= (f \times g) \circ \langle f' \circ fst, g' \circ snd \rangle & \text{definition of } \times \text{ for } f' \text{ and } g' \\
= (f \times g) \circ \langle p, q \rangle & \text{substitute with } p, q \\
= \langle f \circ p, g \circ q \rangle & \text{absorption law} \\
= \langle f \circ f' \circ fst, g \circ g' \circ snd \rangle & \text{substitute back } p, q \\
= \langle (f \circ f') \circ fst, (g \circ g') \circ snd \rangle & \text{association law} \\
= (f \circ f') \times (g \circ g') & \text{reverse of } \times
\end{array}
$$

$\square$

As an instance of bifunctor, define *bimap* for product functor as below:

$$bimap : (A \to C) \to (B \to D) \to (A \times B \to C \times D)$$
$$bimap\ f\ g\ (x, y) = (f\ x,\ g\ y) \tag{4.14}$$

Define the corresponding *bimap* for coproduct functor with `Either` type:

$$bimap : (A \to C) \to (B \to D) \to (Either\ A\ B \to Either\ C\ D)$$
$$bimap\ f\ g\ (left\ x) = left\ (f\ x) \tag{4.15}$$
$$bimap\ f\ g\ (right\ y) = right\ (g\ y)$$

Exercise 4.3

4.3.1 What is the product of two objects in a poset? What is the coproduct?

4.3.2 Prove the absorption law for coproduct and verify the coproduct functor satisfies the composition condition.

Answer: page 163

4.4   Natural Transformation

When Eilenberg and Mac Lane developed category theory in the early 1940s, they wanted to explain why certain "natural" construction are natural and other constructions are not. As the result, categories were invented to support functors, and these were invented to support natural transformations. Mac Lane remarked "I didn't invent categories to study functors; I invented them to study natural transformations." Arrows are used to compare objects; functors are used to compare categories; what will be used to compare functors? It is natural transformation that serves this purpose. Consider the following two functors:

$$Src \underset{G}{\overset{F}{\rightrightarrows}} Trg$$

They connect two categories.[17] How to compare them? Since functors map both objects and arrows, we need compare the mapped objects and mapped arrows. An object A in category Src is mapped to two objects FA and GA in category Trg. We care about the arrow from FA to GA:

$$FA \xrightarrow{\phi_A} GA$$

Besides A, we do the same study to all the objects in Src.

Definition 4.4.1 For a pair of parallel functors F, G, a natural transformation:

$$F \xrightarrow{\phi} G$$

is a family of arrows of Trg indexed by the object A of Src.

$$FA \xrightarrow{\phi_A} GA$$

such that for each arrow $A \xrightarrow{f} B$ of Src, the appropriate square in Trg commutes (as below diagram):

covariant   contravariant

For every arrow f, there is a corresponding square commutes. In covariant case, "commute" means:

$$G(f) \circ \phi_A = \phi_B \circ F(f)$$

Let's use some examples to understand how natural transformation bridges functors.

---

[17] Suppose both are of same variance, covariant, or contravariant.

Example 4.4.1 Define a function *inits*; it enumerates all prefixes of a string or list and behaves like below examples:

```
inits "Mississippi" = ["","M","Mi","Mis","Miss","Missi","Missis",
            "Mississ", "Mississi","Mississip","Mississipp",
            "Mississippi"]

inits [1, 2, 3, 4] = [[],[1],[1,2],[1,2,3],[1,2,3,4]]
```

Or summarized as:

$$inits\ [a_1, a_2, \ldots, a_n] = [[\ ], [a_1], [a_1, a_2], \ldots, [a_1, a_2, \ldots, a_n]] \qquad (4.16)$$

For any set A (or type A) as an object in category **Set**, let $F = $ **List** be the list functor. F sends A to a list type of **List** A, or abbreviated as $[A]$. Let $G = $ **List** $\circ$ **List** be the composite of two list functors. G sends A to a list of lists of **List**(**List**A), or $[[A]]$. The *inits* arrow indexed by A is as below:

$$inits_A : \mathbf{List}A \to \mathbf{List}(\mathbf{List}A)$$

or in square brackets notation:

$$[A] \xrightarrow{inits_A} [[A]]$$

We are going show for any arrow (which is a function in **Set**) $A \xrightarrow{f} B$, the square diagram commutes:

$$
\begin{array}{ccccc}
A & & [A] & \xrightarrow{init_A} & [[A]] \\
\downarrow{\scriptstyle f} & \mathbf{List}(f)\downarrow & & & \downarrow\mathbf{List}(\mathbf{List}(f)) \\
B & & [B] & \xrightarrow[init_B]{} & [[B]]
\end{array}
$$

where:

$$\mathbf{List}(\mathbf{List}(f)) \circ inits_A = inits_B \circ \mathbf{List}(f)$$

Proof We use $fmap$ defined in Eq. (4.6). For n elements of $a_1, a_2, \ldots, a_n$ of type A, let $b_1 = f(a_1), b_2 = f(a_2), \ldots, b_n = f(a_n)$ of type B be their corresponding mapped values:

$\mathbf{List}(\mathbf{List}(f)) \circ init_A\ [a_1, \ldots, a_n]$

$=(fmap_{[[\ ]]}\ f) \circ init_A\ [a_1, \ldots, a_n]$	by Eq. (4.6)
$=(fmap_{[[\ ]]}\ f)\ [[\ ], [a_1], \ldots, [a_1, a_2, \ldots, a_n]]$	by Eq. (4.16)
$=map\ (map\ f)\ [[\ ], [a_1], \ldots, [a_1, a_2, \ldots, a_n]]$	$fmap$ is map for list functor
$=[map\ f\ [\ ], map\ f\ [a_1], \ldots, map\ f\ [a_1, a_2, \ldots, a_n]]$	definition of map
$=[[\ ], [f(a_1)], \ldots, [f(a_1), f(a_2), \ldots, f(a_n)]]$	apply $map\ f$ to every sub-list
$=[[\ ], [b_1], \ldots, [b_1, b_2, \ldots, b_n]]$	definition of f
$=init_B\ [b_1, b_2, \ldots, b_n]$	reverse of $init$
$=init_B\ [f(a_1), f(a_2), \ldots, f(a_n)]$	definition of f
$=init_B \circ (map\ f)\ [a_1, a_2, \ldots, a_n]$	reverse of map
$=init_B \circ (fmap_{[\ ]}\ f)\ [a_1, \ldots, a_n]$	$fmap$ is map for list functor

$=init_B \circ \mathbf{List}(f)\ [a_1, \ldots, a_n]$

Shows $inits$: $\mathbf{List} \rightarrow \mathbf{List} \circ \mathbf{List}$ is a natural transformation.   □

Example 4.4.2 We are going to define a *safeHead* function that safely accesses the first element of any list. Here "safe" means to handle the empty list (Nil) without any exception. To do that, we use the maybe functor defined in Eq. (4.3):

$$
\begin{aligned}
safeHead &: [A] &\rightarrow \mathbf{Maybe}\ A \\
safeHead\ [\ ] &= Nothing \\
safeHead\ (x : xs) &= Just\ x
\end{aligned}
\qquad (4.17)
$$

Let F be the same list functor, but G be the maybe functor. G sends type A (a set) to $\mathbf{Maybe}\ A$. Every A indexes arrow *safeHead* as:

$$[A] \xrightarrow{\ safeHead_A\ } \mathbf{Maybe}\ A$$

We are going to show below square diagram commutes for any arrow $A \xrightarrow{f} B$:

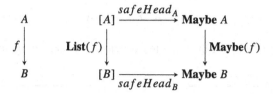

where:

$$\textbf{Maybe}(f) \circ safeHead_A = safeHead_B \circ \textbf{List}(f)$$

Proof First for empty list:

      $\textbf{Maybe}(f) \circ \textbf{safeHead}_A \; [\;]$

$= \textbf{Maybe}(f) \; Nothing$ by Eq. (4.17)

$= Nothing$ by Eq. (4.3)

$= safeHead_B \; [\;]$ reverse of $safeHead$

$= safeHead_B \circ \textbf{List}(f) \; [\;]$ reverse of $fmap$

Second for $x{:}xs$:

      $\textbf{Maybe}(f) \circ safeHead_A \; (x{:}xs)$

$= \textbf{Maybe}(f) \; (Just \; x)$ by Eq. (4.17)

$= Just \; f(x)$ by Eq. (4.3)

$= safeHead_B \; (f(x) : fmap \; f \; xs)$ reverse of $safeHead$

$= safeHead_B \circ \textbf{List}(f) \; (x{:}xs)$ reverse of $fmap$

Proves $safeHead : \textbf{List} \to \textbf{Maybe}$ is a natural transformation.   □

Summarize the two examples: for any object A (a type in programming) in some category, functor **F** sends it to another object **F**A (another type in programming), while the other functor **G** sends A to **G**A. A natural transformation ϕ (a function in programming) indexed by A[18] is in the form:

$$\phi_A : \textbf{F}A \to \textbf{G}A$$

---

[18] Also called as the component at A.

Not only for A, when abstract *all* objects, we obtain a family of arrows (a *polymorphic function*[19] in programming):

$$\phi : \forall A \cdot \mathbf{F}A \to \mathbf{G}A$$

Natural transformation can be written as below[20] in some programming environments:

```
phi :: forall a · F a → G a
```

When we needn't explicitly call out `forall` a, it is simplified as:

```
phi : F a → G a
```

The type of *inits* and *safeHead* are:

```
inits :: [a] → [[a]]

safeHead :: [a] → Maybe a
```

They only rename `phi` to their own names and replace F and G to their own functors respectively.

Natural transformation was developed to compare functors. Analogous to "isomorphism" in abstract algebra, we define when two functors are considered "equal."

Definition 4.4.2 A *natural isomorphism* between two functors $\mathbf{F}$ and $\mathbf{G}$ is a natural transformation:

$$\mathbf{F} \xrightarrow{\phi} \mathbf{G}$$

such that for each source arrow A, the indexed arrow

$$\mathbf{F}A \xrightarrow{\phi_A} \mathbf{G}A$$

is an isomorphism in the target category.

---

[19] Think about the polymorphic function in object-oriented programming and the template function in generic programming. The component at A corresponds to `function<A>(...)` as the instantiated generic function, or `a.function(...)`, where a is a variable of type A.

[20] There is an `ExplicitForAll` option in Haskell, we'll see it again in the next chapter about build/foldr fusion law.

Two functors that are naturally isomorphic are also said to be *naturally equivalent*. For example, function *swap* turns the product of two objects $A \times B$, *swap* into $B \times A$:

$$
\begin{aligned}
swap &: A \times B \to B \times A \\
swap\,(a, b) &= (b, a)
\end{aligned}
\tag{4.18}
$$

swap is a natural transformation. It transforms one bifunctor to another. Both bifunctors happen to be product functors: $\mathbf{F} = \mathbf{G} = \times$. Since they are bifunctors, for every two arrows $A \xrightarrow{f} C$ and $B \xrightarrow{g} D$, we need the following square diagram commutes:

$$
\begin{array}{ccc}
B & \xrightarrow{\ \ g\ \ } & D
\end{array}
$$

$$
\begin{array}{ccc}
A & & A \times B \xrightarrow{\ swap_{A \times B}\ } B \times A \\
{\scriptstyle f}\Big\downarrow & & {\scriptstyle f \times g}\Big\downarrow \qquad\qquad \Big\downarrow{\scriptstyle g \times f} \\
C & & C \times D \xrightarrow[\ swap_{C \times D}\ ]{} D \times C
\end{array}
$$

where:

$$
(g \times f) \circ swap_{A \times B} = swap_{C \times D} \circ (f \times g)
$$

To prove this natural transform condition, one can substitute two products (a, b) and (c, d) to both sides; we leave it as Exercise 4.4.1. We are going to show the indexed arrow, which is $swap_{A \times B}$, is an isomorphism (see Sect. 3.2.3 for the definition). It is because its inverse exists:

$$
swap_{A \times B} \circ swap_{B \times A} = id
$$

Meaning swap twice restores: $(A \times B) \rightsquigarrow (B \times A) \rightsquigarrow (A \times B)$; it is the inverse of itself. *swap* is therefore bijective and a natural isomorphism. It is intuitive that $A \times B$ and $B \times A$ are isomorphic.

Each of the three examples, *inits*, *safeHead*, and *swap*, is both polymorphic function and natural transformation. This is not a coincidence. In fact, all polymorphic functions are natural transformations in functional programming [46].

Exercise 4.4

4.4.1 Prove $swap$ satisfies the natural transformation condition $(g \times f) \circ swap = swap \circ (f \times g)$

4.4.2 Prove that the polymorphic function $length$ defined below is a natural transformation.

$$length : [A] \to Int$$
$$length \,[\,] = 0$$
$$length \,(x : xs) = 1 + length \; xs$$

4.4.3 Natural transformation is composable. Consider two natural transformations $F \xrightarrow{\phi} G$ and $G \xrightarrow{\psi} H$. For any arrow $A \xrightarrow{f} B$, draw the diagram for their composition, and list the commutative condition.

Answer: page 170

4.5   Data Types

Functors allows us to convert primitive data types, like Int, $Bool$, to algebraic data type with structures, like **Maybe** Int, **List** $Bool$, and **Tree** Int; product and coproduct provide ways to further compose them to complex data types. We show how to realize recursive data types and computation on them in this section.

4.5.1   Initial and Terminal Object

We start from the two simple data types, *initial object* and *final object*. They are symmetric like the left and right hands. They are simple, but not easy.

Definition 4.5.1 In category C, if there is a special object S, such that there is a unique arrow for every object A.

$$S \longrightarrow A \qquad A \longrightarrow S$$

In other words, S has a unique arrow that

points to every other object   pointed from every other object

We call this special object S

initial object   final object

Sometimes, a final object is said to be *terminal*.

Traditionally, people use 0 to represent initial object and use 1 for final object. The unique arrow is written as:

$$0 \longrightarrow A \qquad A \longrightarrow 1$$

Why are the initial and final objects symmetric? Suppose S is the initial object of category C, by reversing the direction for all the arrows, S becomes the final object of category C^{op}. There may or may not be initial or final object in a category. A category can have one without the other or have neither. If it has both, then these objects may or may not be the same. The initial object or final object of a category is unique up to isomorphism (or called isomorphic uniqueness). We only prove the uniqueness for initial object; the proof for final object is symmetric.

Proof Suppose there is another initial object 0' besides 0. Because 0 is initial object, there exists an arrow f from 0 to 0' by definition, while there is also an arrow g from 0' to 0 because 0' is also initial object. Their composition $f \circ g$ points back to 0, and $g \circ f$ points back to 0'. By the identity axiom of category, there are identity arrows id_0 self-pointed to 0 and $id_{0'}$ self-pointed to 0', respectively, as below diagram:

However, the identity arrow is unique for each object (see Exercise 4.1.1); we have:

$$id_0 = f \circ g \quad \text{and} \quad id_{0'} = g \circ f$$

The uniqueness of identity arrow asserts 0' is isomorphic to 0, i.e., the initial object is isomorphic unique.   □

This is the reason we usually use the word *the* initial object rather than an initial object and so as for *the* final object. Specially, if the initial object is also the final object, we call it *zero object* or *null object*. A category may not necessarily have the zero object.

Below are examples of data types that correspond to the initial and final objects.

Example 4.5.1 In a poset (partial order set), arrows are ordering relations. If there is a minimum element, then it is the initial object. Similarly, if there is a maximum element, then it is the final object. Use the Duck Donald's family, for example, in poset of {Huey, Thelma Duck, Quackmore Duck, Grandma Duck} under ancestral ordering, Donald's nephew Huey is the initial object, and Grandma Duck is the final object. For another example of prime numbers $\{2, 3, 5, 7, \dots\}$ under normal $\leq$ ordering, 2, as the minimum prime, is the initial object, but there is no final object. For another example, the poset of real numbers R under normal $\leq$, there is neither the minimum nor the maximum, hence no initial or final object. Go back to the poset of the complete Duck Donald's family tree in Fig. 4.3, as there is no common ancestor for all members, so it has no initial or final object.

Example 4.5.2 For the category **Grp** of all groups, arrows are morphisms between groups. A morphism $f : H \to G$ must send the identity element e_H of group H to the identity element e_G of another group G, that is $e_H \rightsquigarrow e_G$. The trivial group $\{e\}$ only contains the identity element (see Sect. 3.2.7). Let H be $\{e\}$ and G be any group. There is an arrow $h : \{e\} \to G$ that sends $e \rightsquigarrow e_G$. Note h is unique because e is the only element. This shows $\{e\}$ is the initial object

$$\{e\} \longrightarrow G$$

Conversely, from any group G, there is a unique arrow $g : G \to \{e\}$ that sends $x \rightsquigarrow e$ for every element x in G. This shows $\{e\}$ is the final objects:

$$G \longrightarrow \{e\}$$

Since $\{e\}$ is both initial and final objects, it is the zero object. Particularly, if compose the two arrows:

$$G \longrightarrow \{e\} \longrightarrow G'$$

the result is zero arrow; it bridges three groups G, $\{e\}$, and G' together. All elements in G are sent to e and then sent to $e_{G'}$, as shown in Fig. 4.8. $\{e\}$ to groups is analogous to zero to numbers; this is where the name "zero" object comes from.

It does not matter what we call the only element in the trivial group. It can be e, can be 0 (group **Z** under $+$), can be 1 (group $\mathbf{Z}_n$ under $\times$ modulo n), can be I (the identity matrix), can be (1) (the identity permutation), can be id, etc. They are all isomorphic (equivalent) to the trivial group of order 1.

Example 4.5.3 For the category **Set** of all sets with total functions as arrows, it's easy to find the final object, which is the singleton set: $\{\bigstar\}$. Same as above example,

Fig. 4.8 Zero object

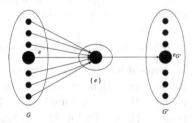

for any set S, there is a unique function $g : S \to \{\bigstar\}$ sends $x \rightsquigarrow \bigstar$ for every x in S, or written as $g(x) = \bigstar$. It is the arrow to the final object:

$$S \longrightarrow \{\bigstar\}$$

However, how does it map the empty set $\varnothing^{21}$ to $\{\bigstar\}$? (This is not a problem for groups, because group can't be empty; a group has at least one element, the identity.) In fact, $\varnothing$ is the initial object of this category, because we can define an arrow from it to any set S:

$$\varnothing \xrightarrow{\ f\ } S$$

Please take a while to think about it. By identity axiom of category, every object has id arrow, so as the empty set: $\varnothing \xrightarrow{\ id\ } \varnothing$. The empty set is the object that has unique arrow to every set (include $\varnothing$ itself). The only thing special is that f has no argument. We cannot write $f(x) = y$ for there is no x in $\varnothing$ at all.

To answer the above question, since there is unique arrow from $\varnothing$ to any set, there is unique arrow from $\varnothing$ to $\{\bigstar\}$. In summary, there is unique arrow from every set to $\{\bigstar\}$, therefore, $\{\bigstar\}$ is the final object of *Set*.

Why can't we make $\{\bigstar\}$ to be the initial object like $\{e\}$ to groups? The reason is because the arrow $\{\bigstar\} \to S$ may not be unique for nonempty set $S = \{x, y, z, \dots\}$. We can define an arrow $\vec{x} : \{\bigstar\} \to S$, that sends $\bigstar \rightsquigarrow x$; in the same way, we can also define other arrows $\vec{y} : \bigstar \rightsquigarrow y$ or $\vec{z} : \bigstar \rightsquigarrow z$ or $\dots$ The family of arrows

$$\{\bigstar\} \longrightarrow S$$

has particular meaning: select an element from S. For example $\vec{x}$ selects x out. For this reason, we call it *selection arrow*.

We intend to choose the symbol $\{\bigstar\}$ to emphasize: it does not matter what the element is as far as it is a singleton set. All singleton sets are isomorphic.

Example 4.5.4   The type system in programming can be viewed as category *Set*. A data type, for example, *Int*, is the set of all integers; Boolean type is the set of $\{True, False\}$. We know $\{\bigstar\}$ is the final object in *Set*. What data type does it correspond to? Since the final object is unique up to isomorphism, any data type with only one value is the final object in programming.

We can define a dedicated data type, named as "()", with only one element, also named as "()".

```
data  ()  =  ()
```

---

[21] André Weil said he was responsible for the null set symbol $\varnothing$ and that it came from the Norwegian alphabet, which he alone among the Bourbaki group was familiar with.

It is isomorphic to any singleton set, hence $\{\bigstar\} = \{()\}$. We will see the advantage of choosing "()" notation later. Define the arrow (total function in programming) from any data type (set) to this final object as below:

```
unit :: a → ()
unit _ = ()
```

There are other ways to define the final object, for example, singleton of any data type that is isomorphic to $\{()\}$.[22]

```
data Singleton = S

proj :: a → Singleton
proj _ = S
```

It looks like a constant function, which maps any value to a constant. However, not all constant functions point to the final object. For example, the following two:

```
yes :: a → Bool
yes _ = True

no :: a → Bool
no _ = False
```

As the *Bool* data type contains two values, any other data type has two arrows: *yes* and *no*, point to *Bool*. It does not satisfy the uniqueness of arrow to the final object.

In *Set* category, the initial object is the empty set $\varnothing$. What is the corresponded data type in programming? View the data type as set, the empty set is a data type without any values. In programming, it means declare a data type without defining it. Below example program declares Void, but leaves it without any values:

```
data Void
```

Thus Void represents empty set. However, the initial object must have a unique arrow to any object. It requires to define a function from Void to all data types.

```
absurd :: Void → a
absurd _ = undefined
```

---

[22] Many programming environments, like C++, Java, Scala, etc. support defining singleton object. We can also treat the force type casting as an arrow, then it also forms the final object.

The implementation does not matter here, because there does not exist a real argument to call this function. It is quite OK to implement it as below:

```
absurd :: Void → a
absurd a = case a of {}
```

One may define other empty types (by declare the type without defining any values), but they are all isomorphic. Below example program establishes two isomorphisms between Empty and Void. It is easy to verify $iso \cdot iso' = id$.

```
data Empty

f :: Empty → a
f _ = undefined

iso :: Void → Empty
iso _ = undefined

iso' :: Empty → Void
iso' _ = undefined
```

For **Set**, we define selection arrow from the final object to a nonempty set. What does it correspond to in programming? View type as set, it means to select a specific value of the type. Below example program selects 0 and 1 from Int, respectively:

```
zero :: () → Int
zero () = 0

one :: () → Int
one () = 1
```

We see the advantage of "()" notation. It looks like passing null argument to functions: $zero$ () returns 0, and one () returns 1.

Exercise 4.5

4.5.1 In the example of poset, if there exists the minimum (or the maximum) element, then the minimum (or the maximum) is the initial object (or the final object). Consider the category **Poset** of all posets, if there exists the initial object, what is it? If there exists the final object, what is it?

4.5.2 In Peano category **Pno** (see Exercise 4.1.3), what is the initial object in form of (A, f, z)? What is the final object?

Answer: page 175

4.5.2   *Exponentials*

The initial and final objects are analogues to 0 and 1; the product and coproduct are analogous to × and +. If find the analogue to exponential, then we will be empowered with the basic arithmetic in the abstract world of categories and further develop the analogue to polynomials. We start from examples.

Example 4.5.5 There are infinitely many functions (as arrows) from *Bool* to *Int*. They form a set denoted by the type *Bool* → *Int*. Below function for example, is in this set:

$$ord : Bool \rightarrow Int$$
$$ord\ False = 0 \qquad\qquad (4.19)$$
$$ord\ True\ = 1$$

It labels false and true as a pair of numbers (0, 1). From this instance, we abstract all functions in this set *Bool* → *Int* as below:

$$f : Bool \rightarrow Int$$
$$f\ False = \ldots$$
$$f\ True\ = \ldots$$

No matter how f varies, the result is always a pair of integers (a, b). It's good to replace ... with a and b:

$$f : Bool \rightarrow Int$$
$$f\ False = a \qquad\qquad (4.20)$$
$$f\ True\ = b$$

The set of f arrows is isomorphic to the set of integer pairs (a, b):

$$
\begin{aligned}
Bool \xrightarrow{\ f\ } Int &= \{(f\ False, f\ True)\} & \\
&= \{(a, b)\} & \text{by Eq. (4.20)} \\
&= Int \times Int & \text{product of } Int \\
&= Int^2 & \text{exponential notation} \\
&= Int^{Bool} & Bool \cong 2
\end{aligned}
$$

The type (set) of arrows *Bool* → *Int* corresponds to an exponential Int^{Bool}. Why can we use *Bool* to substitute 2 such that $Int^2 = Int^{Bool}$? The reason is isomorphic. Use size to name set of natural numbers: $\mathbf{0} = \{\}, \mathbf{1} = \{0\}, \mathbf{2} = \{0, 1\}, \ldots$, we can establish an isomorphism between *Bool* and **2** under *ord* function (Eq. (4.19)):

$$Bool = \{False, True\} \xrightarrow{ord} \{0, 1\} = \mathbf{2}$$

Example 4.5.6 The set of all the functions $Char \rightarrow Bool$. There are finite many functions in this set. For example isDigit(c) tests whether a character is digit or not. Below is one of the implementations (though not practical):

```
isDigit : Char → Bool
...
isDigit '0' = True
isDigit '1' = True
...
isDigit '9' = True
isDigit 'a' = False
isDigit 'b' = False
...
```

Although naive, it reflects a fact: if there are 256 different chars (ASCII code for example), the possible values of isDigit form a 256-tuple of (False, ..., True, ... True, False, ...), where the positions corresponding to digits are True, the rest are False. Among the varies of functions of type $Char \rightarrow Bool$, like isUpper, isLower, isWhitespace, and etc., everyone corresponds to a specific tuple. For example, in the tuple of isUpper, positions for the upper case characters are True, the rest are False. Although there are infinitely many ways to define a function from $Char$ to $Bool$, from the result perspective, there are only 2^{256} distinct functions essentially. The type of arrows $Char \rightarrow Bool$ is isomorphic to $Bool^{Char}$, which is the set of 256-tuple of Boolean values.

In general, we express the set of functions $f : B \rightarrow C$ in exponential form of C^B. With such a f in C^B and some b in B, define a special *apply* function[23] that applies f to b:

$$apply(f, b) = f(b)$$

Where is symbol A? We actually need A for other purpose: a function of two variables (or binary function), for example, $f(x, y) = z$ defined for natural numbers, like $f(x, y) = x^2 + 2y + 1$. Treat the input to f as a whole thing, i.e., a pair of (x, y); the function sends the pair to some number: $f : (x, y) \rightsquigarrow z$. We know the concept of pair is product, so the type of the function is from a product to a number:

$$f : \mathbb{N} \times \mathbb{N} \rightarrow \mathbb{N}$$

---

[23] Some authors use *eval* to highlight the evaluation process ([48] pp. 111–112, and [49]). We adopt *apply* ([5] pp. 72) to follow the tradition in Lisp.

If only pass x in, the Curried result (something like $f(x, \bullet)$, see Sect. 2.3.1) is another function. This new function sends a natural number to another number; its type is $\mathbb{N} \to \mathbb{N}$. Express in exponentials as $\mathbb{N}^{\mathbb{N}}$. This is a (infinite) set of (infinite long) tuples; each element in the tuple is a natural number.

In summary, for every function of two variables $g : A \times B \to C$, when pass some a in, it becomes a Curried function $g(a, \bullet) : B \to C$ belonging to C^B. In other words, there is a unique unary function $A \xrightarrow{\lambda g} C^B$ sending a to $B \xrightarrow{g(a, \bullet)} C$. We call λg the exponential transpose of g.[24] Further, applying $\lambda g(a)$ to b is equivalent to call $g(a, b)$:

$$apply(\lambda g(a), b) = g(a, b)$$

Function *apply* takes two arguments: (1) function $\lambda g(a)$ of type C^B; and (2) b of type B. The result is $g(a, b)$ of type C. The type of *apply* is:

$$C^B \times B \xrightarrow{apply} C$$

Definition 4.5.2 (Exponentials) If a category C has final object and supports product, then an *exponential object* is a pair of object and arrow

$$(C^B, apply)$$

For any object A and arrow $A \times B \xrightarrow{g} C$, there is a unique transpose arrow

$$A \xrightarrow{\lambda g} C^B$$

such that below diagram commutes

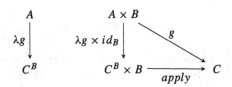

where:

$$apply \circ (\lambda g \times id_B) = g \tag{4.21}$$

---

[24] Some authors use $\bar{g}$ for exponential transpose. We name it λg to indicate λ-expression: $a \mapsto (\bullet \mapsto \cdot g(a, \bullet))$.

Define *curry* that converts a binary function (function on product) to Curried function:

$$curry : (A \times B \to C) \to A \to (C^B)$$
$$curry\ g = \lambda g \tag{4.22}$$

or expand it to λ-expression: $curry\ g = (a \mapsto (b \mapsto g(a, b)))$ (listing 4.6 gives its example implementation). Therefore, *curry* g is the exponential transpose of g. Substitute it into Eq. (4.21) of above diagram, and omit the subscript of id:

$$apply \circ (curry\ g \times id) = g \tag{4.23}$$

We obtain a universal property:

$$f = curry\ g \quad \Longleftrightarrow \quad apply \circ (f \times id) = g$$

It explains why the exponential transpose arrow is unique. Assume there exists another arrow $A \xrightarrow{h} C^B$, such that $apply \circ (h \times id) = g$. The universal property asserts $h = curry\ g$.

By fixing objects B and C, we can form a category **Exp**. Objects are binary functions like $A \times B \to C$, for example, h and k in below diagram:

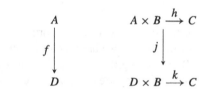

Arrows are defined as this, for $A \xrightarrow{f} D$ in category **C**, the corresponding arrow in **Exp** is $h \xrightarrow{j} k$. The arrows commute if and only if below diagram commutes when combining object C on the right:

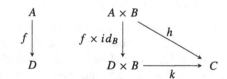

where $k \circ (f \times id_B) = h$. There exists the final object in **Exp**, which is exact $C^B \times B \xrightarrow{apply} C$, because from any object $A \times B \xrightarrow{g} C$, there is the *apply* arrow: $\lambda g = curry\ g$. While the self-pointed arrow to the final object must be id, hence we obtain the *reflection law*:

$$curry\ apply = id \tag{4.24}$$

Exercise 4.6

4.6.1 Verify that **Exp** is a category. What is the *id* arrow? and how to compose arrows?

4.6.2 What is the subscript of *id* in the reflection law $curry\ apply = id_?$? Find another way to prove the reflection law.

4.6.3 Define

$$(curry\ f) \circ g = curry(f \circ (g \times id))$$

as the fusion law for Currying. Prove this law and give its diagram.

Answer: page 180

4.5.3   Object Arithmetic

With the initial object as 0, the final object as 1, coproduct as add, product as multiplication, and arrow as exponential, we are ready to realize object arithmetic and construct polynomials. Before doing that, we need be aware of the scope and boundaries of this abstraction. Some categories may not have the initial, final object, or exponentials. If a category has finite product, for any object A and B, there is A^B; it is called *Cartesian closed*. A Cartesian closed category satisfy below three conditions:

(1) A final object: 1;
(2) Every two objects have product ($\times$);
(3) Every two objects have exponential (A^B).

We can either consider the final object 1 as the 0th power of an object $A^0 = 1$ (see next section) or the product of zero objects. Fortunately, programming, category of sets and total functions, is Cartesian closed. A Cartesian closed category can model the simply typed λ calculus (see 2.5) and serve as the foundation of all the programming languages with types ([47] pp. 148).

If a Cartesian closed category also has the initial object and supports coproduct and distribution law of product over coproduct, then it is *Bicartesian closed*. A Bicartesian closed category satisfies the following conditions in additional:

(4) The initial object: 0;
(5) Every two objects have coproduct ($+$);

(6) Product can be distributed from both sides to coproduct:

$$A \times (B + C) = A \times B + A \times C$$
$$(B + C) \times A = B \times A + C \times A$$

We are going to show the basic object arithmetic in programming, namely, equational theory.

0th Power

$$A^0 = 1$$

0 and 1 are the initial and final objects. The 0th power of A represents the type (set) of all arrows $0 \to A$. As the initial object, 0 has the unique arrow to any object A. Hence the set of $0 \to A$ contains only one arrow. This singleton set is isomorphic to $\{\bigstar\}$, which is exact the final object 1. The equal symbol in the below reasoning is up to isomorphism.

$$A^0 = 0 \to A \qquad \text{exponential}$$
$$= \{\bigstar\} \qquad \text{unique arrow from the initial object}$$
$$= 1 \qquad \{\bigstar\} \text{ is the final object}$$

$\square$

It reflects our common sense in number arithmetic that the 0th power of any number is 1.

Powers of 1

$$1^A = 1$$

1 is the final object. The exponential 1^A represents the type (set) of all arrows from A to the final object, which is $A \to 1$. Because there is unique arrow from any object A to the final object, the set $A \to 1$ contains only one arrow. This singleton set is isomorphic to $\{\bigstar\}$, which is exact the final object.

$$1^A = A \to 1 \qquad \text{exponential}$$
$$= \{\bigstar\} \qquad \text{unique arrow to the final object}$$
$$= 1 \qquad \{\bigstar\} \text{ is the final object}$$

$\square$

First Power

$$A^1 = A$$

This is exact the "selection arrow" (see Sect. 4.5.3). 1 is the final object. The exponential A^1 represents the type (set) of arrows from 1 to A, which is $1 \to A$. For nonempty set A, define a function from the final object 1 to some element a in A:

$$f_a : 1 \rightsquigarrow a$$

Such function selects a out of A, called selection function. On one hand, the set of all selection functions $\{f_a : 1 \rightsquigarrow a | a \in A\}$ is the set of arrows $1 \to A$ exactly; on the other hand, the set $\{f_a\}$ has bijective (one-to-one) mapping to $\{a\} = A$; hence they are isomorphic:

$$A^1 = 1 \to A \qquad\qquad \text{exponential}$$
$$= \{f_a : 1 \rightsquigarrow a | a \in A\} \qquad \text{the set of maps from 1 to } A$$
$$= \{a | a \in A\} = A \qquad\qquad \text{isomorphic}$$

$\square$

Exponentials of Sum

$$A^{B+C} = A^B \times A^C$$

The exponential A^{B+C} represents the set of arrows from the coproduct $B + C$ to A, which is $(B + C) \to A$. Denote the coproduct by **Either** B C (see Sect. 4.3.1). For any function of type **Either** B $C \to A$, we implement it in the form:[25]

$$f : \textbf{Either } B\ C \to A$$
$$f\ (Left\ b) \quad = \ldots$$
$$f\ (Right\ c) \quad = \ldots$$

No matter how f varies, the implementation always contains two branches: either from $b \rightsquigarrow a_1$ or from $c \rightsquigarrow a_2$ for a_1, a_2 in A. Essentially, every function on coproduct is equivalent to a pair of maps ($b \rightsquigarrow a_1, c \rightsquigarrow a_2$), and it is exact the

---

[25] Alternatively, we reason it with tags:

$$f : B + C \to A$$
$$f(b, 0) = \ldots$$
$$f(c, 1) = \ldots$$

product of $B \rightarrow A$ and $C \rightarrow A$. Therefore, $(B+C) \rightarrow A = (B \rightarrow A) \times (C \rightarrow A)$. Write them in exponentials: $B \rightarrow A$ is A^B, and $C \rightarrow A$ is A^C; it explains the exponentials of sum:

$$
\begin{aligned}
A^{B+C} &= (B + C) \rightarrow A && \text{exponentials} \\
&= \{(b \rightsquigarrow a_1, c \rightsquigarrow a_2)|a_1, a_2 \in A, b \in B, c \in C\} && \text{pair of arrows} \\
&= (B \rightarrow A) \times (C \rightarrow A) && \text{pairs are product} \\
&= A^B \times A^C && \text{in exponentials}
\end{aligned}
$$

□

Exponentials of Exponentials

$$(A^B)^C = A^{B \times C}$$

The right side exponential $A^{B \times C}$ is the set of all binary functions of $B \times C \xrightarrow{g} A$. It is obviously natural isomorphic under swapping to $C \times B \xrightarrow{g \circ swap} A$ (see Eq. (4.18) as natural transformation). The Curried form $curry(g \circ swap)$ is of type $C \rightarrow A^B$. Written in exponentials, we obtain $(A^B)^C$.

$$
\begin{aligned}
A^{B \times C} &= B \times C \xrightarrow{g} A && \text{exponentials} \\
&= C \times B \xrightarrow{g \circ swap} A && \text{natural isomorphic} \\
&= C \xrightarrow{curry(g \circ swap)} A^B && \text{Currying} \\
&= (A^B)^C && \text{in exponentials}
\end{aligned}
$$

□

Exponentials of Product

$$(A \times B)^C = A^C \times B^C$$

The exponential $(A \times B)^C$ is the set of arrows of $C \rightarrow A \times B$. It's equivalent to a set of functions each returning a pair of values $\{c \rightsquigarrow (a, b)\}$, where $c \in C, a \in A, b \in B$. This set is obviously isomorphic to $\{(c \rightsquigarrow a, c \rightsquigarrow b)\}$, which is the product of arrows $C \rightarrow A$ and $C \rightarrow B$. Written the two arrows in exponentials:

$$
\begin{aligned}
(A \times B)^C &= C \rightarrow A \times B && \text{exponential} \\
&= \{c \mapsto (a, b)|a \in A, b \in B, c \in C\} && \text{arrows} \\
&= \{(c \mapsto a, c \mapsto b)\} && \text{pairs of arrows}
\end{aligned}
$$

$$= (C \to A) \times (C \to B) \qquad\qquad \text{product of arrows}$$

$$= A^C \times B^C \qquad\qquad\qquad \text{in exponentials}$$

$$\square$$

4.5.4   Polynomial Functors

Extend the object arithmetic to functors, such that the arithmetic applies to both objects and arrows. A *polynomial functor* is built from constant functors (example 4.2.2), product functors, and coproduct functors (Eq. (4.12)) recursively.

(1) The identity functor id and constant functor $\mathbf{K}_A$ are polynomial functors;
(2) if $\mathbf{F}$ and $\mathbf{G}$ are polynomial functors, then their composition $\mathbf{FG}$, sum $\mathbf{F} + \mathbf{G}$, and product $\mathbf{F} \times \mathbf{G}$ are also polynomial functors. Where the sum and product are defined as below:

$$(\mathbf{F} + \mathbf{G})h = \mathbf{F}h + \mathbf{G}h$$
$$(\mathbf{F} \times \mathbf{G})h = \mathbf{F}h \times \mathbf{G}h$$

For example, if functor $\mathbf{F}$ is defined as below for objects and arrows:

$$\begin{cases} \mathbf{F}X = A + X \times A & \text{object} \\ \mathbf{F}h = id_A + h \times id_A & \text{arrow} \end{cases}$$

where A is some object, then $\mathbf{F}$ is a polynomial functor; express as a polynomial:

$$\mathbf{F} = \mathbf{K}_A + (id \times \mathbf{K}_A)$$

4.5.5   F-Algebra

In order to construct complex algebraic data types that support recursive structure, we need another brick, the F-algebra. Many abstract algebraic objects, like monoid, group, ring, and etc., have structures. It is the relation among structures that makes them different from concrete numbers and shapes. When understanding the structures and their relations, we actually understand all things of the same kind; not only numbers, points, lines, and planes but also "tables, chairs, and beer mugs" said by Hilbert. Start from examples.

Example 4.5.7 (Monoid) A monoid is a set M defined with the identity element (denoted by 1_M) under associative binary operation (denoted by $\oplus$). It follows two monoid axioms:

$$\begin{cases} \text{associativity:} & (x \oplus y) \oplus z = x \oplus (y \oplus z) \text{ for all } x, y, z \text{ in } M \\ \text{identity element:} & x \oplus 1_M = 1_M \oplus x = x \qquad \text{for all } x \text{ in } M \end{cases} \tag{4.25}$$

Step 1. Represent $\oplus$ as a binary function, 1_M as a selection function, define:

$$\begin{cases} m\,(x, y) & = x \oplus y \\ e\,() & = 1_M \end{cases}$$

The corresponding types are:

	type	exponential
binary function:	$M \times M \to M$	$M^{M \times M}$
selection:	$1 \to M$	M^1

The first arrow m sends the product of two elements to another element in the monoid; it is closed. The second arrow e is the selection function. 1 is the final object. We intend to use "()" denote the final object, as $1 = \{\bigstar\} = \{()\}$ up to isomorphism. As the advantage, $e\,()$ looks like a function call without any arguments, actually, it accepts the final object, a singleton set, as the argument. Next substitute the monoid axioms in Eq. (4.25) with m and e:

$$\begin{cases} \text{associativity:} & m(m(x, y), z) = m(x, m(y, z)) \text{ for all } x, y, z \text{ in } M \\ \text{identity element:} & m(x, e()) = m(e(), x) = x \qquad \text{for all } x \text{ in } M \end{cases}$$

Step 2. Remove all the concrete elements like x, y, z, and object M. We obtain the monoid axioms purely in function (arrow) form:

$$\begin{cases} \text{associativity:} & m \circ (m, id) = m \circ (id, m) \\ \text{identity element:} & m \circ (id, e) = m \circ (e, id) = id \end{cases}$$

The two axioms asserts diagrams in Fig. 4.9 commute.

Fig. 4.9 Diagrams of monoid axioms. (**a**) Associativity. (**b**) Identity element

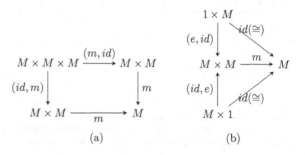

(a)   (b)

We thus represent any monoid with a tuple (M, m, e), where M is the set; m is the binary function; and e is the function selecting the identity element out.

Step 3. The tuple (M, m, e) specifies the set of the monoid, the binary operation, and the identity element. It completely defines the algebraic structure of monoid. Given a set M, all possible combinations of m and e that form the monoid are products of two exponentials:

$$M^{M \times M} \times M^1 = M^{M \times M + 1} \qquad\qquad \text{exponentials of sum}$$

Next turn the right side exponentials into arrow form. For all monoids defined on set M, the algebraic operations must be one of the following:

$$\alpha : 1 + M \times M \to M \qquad\qquad \text{sum (coproduct)}$$

$$() \rightsquigarrow 1_M \qquad\qquad \text{identity element}$$

$$(x, y) \rightsquigarrow x \oplus y \qquad\qquad \text{binary operation}$$

We represent this relation as the sum (coproduct) $\alpha = e + m$, which is a polynomial functor $\mathbf{F}M = 1 + M \times M$. In summary, the algebraic structure of monoid consists of three parts:

(1) Object M is the set that carries the algebraic structure of the monoid, called *carrier object*;
(2) Functor $\mathbf{F}$ defines the algebraic operations. It is a polynomial functor $\mathbf{F}M = 1 + M \times M$;
(3) Arrow $\mathbf{F}M \xrightarrow{\alpha} M$. It is the sum (coproduct) of the identity arrow e and the binary operation arrow m, i.e., $\alpha = e + m$.

We say it defines the F-algebra (M, α) of monoid.

In programming, define the type of (all) F-algebra arrows:

```
type Algebra f a = f a → a
```

It is an alias to the arrow $\mathbf{F}A \to A$ essentially, namely, `Algebra F, A`. For monoid, we also need define a functor:[26]

```
data MonoidF a = MEmptyF | MAppendF a a
```

---

[26] There is a definition of monoid in Haskell standard library (see listing 4.7). However, it is the definition of the monoid algebraic structure, but not the definition of monoid functor.

If a is string, below example program implements the arrow of the F-algebra (**String**, evals) as below:

```
evals :: Algebra MonoidF String
evals MEmpty = e ()
evals (MAppendF s1 s2) = m s1 s2

e :: () → String
e () = ""

m :: String → String → String
m = (+)
```

or simplify it by embedding e and m in evals:

```
evals :: Algebra MonoidF String
evals MEmpty = ""
evals (MAppendF s1 s2) = s1 + s2
```

Function evals is one of the implementations to the arrow $\alpha : \mathbf{F}A \to A$, where **F** is MonoidF and A is **String**. There can be other implementations as far as they satisfy the monoid axioms.

Example 4.5.8 (Group) Groups need an additional axiom of inverse compare to monoid. For group G, let 1_G denote the identity element; "·" denote the binary operator; and x^{-1} denote the inverse of x.

$$\begin{cases} \text{associativity:} & (x \cdot y) \cdot z = x \cdot (y \cdot z) & \text{for all } x, y, z \text{ in } G \\ \text{identity element:} x \cdot 1_G = 1_G \cdot x = x & \text{for all } x \text{ in } G \\ \text{inverse element:} & x \cdot x^{-1} = x^{-1} \cdot x = 1_G & \text{for all } x \text{ in } G \end{cases} \quad (4.26)$$

Step 1. Represent the binary operation, the identity element, and the reverse as functions:

$$\begin{cases} m\,(x, y) & = x \cdot y \\ e\,() & = 1_G \\ i\,x & = x^{-1} \end{cases}$$

The corresponding function types and exponentials are:

	type	exponential
binary operation:	$G \times G \to G$	$G^{G \times G}$
selection:	$1 \to G$	G^1
reverse:	$G \times G$	G^G

The first two arrows are same as monoid. The third one i inverses x to x^{-1} in the group. Next substitute 1_G, "$\cdot$", and the inverse in the group axioms (Eq. (4.26)) with e, m, and i.

$$\begin{cases} \text{associativity:} \quad m(m(x, y), z) = m(x, m(y, z)) \text{ for all } x, y, z \text{ in } G \\ \text{identity element:} m(x, e()) = m(e(), x) = x \quad \text{ for all } x \text{ in } G \\ \text{reverse element:} \ m(x, i(x)) = m(i(x), x) = e() \ \text{ for all } x \text{ in } G \end{cases}$$

Step 2. Remove the concrete elements x, y, z, and G; represent the group axioms purely in arrows:

$$\begin{cases} \text{associativity:} \quad m \circ (m, id) = m \circ (id, m) \\ \text{identity element:} \quad m \circ (id, e) = m \circ (e, id) = id \\ \text{reverse element:} \quad m \circ (id, i) = m \circ (i, id) = e \end{cases}$$

Now every group is represented in a tuple (G, m, e, i).

Step 3. Use the coproduct of m, e, i as sum $\alpha = e + m + i$. The polynomial functor $\mathbf{F}G = 1 + G + G \times G$ specifies the algebraic operations on G.

$$\alpha : 1 + G + G \times G \to G \qquad\qquad \text{sum}$$

$$() \rightsquigarrow 1_G \qquad\qquad \text{identity element}$$

$$x \rightsquigarrow x^{-1} \qquad\qquad \text{inverse element}$$

$$(x, y) \rightsquigarrow x \cdot y \qquad\qquad \text{binary operation}$$

In summary, the algebraic structure of group consists of three parts:

(1) The carrier object G, which is the set furnished with the algebraic structure;
(2) Polynomial functor $\mathbf{F}G = 1 + G + G \times G$, which defined the algebraic operations on the group;
(3) Arrow $\mathbf{F}G \xrightarrow{\alpha = e + m + i} G$. It is the sum of the identity element arrow e, the binary operation arrow m, and the inverse arrow i.

We say it defines the F-algebra (G, α) of group.

It's ready to give the definition of F-algebra. There are many dual concepts appear in pairs. We benefit from them like buy one get one. When define F-algebra, we obtain F-coalgebra at the same time. We will postpone the examples of F-coalgebra in Chap. 6 about infinity.

Definition 4.5.3 Let C be a category; $C \xrightarrow{\mathbf{F}} C$ is an endo-functor of category C. For the object A and morphism α in this category, arrow:

$$FA \xrightarrow{\alpha} A \qquad\qquad A \xrightarrow{\alpha} FA$$

forms a pair (A, α). It is called

F-algebra   F-coalgebra

where A is the carrier object.

Treat F-algebra (A, α) (or F-coalgebra (A, α)) as object. When the context is clear, we may simplify the notation as a pair (A, α). The arrows between such objects are defined as below.

Definition 4.5.4 F-morphism is the arrow between F-algebra (or F-coalgebra) objects:

$$(A, \alpha) \longrightarrow (B, \beta)$$

If the arrow $A \xrightarrow{f} B$ between the carrier objects make the below diagram commutes:

FA $\xrightarrow{\alpha}$ A   A $\xrightarrow{\alpha}$ FA

$\mathbf{F}(f) \downarrow$ $\downarrow f$ $f \downarrow$ $\downarrow \mathbf{F}(f)$

FB $\xrightarrow{\beta}$ B   B $\xrightarrow{\beta}$ FB

where:

$$f \circ \alpha = \beta \circ \mathbf{F}(f) \qquad \beta \circ f = \mathbf{F}(f) \circ \alpha$$

F-algebra and F-morphism and F-coalgebra and F-morphism form

F-algebra category $Alg(\mathbf{F})$ F-coalgebra category $CoAlg(\mathbf{F})$

respectively.

Exercise 4.7

4.7.1 Draw the diagram to illustrate the group axiom of inverse element.
4.7.2 Let p be a prime number. Use the F-algebra to define the α arrow for the multiplicative group for integers modulo p (refer to Chap. 3 for the definition of this group).
4.7.3 Define F-algebra for ring (see Sect. 3.3.1 for the definition of ring).
4.7.4 What is the id arrow for F-algebra category? How to compose arrows?

Answer: page 189

Recursion and Fixed Point

Apply F-algebra to Peano's axioms, we can describe all things isomorphic to natural numbers. In order to model this *kind* of algebraic structure in F-algebra, we need three things: (1) functor, (2) carrier object, and (3) α arrow. Use Fibonacci numbers, for example, let the functor be **NatF** as it's isomorphic to natural numbers; the carrier object A is the set of integer pairs of (Int, Int); the initial pair is $(1, 1)$; the successor function is $h(m, n) = (n, m + n)$, so the arrow $\alpha = e + h$, where $e() = (1, 1)$. From Peano's axioms, define the functor of natural numbers as below:

```
data NatF A = ZeroF | SuccF A
```

This is a polynomial functor **NatF** $A = 1 + A$. Here is an interesting question: let $A' = $ **NatF** A, substitute it into functor **NatF**, then what will the **NatF** A' look like? It will apply the functor twice **NatF**(**NatF** A). Repeat applying the functor three times gives **NatF**(**NatF**(**NatF**A)). Actually, we may repeat infinitely many times and name the type **NatF**(**NatF**$(\dots)$) as **Nat**.

$$\text{data } \textbf{Nat} = \textbf{NatF}(\textbf{NatF}(\dots)) \qquad \text{infinitely many times}$$
$$= ZeroF \mid SuccF \ (SuccF(\dots)) \quad \text{infinitely many times of } SuccF$$
$$= ZeroF \mid SuccF \ \textbf{Nat} \qquad \text{named as } \textbf{Nat}$$
$$= Zero \mid Succ \ \textbf{Nat} \qquad \text{rename}$$

It is exact the definition of natural numbers (compare with 5 in Chap. 1). List `Nat` and `NatF A` together:

```
data Nat = Zero | Succ Nat
```

```
data NatF A = ZeroF | SuccF A
```

Nat as a functor is recursive, while **NatF** is not. Extend the fixed point we've seen (Eq. (2.12)) in Chap. 2 to functors:

$$\textbf{Fix F} = \textbf{F (Fix F)} \qquad (4.27)$$

The *fixed point* of endo-functor **F** is **Fix F**. On one hand, applying **F** to a fixed point still gives the fixed point; on the other hand, applying an endo-functor to itself infinitely many times gives its fixed point. Hence **Nat** is the fixed point of the functor **NatF**, which means **Nat** = **Fix NatF**.

Exercise 4.8

4.8.1 Someone write the natural number like functor as the below recursive form. What do you think about it?

```
data NatF A = ZeroF | SuccF (NatF A)
```

4.8.2 We can define an α arrow for **NatF** $Int \rightarrow Int$, named $eval$:

$$eval : \textbf{NatF}\ Int \rightarrow Int$$
$$eval\ ZeroF \quad = 0$$
$$eval\ (SuccF\ n) = n + 1$$

We can recursively substitute $A' = \textbf{NatF}\ A$ to functor **NatF** by n times. We denote the functor obtained as $\textbf{NatF}^n\ A$. Can we define the following α arrow?

$$eval : \textbf{NatF}^n\ Int \rightarrow Int$$

Answer: page 190

Initial Algebra and Catamorphism

If there exists initial object in the category of F-algebra $Alg(F)$, what are the properties for it? There must be unique arrow from the initial object to other objects, what kind of relationship does it represent? This section deals with these questions. Using abstract natural numbers (things being isomorphic to $\mathbb{N}$), for example, in category $Alg(\textbf{NatF})$, denote object as (A, α), where A is the carrier object and $\textbf{NatF}\ A \xrightarrow{\alpha} A$ is the arrow. Fibonacci numbers form an F-algebra object $(Int \times Int, fib)$ in this category, where carrier object is the product of integers; and the arrow is fib. Define this arrow $\textbf{NatF}\ (Int \times Int) \xrightarrow{fib} (Int \times Int)$ as below:

$$fib : \textbf{NatF}\ (Int \times Int) \rightarrow (Int \times Int)$$
$$fib\ ZeroF \qquad\qquad = (1, 1) \qquad\qquad\qquad (4.28)$$
$$fib\ (SuccF\ (m, n)) \quad = (n, m + n)$$

Alternatively, we can write it as sum (coproduct) $fib = [e, h]$, where e always returns $(1, 1)$ and $h(m, n) = (n, m + n)$.

Let the initial object of category $Alg(\textbf{NatF})$ be (I, i). There is unique arrow from this initial object to any object (A, α). It implies there exists arrow $I \xrightarrow{f} A$, such that below diagram commutes:[27]

$$
\begin{array}{ccc}
\textbf{NatF}\ A & \xrightarrow{\ \alpha\ } & A \\[4pt]
\textbf{NatF}(f) \Big\uparrow & & \Big\uparrow f \\[4pt]
\textbf{NatF}\ I & \xrightarrow[\ i\ ]{} & I
\end{array}
$$

[27] In notation (I, i), the upper case I stands for the **I**nitial carrier object, and the lower case i stands for the initial arrow.

Since there is unique arrow from the initial object to *every* object, there is arrow from (I, i) to object (**NatF** I, **NatF**(i)), which is generated by applying functor **NatF** to carrier object I and arrow i. Alternatively, we can view it as substituting $A = $ **NatF** I in above diagram. This F-algebra is furnished with three things:

(1) The functor **NatF**;
(2) The carrier object **NatF** I; obtained from the initial carrier object;
(3) Arrow **NatF**(**NatF** I) → **NatF** I. As functor **NatF** applies to both object and arrow, it "lifts" the initial arrow i to **NatF**(i).

For the arrow from the initial object (I, i) to F-algebra (**NatF** I, **NatF**(i)), there exists an arrow $I \xrightarrow{\ j\ }$ **NatF** I, such that below diagram commutes:

$$
\begin{array}{ccc}
\textbf{NatF}(\textbf{NatF}\ I) & \xrightarrow{\ \textbf{NatF}(i)\ } & \textbf{NatF}\ I \\
\Big\uparrow{\scriptstyle \textbf{NatF}(j)} & & \Big\uparrow{\scriptstyle j} \\
\textbf{NatF}\ I & \xrightarrow[\ i\ ]{} & I
\end{array}
$$

We compose two arrows: **NatF**(i) and i as below.

$$\textbf{NatF}(\textbf{NatF}\ I) \xrightarrow{\ \textbf{NatF}(i)\ } \textbf{NatF}\ I \xrightarrow{\ i\ } I$$

To make it obvious, draw the second arrow vertically:

$$
\begin{array}{ccc}
\textbf{NatF}(\textbf{NatF}I) & \xrightarrow{\ \textbf{NatF}(i)\ } & \textbf{NatF}I \\
& & \Big\downarrow{\scriptstyle i} \\
& & I
\end{array}
$$

Compare with the above diagram, i is inverse of j. Their composition $i \circ j$ is self-pointed to I itself; hence, it is the unique endo-arrow of the initial object id. Conversely, the arrow from the initial object to **NatF** I is also unique.

$$i \circ j = id = j \circ i$$

In summary, **NatF** I is isomorphic to I: **NatF** $I = I$. This is exactly the concept of fixed point, because the fixed point of **NatF** is **Nat**.

$$\textbf{NatF Nat} = \textbf{Nat}$$

It tells us $I = $ **Nat** and $i = [zero, succ]$ (or $zero + succ$), the initial object (**Nat**, $[zero, succ]$) is exact natural numbers under Peano's arithmetic. This result reflects Dedekind's finding: for any algebraic structure $(A, [c, f])$

satisfying Peano's axioms, there is unique morphism from natural numbers
(**Nat**, $[zero, succ]$), which sends 0 to c, and n to $f^n(c) = f(f(\dots f(c))\dots)$
(see Exercise 4.1.3).

Definition 4.5.5 If there exists the initial object in the category of F-algebra
$Alg(F)$, then the initial object is the *initial algebra*.

For example, natural numbers (**Nat**, $[zero, succ]$) is the initial algebra of
Alg(**NatF**). Lambek, in 1968, pointed out i was an isomorphism and named the
initial algebra (I, i) the fixed point of functor **F**. This result is called *Lambek
theorem* nowadays [51]. In the category of sets under total functions, many functors,
including polymorphic functors, have the initial algebra [50]. We skip its proof for
existence. Given a functor **F**, we get the initial algebra through its fixed point.

Joachim Lambek (1922–2014)

If there exists the initial algebra (I, i), then there is unique morphism to any
algebra (A, f). Denote the morphism from carrier object I to A by $(\!|f|\!)$, so that
below diagram commutes:

$$
\begin{array}{ccc}
FI & \xrightarrow{\;i\;} & I \\
{\scriptstyle \mathbf{F}(\!|f|\!)}\Big\downarrow & & \Big\downarrow{\scriptstyle (\!|f|\!)} \\
FA & \xrightarrow[\;f\;]{} & A
\end{array}
$$

where:

$$h = (\!|f|\!), \text{ if and only if } h \circ i = f \circ \mathbf{F}(h)$$

We call the arrow $(\!|f|\!)$ *catamorphism*. It comes from Greek word $\kappa\alpha\kappa\alpha$, means
downward. The brackets "$(\!|\ |\!)$" look like a pair of bananas, namely, "banana
brackets." Catamorphism is powerful; it can convert a normal non-recursive function

f to $(\!(f)\!)$ on recursive structure; hence, it builds complex recursive computation. For example, although the natural number functor **NatF** is not recursive, the initial algebra of **Nat** is recursive:

```
data NatF A = ZeroF | SuccF A
```

```
data Nat = Zero | Succ Nat
```

Although arrow **NatF**$A \xrightarrow{f} A$ is not recursive, a catamorphism, named *cata*, builds an arrow **Nat** $\to A$ from f. This new arrow applies computation to the recursive **Nat**.

$$(\mathbf{NatF}A \xrightarrow{f} A) \xrightarrow{cata} (\mathbf{Nat} \to A)$$

cata f as a function of **Nat** $\to A$, should handle both *Zero* and (*Succ n*) as values of **Nat**:

$$
\begin{aligned}
&cata : (\mathbf{NatF}\ A \to A) \to \mathbf{Nat} \to A \\
&cata\ f\ Zero \qquad\quad = f\ ZeroF \\
&cata\ f\ (Succ\ n) \qquad = f\ (SuccF\ (cata\ f\ n))
\end{aligned}
\qquad (4.29)
$$

The first clause sends *Zero* $\rightsquigarrow f\ ZeroF$; for the recursive case of (*Succ n*), it first recursively evaluates *cata f n* to get a value *a* of type *A*; then convert it to *SuccF a*, which is a value of type **NatF** *A*; finally passes it to f. This catamorphism of natural numbers is generic to any carrier object *A*. Here are two examples. The first example converts **Nat** back to *Int*.

```
toInt :: Nat → Int
toInt = cata eval where
  eval :: NatF Int → Int
  eval ZeroF = 0
  eval (SuccF x) = x + 1
```

For example, `toInt zero` gives 0, and `toInt (Succ (Succ (Succ Zero)))` gives 3. Define its inverse function as below:

```
fromInt :: Int → Nat
fromInt 0 = Zero
fromInt n = Succ (fromInt (n-1))
```

For any integer, n = (`toInt` ∘ `fromInt`) n holds. This example looks trivial. Here is another example about Fibonacci numbers.

```
toFib :: Nat → (Integer, Integer)
toFib = cata fibAlg where
  fibAlg :: NatF (Integer, Integer) → (Integer, Integer)
  fibAlg ZeroF = (1, 1)
  fibAlg (SuccF (m, n)) = (n, m + n)
```

We intentionally rename fib to `fibAlg` to call out our purpose: catamorphism from the non-recursive algebraic relation to recursively computation on Fibonacci numbers. `toFib Zero` gives pair $(1, 1)$, and `toFib (Succ (Succ (Succ Zero)))` gives pair $(3, 5)$. Below helper function calculates the n-th Fibonacci number.

$$fibAt = fst \circ toFib \circ fromInt$$

Proposition 4.5.1 *For any algebraic structure* $(A, c+f)$ *satisfying Peano's axioms (isomorphic to natural numbers), there is a catamorphism from the initial algebra* (**Nat**, $zero + succ$) *that generates a recursive computation tool* $(\!|c + f|\!)$.

Proof The catamorphism makes below diagram commute (sum $f + g$ is coproduct $[f, g]$):

$$
\begin{array}{ccc}
\mathbf{NatFNat} & \xrightarrow{[zero,\, succ]} & \mathbf{Nat} \\
\mathbf{NatF}(h) \downarrow & & \downarrow h \\
\mathbf{NatF}A & \xrightarrow[{[c,\, f]}]{} & A
\end{array}
$$

$$
\begin{aligned}
h \circ [zero, succ] &= [c, f] \circ \mathbf{NatF}(h) & \text{diagram commute} \\
&= [c, f] \circ (id + h) & \text{polynomial functor} \\
&= [c \circ id, f \circ h] & \text{absorption law equation (4.13)} \\
&= [c, f \circ h] & c \circ id = c \\
&= [h \circ zero, h \circ succ] & \text{fusion law equation (4.9) on the left}
\end{aligned}
$$

$$
\Rightarrow \begin{cases} h \circ zero = c \\ h \circ succ = f \circ h \end{cases}
$$

$$
\Rightarrow \begin{cases} h\, Zero = c \\ h\, (Succ\, n) = f(h(n)) \end{cases}
$$

$$
\Rightarrow \begin{cases} h(0) = c \\ h(n + 1) = f(h(n)) \end{cases}
$$

This is exact the definition of folding for natural numbers (Eq. (1.8)): $h = foldn(c, f)$.   □

The catamorphism of natural numbers $(\!|c + f|\!) = foldn(c, f)$. For Fibonacci numbers, it is:

$$(\!|fibAlg|\!) = (\!|start + next|\!) = foldn(start, next)$$

It seems we go back to Chap. 1 after a long journey. It's actually a spiral of understanding. Chapter 1 shows this result from induction and abstraction; Here we apply abstract pattern to a concrete problem and obtain the same result.

Algebraic Data Type

We are going to define more algebraic data types with initial F-algebra, like list and binary tree.

Example 4.5.9 The definition of list given in Chap. 1 is as below.

```
data List A = Nil | Cons A (List A)
```

The corresponding non-recursive functor is:

```
data ListF A B = NilF | ConsF A B
```

List is the fixed point of functor **ListF**. To verify it, let $B' = $ **ListF** A B, then recursively apply to itself infinitely many times. Denote the result by **Fix** (**ListF** A).

$$
\begin{aligned}
\textbf{Fix (ListF } A) &= \textbf{ListF } A \textbf{ (Fix (ListF } A)) && \text{definition of fixed point} \\
&= \textbf{ListF } A \textbf{ (ListF } A \, (\ldots)) && \text{expand} \\
&= NilF \mid ConsF \; A \textbf{ (ListF } A \, (\ldots)) && \text{definition of } \textbf{ListF } A \\
&= NilF \mid ConsF \; A \textbf{ (Fix (ListF } A)) && \text{reverse of fixed point}
\end{aligned}
$$

Compare with the definition of **List** A:

$$\textbf{List } A = \textbf{Fix (ListF } A)$$

Fix A in functor **ListF**, for any carrier object B, define arrow

$$\textbf{ListF } A \; B \xrightarrow{\; f \;} B$$

It forms F-algebra for list. The initial algebra is (**List** A, $[nil, cons]$); the catamorphism is as below diagram:

$$
\begin{CD}
\textbf{ListF } A \textbf{ (List } A) @>{[nil,\,cons]}>> \textbf{(List } A) \\
@V{(\textbf{ListF } A)(h)}VV @VV{h}V \\
\textbf{ListF } A \; B @>>{[c,\,f]}> B
\end{CD}
$$

Given a non-recursive function f, define catamorphism to convert it to computation on recursive list.

```
cata :: (ListF a b → b) → (List a → b)
cata f Nil = f NilF
cata f (Cons x xs) = f (ConsF x (cata f xs))
```

As its application, below example program calculates the length of a list:

```
len :: (List a) → Int
len = cata lenAlg where
  lenAlg :: ListF a Int → Int
  lenAlg NilF = 0
  lenAlg (ConsF _ n) = n + 1
```

Hence `len Zero` gives 0, and `len (Cons 1 (Cons 1 Zero))` gives 2. Below function converts list in bracket notation to *List*.

```
fromList :: [a] → List a
fromList [] = Nil
fromList (x:xs) = Cons x (fromList xs)
```

Chain them together, for example, `len (fromList [1, 1, 2, 3, 5, 8])` calculates the length of a list. Different algebraic rule f gives different computation. Below example sums up the elements in a list:

```
sum :: (Num a) ⇒ (List a) → a
sum = cata sumAlg where
  sumAlg :: (Num a) ⇒ ListF a a → a
  sumAlg NilF = 0
  sumAlg (ConsF x y) = x + y
```

We next show the catamorphism on list is essentially folding computation $foldr$. List functor is a polynomial functor. Fix A, it sends carrier object B to **ListF** A $B = 1 + A \times B$ and sends arrow h to $(\textbf{ListF } A)(h) = id + (id \times h)$. The above diagram commutes.

$$\begin{aligned}
h \circ [nil, cons] &= [c, f] \circ (\textbf{ListF } A)(h) && \text{diagram commute} \\
&= [c, f] \circ (id + (id \times h)) && \text{polynomial functor} \\
&= [c \circ id, f \circ (id \times h)] && \text{absorption law of coproduct} \\
&= [c, f \circ (id \times h)] && \\
&= [h \circ nil, h \circ cons] && \text{fusion law of coproduct on the left}
\end{aligned}$$

$$\Rightarrow \begin{cases} h \circ nil = c \\ h \circ cons = f \circ (id \times h) \end{cases}$$

$$\Rightarrow \begin{cases} h \ Nil = c \\ h \ (Cons \ a \ x) = f(a, h(x)) \end{cases}$$

This is exactly the definition of list folding: $h = foldr(c, f)$.   □

The catamorphism of list F-algebra $(\!|c + f|\!) = foldr(c, f)$. For example, $foldr(0, +)$ sums up the elements of a list, and $foldr(0, (a, n) \mapsto n+1)$ computes the length of a list.

Example 4.5.10 In Chap. 2, we defined binary tree as below:

```
data Tree A = Nil | Br A (Tree A) (Tree A)
```

To describe all things isomorphic to binary tree with F-algebra, define binary tree functor as below.

```
data TreeF A B = NilF | BrF A B B
```

B is the carrier object; the initial algebra is (**Tree** A, $[nil, br]$). We leave its proof as Exercise 4.9.1. Below diagram shows the catamorphism of binary tree:

$$
\begin{array}{ccc}
\textbf{TreeF}\ A\ (\textbf{Tree}\ A) & \xrightarrow{\ [nil,\ br]\ } & (\textbf{Tree}\ A) \\
{\scriptstyle (\textbf{TreeF}\ A)(h)}\Big\downarrow & & \Big\downarrow {\scriptstyle h} \\
\textbf{TreeF}\ A\ B & \xrightarrow[\ [c,\ f]\ ]{} & B
\end{array}
$$

The catamorphism of binary tree accepts an arrow of F-algebra **TreeF** $A \ B \xrightarrow{f} B$, and generates a function of **Tree** $A \rightarrow B$:

$$
\begin{aligned}
cata &: (\textbf{TreeF}\ A\ B \rightarrow B) \rightarrow (\textbf{Tree}\ A \rightarrow B) \\
cata\ f\ Nil &= f\ NilF \\
cata\ f\ (Br\ k\ l\ r) &= f\ (BrF\ k\ (cata\ f\ l)\ (cata\ f\ r))
\end{aligned}
\qquad (4.30)
$$

Below example defines the sum algebra, and then the catamorphism applies it to sum up all elements of the tree recursively:

$$
\begin{aligned}
sum &: \textbf{Tree}\ A \rightarrow A \\
sum &= cata \circ sumAlg
\end{aligned}
$$

where:

$$sum Alg : \textbf{TreeF } A \ B \rightarrow B$$
$$sum Alg \ Nil F \qquad = 0_B$$
$$sum Alg \ (Br \ k \ l \ r) \quad = k + l + r$$

Proof We next show the catamorphism is the essentially the folding operation $foldt$ of binary tree. Fix object A, the binary tree functor **TreeF** $A \ B$, is a polynomial functor. It sends carrier object B to **TreeF** $A \ B = 1 + (A \times B \times B)$ and sends arrow h to (**TreeF** $A)(h) = id + (id_A \times h \times h)$. Since the diagram commutes, we have:

$$h \circ [nil, br] = [c, f] \circ (\textbf{TreeF } A)(h) \qquad \text{diagram commute}$$
$$= [c, f] \circ (id + (id_A \times h \times h)) \quad \text{polynomial functor}$$
$$= [c \circ id, f \circ (id_A \times h \times h)] \qquad \text{absorption law of coproduct}$$
$$= [c, f \circ (id_A \times h \times h)]$$
$$= [h \circ nil, h \circ br] \qquad \text{fusion law of coproduct on the left}$$

$$\Rightarrow \begin{cases} h \circ nil = c \\ h \circ br = f \circ (id_A \times h \times h) \end{cases}$$

$$\Rightarrow \begin{cases} h \ Nil = c \\ h \ (Br \ k \ l \ r) = f(k, h(l), h(r)) \end{cases}$$

It is exact the definition of folding of binary tree $h = foldt(c, f)$, where:

$$foldt \ c \ h \ Nil \qquad = c$$
$$foldt \ c \ h \ (Br \ k \ l \ r) = h(k, foldt \ c \ h \ l, foldt \ c \ h \ r)$$

$\square$

For example, $foldt(0_B, (klr \mapsto k + l + r))$ sums all elements of a binary tree.

Exercise 4.9

4.9.1 For the binary tree functor **TreeF** $A \ B$, fix A, use the fixed point to show that (**Tree** A, $[nil, branch]$) is the initial algebra

Answer: page 199

4.6   Fold

We've seen the basics of category theory, including category, functor, natural transformation, product and coproduct, initial object and final object, exponentials, and F-algebra. They can construct complex algebraic structures. By the end of this chapter, we show how the generic folding operation is realized in the language of category theory [52]:

```
foldr f z t = appEndo (foldMap (Endo o f) t) z
```

The reason why it is defined like this is abstraction, to unlimit $foldr$ to list, expand it to any foldable structures. $foldr$ for list is defined as below:

```
foldr :: (a → b → b) → b → [a] → b
foldr f z [] = z
foldr f z (x:xs) = f x (foldr f z xs)
```

where f is a binary operation. Denote it by infix operator $\oplus$, and explicitly denote z, the identity element by e. For example, $foldr\ (\oplus)\ e\ [a, b, c]$ is expanded as:

$$a \oplus (b \oplus (c \oplus e))$$

It reminds us of monoid; $foldr$ essentially repeats the binary operation on top of monoid. Atop the monoid definition of the identity element and binary operation, we add an additional operation to "sum up" a list of elements with the $\oplus$ and e:

$$concat_M : [M] \to M$$
$$concat_M = foldr\ (\oplus)\ e$$

It is named as mconcat in some programming environment. For any monoid M, function $concat_M$ processes a list of elements of M, folding them together through the binary operation and the identity element. For example, string is an instance of monoid, the identity element is the empty string, and the binary operation is string concatenation. Hence $concat_M$ ["Hello", "String" "Monoid"] gives the following result:

"Hello" ++ ("String" ++ ("Monoid" ++ "")) = "HelloStringMonoid"

We are able to sum up any monoid elements. Can we make it more generic? Consider there is a list of elements that don't belong to monoid. We can still sum them if there is a way to convert them to monoid elements. List functor behaves exactly as we expect, to convert a list of some types to a list of monoids. In other words, we can use the list functor to "lift" the arrow $A \xrightarrow{g} M$ to **List**(g) for sum:

$$[A] \xrightarrow{fmap\ g} [M] \xrightarrow{concat_M} M$$

$$\mathbf{List} \uparrow \qquad\qquad \uparrow \mathbf{List}$$

$$A \xrightarrow{\quad g \quad} M$$

$$foldMap : (A \to M) \to [A] \to M$$
$$foldMap\ g = concat_M \circ fmap\ g$$

However, there is still limitation. For the binary operation $f : A \to B \to B$ passed to $foldr$, if B is not monoid, we can't do folding. To solve it, consider the Curried $f : A \to (B \to B)$. If we treat the arrow $B \to B$ as object, they form a monoid, in which the identity element is the id arrow, and the binary operation is arrow composition. To make it clear, we wrap the $B \to B$ arrow as a type (set) through a functor:

```
newtype Endo B = Endo (B → B)
```

It's called "endo" functor as it point from B to B itself. Besides, define a function of **Endo** $B \xrightarrow{appEndo} B$:

$$appEndo(Endo\ a) = a$$

Now we can declare **Endo** is a monoid, with id arrow as the identity element and function composition as the binary operation.

```
instance Monoid Endo where
    mempty = Endo id
    Endo f `mappend` Endo g = Endo (f ∘ g)
```

Given any binary function f, we can use $foldMap$ to fold on $Endo$ monoid:

$$foldCompose : (A \to (B \to B)) \to [A] \to \mathbf{Endo}B$$
$$foldCompose\ f = foldMap\ (Endo \circ f)$$

For example, $foldCompose\ f\ [a, b, c]$ expands as below:

$$Endo(f\ a) \oplus (Endo(f\ b) \oplus (Endo(f\ c) \oplus Endo(id)))$$
$$= Endo(f\ a \oplus (f\ b \oplus (f\ c \oplus id)))$$

Here is a concrete example:

$$foldCompose\ (+)\ [1, 2, 3]$$

$$\Rightarrow foldMap\ (Endo \circ (+))\ [1, 2, 3]$$

$$\Rightarrow concat_M\ (fmap\ (Endo \circ (+)))\ [1, 2, 3]$$

$$\Rightarrow concat_M\ (fmap\ Endo\ [(+1), (+2), (+3)])$$

$$\Rightarrow concat_M\ [Endo\ (+1),\ Endo\ (+2),\ Endo\ (+3)]$$

$$\Rightarrow Endo\ ((+1) \circ (+2) \circ (+3))$$

$$\Rightarrow Endo\ (+6)$$

As the last step, we need extract the result from the $Endo$ object. For this example, we use $appEndo$ to extract $(+6)$ and then apply it to the initial value z passed to $foldr$:

$$foldr\ f\ z\ xs = appEndo\ (foldCompose\ f\ xs)\ z$$

$$= appEndo\ (foldMap\ (Endo \circ f)\ xs)\ z$$

This is the complete definition of $foldr$ in the language of categories. We can further define a dedicated type $Foldable$, such that for any data structure, user can either realize $foldMap$ or $foldr$ (see listing 4.8).

We cannot introduce the category theory just in one chapter. We only scratch the surface of the iceberg. The key idea is abstraction. The dual concepts like initial and final objects, product and coproduct, can be further abstracted to the higher level dual things, limits, and colimits. We haven't introduced the adjunctions nor Yoneda lemma. We have not touched the popular monad concept in programming. We hope this chapter could serve as a path to the category world.

Appendix: Example Programs

Listing 4.1 Definition of functors

```
class  Functor f  where
    fmap           :: (a → b) → f a → f b
```

Listing 4.2 The Maybe functor

```
instance  Functor Maybe  where
    fmap _ Nothing      = Nothing
    fmap f (Just a)     = Just (f a)
```

Look up in binary search tree with Maybe functor, and convert the result to binary format:

Listing 4.3 lookup binary search tree

```
lookup Nil _ = Nothing
lookup (Node l k r) x | x < k = lookup l x
                      | x > k = lookup r x
                      | otherwise = Just k

lookupBin = (fmap binary) o lookup
```

Listing 4.4 Definition of bifunctor

```
class Bifunctor f where
  bimap :: (a → c) → (b → d) → f a b → f c d
```

Listing 4.5 Product and coproduct functors

```
instance Bifunctor (,) where
  bimap f g (x, y) = (f x, g y)

instance Bifunctor Either where
  bimap f _ (Left a) = Left (f a)
  bimap _ g (Right b) = Right (g b)
```

Listing 4.6 Definition of *curry* and its reverse *uncurry*

```
curry       :: ((a, b) → c) → a → b → c
curry f x y = f (x, y)

uncurry     :: (a → b → c) → ((a, b) → c)
uncurry f (x, y) = f x y
```

Definition of monoid. It requires any monoid a be a semigroup as precondition:

Listing 4.7 Monoid

```
class Semigroup a ⇒ Monoid a where
    mempty  :: a

    mappend :: a → a → a
    mappend = (<>)

    mconcat :: [a] → a
    mconcat = foldr mappend mempty
```

Listing 4.8 Definition of foldable

```
newtype Endo a = Endo { appEndo :: a → a }

class Foldable t where
    foldr :: (a → b → b) → b → t a → b
    foldr f z t = appEndo (foldMap (Endo ∘ f) t) z

    foldMap :: Monoid m ⇒ (a → m) → t a → m
    foldMap f = foldr (mappend ∘ f) mempty
```

Chapter 5
Fusion

> ... *mathematical knowledge ... is, in fact, merely verbal knowledge. "3" means "2+1", and "4" means "3+1". Hence it follows (though the proof is long) that "4" means the same as "2+2". Thus mathematical knowledge ceases to be mysterious.*
>
> *Bertrand Russell*

Penrose triangle

I still remember my math class at school. My teacher often wrote down a long expression with many alphabetic symbols in the blackboard and then asked the class to simplify it. Some student volunteered to do it in front of the class. Merged terms, factorized, ... varies of tricks. It is like a magic process that often led to unbelievable simple result. Sometimes the guy was stuck or trapped in loops and finally saved by my teacher. Chalk and blackboard, such experience was unforgettable, just like it happened yesterday. I was so impressed to the mythical power of reasoning. The magic is that I needn't care about the concrete meanings when doing the deduction, like building bricks, from different parts to an interesting toy. The formulas and theorems are combined together and finally build an interesting result. When meet $a^2 + 2ab + b^2$, turn it into $(a + b)^2$, just like mating two bricks. We needn't force ourselves to remind the geometric meanings for this formula when manipulating it.

We use two examples in this chapter to demonstrate how to do deduction in programming. For each example, we'll explain the intuitive concrete meanings and

© China Machine Press 2024
X. Liu, *Mathematics in Programming*,
https://doi.org/10.1007/978-981-97-2432-1_5

Fig. 5.1 Geometric
illustration for
$(a + b)^2 = a^2 + 2ab + b^2$

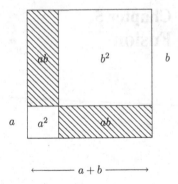

give the purely formal deduction. Just like $(a + b)^2$, on one hand, we explain it as
the total area from two different squares and two equal rectangles (Fig. 5.1); on the
other hand, we deduce it step by step.

$$(a + b)^2 = (a + b)(a + b) \qquad \text{definition of square}$$
$$= a(a + b) + b(a + b) \qquad \text{distribution law}$$
$$= a^2 + ab + ba + b^2 \qquad \text{distribution law again}$$
$$= a^2 + 2ab + b^2 \qquad \text{merge } ab \text{ and } ba$$

5.1 Foldr/build Fusion

The first example is foldr/build fusion law. In 2015, a main stream programming
language Java adopted lambda expression and a set of functional tools in its
1.8 version. Some programmers soon found the chained function calls, which
brought elegance and expressiveness, were at the expense of performance if used
carelessly. The chained functions might generate excessive intermediate results,
while these intermediate results were not necessary simple values, but list, container,
or collection of complex structures. They were thrown away after being consumed
by the next functions and then duplicated again and passed on. Such pattern
of produce—consume—thrown away, repeated along the function chain, caused
computation overhead. For example, below function tests whether every element
of a list satisfies a given prediction [53]:

$$all(p, xs) = and(map(p, xs))$$

$all(prime, [2, 3, 5, 7, 11, 13, 17, 19, \ldots])$ tells if a list contains all primes. However, the performance of this implementation is poor. First, $map(prime, xs)$
generates a list of Boolean values [True, True, ...] of the same length as xs;

indicating the corresponding number is prime or not. Then it passes the list to *and* function; examines if there exists any False value. Finally, both xs and the list of Boolean are thrown away; it only returns a Boolean value as the result. Alternatively, here is another implementation that avoids generating the intermediate Boolean list:

$$all(p, xs) = h(xs)$$

$$\text{where} \begin{cases} h([\,]) & = True \\ h(x:xs) & = p(x) \wedge h(xs) \end{cases}$$

This implementation, though does not generate intermediate list, is neither intuitive nor elegant compare to $and(map(p, xs))$. Is there any way that shares the advantage of both methods? We know some transformations satisfying such requirement, for example,

$$map\ sqrt\ (map\ abs\ xs) = map\ (sqrt \circ abs)\ xs$$

The left side first generates a list of absolute values and then evaluates square root for every number. It is equivalent to the right side, which takes absolute value, followed by evaluating the square root for every number in that list. Abstract to the following rule:

$$map\ f\ (map\ g\ xs) = map\ (f \circ g)\ xs \tag{5.1}$$

However, there are too many rules to list. It's not practical to apply them to a complex program. Gill, Launchbury, and Peyton Jones developed a method in 1993; starting from the basic build and folding operations, they found the pattern to optimize the chained functions.

5.1.1 Foldr

foldr stands for folding from right (Eq. (1.15)):

$$foldr \oplus z\,[\,] = z$$
$$foldr \oplus z\,(x:xs) = x \oplus (foldr \oplus z\,xs)$$

For nonempty list, it expands to:

$$foldr \oplus z\,[x_1, x_2, \ldots, x_n] = x_1 \oplus (x_2 \oplus (\ldots (x_n \oplus z))\ldots) \tag{5.2}$$

Many list computations are essentially folding, for example:

(1) Sum:

$$sum = foldr + 0$$

(2) The *and* function that applies logic-and for all Boolean values in a list:

$$and = foldr \wedge True$$

This is because:

$$and\ [x_1, x_2, \dots, x_n] = x_1 \wedge (x_2 \wedge (\dots (x_n \wedge True))\dots)$$

(3) Test if an element belongs to a list:

$$elem\ x\ xs = foldr\ (a\ b \mapsto (a = x) \vee b)\ False\ xs$$

(4) Map (for each map):

$$map\ f\ xs = foldr\ (x\ ys \mapsto f(x) : ys)\ [\,]\ xs$$
$$= foldr\ ((:) \circ f)\ [\,]\ xs$$

(5) Filter with a prediction:

$$filter\ p\ xs = foldr\ \left(x\ ys \mapsto \begin{cases} p(x) : & x:ys \\ otherwise : & ys \end{cases} \right)\ [\,]\ xs \qquad (5.3)$$

(6) Concatenate two lists:

$$xs + \!\!\!+ ys = foldr\ (:)\ ys\ xs \qquad (5.4)$$

This is because:

$$[x_1, x_2, \dots, x_n] + \!\!\!+ ys = x_1 : (x_2 : (\dots (x_n:ys))\dots)$$

(7) Concatenate multiple lists:

$$concat\ xss = foldr + \!\!\!+ [\,]\ xss$$

Actually all list computations can be defined by folding (we've shown *foldr* is the initial algebra for list in Example 4.5.9). If we simplify folding, then we will simplify all list computations.

5.1.2 Foldr/build Fusion Law

What if fold from right by the cons operation (:) starting from the empty list [ ]?

$$foldr \ (:) \ [\,] \ [x_1, x_2, \ldots, x_n] = x_1 : (x_2 : (\ldots (x_n : [\,])) \ldots) \tag{5.5}$$

It ends up with the list itself. One may remind the fixed point in Eq. (4.27); we'll go back this in later section. Basically, we generate a list from a binary operation g (e.g., (:)) and a starting value (e.g., [ ]). Define the list construction process as *build*:

$$build(g) = g((:), [\,]) \tag{5.6}$$

When fold the list with another binary operation f and start value z, the result is equivalent to call g by substituting [ ] by z, and (:) by f.

$$foldr(f, z, build(g)) = g(f, z) \tag{5.7}$$

Written in pointless form (without parentheses and named arguments):

$$foldr \ f \ z \ (build \ g) = g \ f \ z \tag{5.8}$$

This is the *foldr/build fusion law* for list. For example, to sum up all the integers from a to b, which is $sum([a, a+1, \ldots, b-1, b])$, we generate all integers from a to b as below:

$$range(a, b) = \begin{cases} a > b : & [\,] \\ \text{otherwise}: & a : range(a+1, b) \end{cases}$$

For example, $range(1, 5)$ builds the list $[1, 2, 3, 4, 5]$. Sum up the enumerated numbers to give the answer.

$$sum(range(a, b))$$

Next, extract the binary operation (:) and the start value [ ] out as parameters, named as $\oplus$ and z, respectively:

$$range'(a, b, \oplus, z) = \begin{cases} a > b : & z \\ \text{otherwise}: & a \oplus range'(a+1, b, \oplus, z) \end{cases}$$

Further "Curry" the last two arguments:

$$range' \ a \ b = \oplus z \mapsto \begin{cases} a > b : & z \\ \text{otherwise}: & a \oplus (range' \ (a+1) \ b \oplus z) \end{cases}$$

We redefine *range* with *range'* and *build*:

$$range(a, b) = build(range'(a, b))$$

Finally, simplify *sum* with the fusion law:

$$
\begin{aligned}
sum(range(a, b)) &= sum(build(range'(a, b))) & \text{substitute} \\
&= foldr\ (+)\ 0\ (build\ (range'\ a\ b)) & \text{define sum with foldr} \\
&= range'\ a\ b\ (+)\ 0 & \text{fusion law}
\end{aligned}
$$

This optimization sums up numbers without generating intermediate list. Below shows how it expands:

$$
range'\ a\ b\ (+)\ 0 = \begin{cases} a > b : & 0 \\ \text{otherwise} : & a + (range'\ (a + 1)\ b\ (+)\ 0) \end{cases}
$$

5.1.3 Build

To leverage the fusion law conveniently, we need to rewrite the common functions that generate list in form of *build* · · · *foldr*. Such that when composite computations, we can apply the fusion law to *foldr* · · · (*build* · · · *foldr*) for reduction.

(1) The simplest one generates the empty list:

$$[\ ] = build\ (f\ z \mapsto z)$$

Proof We show it by substituting to the definition of *build*:

$$
\begin{aligned}
build\ (f\ z \mapsto z) &= (f\ z \mapsto z)\ (:)\ [\ ] & \text{by Eq. (5.6)} \\
&= (:)\ [\ ] \mapsto [\ ] & \beta - \text{reduction, see Sect. 2.3.3} \\
&= [\ ]
\end{aligned}
$$

(2) *cons* operation (:). It links an element to a list:

$$x{:}xs = build\ (f\ z \mapsto f\ x\ (foldr\ f\ z\ xs))$$

Proof

$$build\ (f\ z \mapsto f\ x\ (foldr\ f\ z\ xs))$$

$$=(f\ z \mapsto f\ x\ (foldr\ f\ z\ xs))\ (:)\ [\ ]\qquad \text{definition of } build$$

$$=x:(foldr\ (:)\ [\ ]\ xs)\qquad\qquad\quad \beta-\text{reduction}$$

$$=x:xs\qquad\qquad\qquad\qquad\qquad \text{by Eq. (5.5), the fixed point}$$

(3) List concatenation:

$$xs \mathbin{+\!\!+} ys = build\ (f\ z \mapsto foldr\ f\ (foldr\ f\ z\ ys)\ xs) \qquad (5.9)$$

Proof

$$build\ (f\ z \mapsto foldr\ f\ (foldr\ f\ z\ ys)xs)$$

$$=(f\ z \mapsto foldr\ f\ (foldr\ f\ z\ ys)\ xs)\ (:)\ [\ ]\quad \text{definition of } build$$

$$=foldr\ (:)\ (foldr\ (:)\ [\ ]\ ys)\ xs\qquad\quad \beta-\text{reduction}$$

$$=foldr\ (:)\ ys\ xs\qquad\qquad\qquad\qquad \text{fixed point for inner } foldr$$

$$=xs \mathbin{+\!\!+} ys\qquad\qquad\qquad\qquad\qquad \text{by Eq. (5.4), list concatenation}$$

For the remaining examples, we leave the proof as Exercise 5.1.2.

4. Concatenate multiple lists:

$$concat\ xss = build\ (f\ z \mapsto foldr\ (xs\ x \mapsto foldr\ f\ x\ xs)\ z\ xss) \qquad (5.10)$$

5. For each map:

$$map\ f\ xs = build\ (\oplus z \mapsto foldr\ (y\ ys \mapsto (f\ y) \oplus ys)\ z\ xs) \qquad (5.11)$$

6. Filter:

$$filter\ p\ xs = build\ (\oplus z \mapsto foldr\ \left(x\ xs' \mapsto \begin{cases} p(x): & x \oplus xs' \\ \text{otherwise}: & xs' \end{cases}\right)\ z\ xs)$$

7. Generate infinite long list of an element:

$$repeat\ x = build\ (\oplus z \mapsto let\ xs = x \oplus xs\ in\ xs) \qquad (5.12)$$

It's not a typo in $xs = x \oplus xs$, where xs is an infinite list of x; this relation says $x:[x, x, \ldots] = [x, x, \ldots]$. We'll explain infinite list in the next chapter.

5.1.4 Reduction with the Fusion Law

Empowered by the fusion law, we are ready to simplify list computations. Go back to the example of $all(p, xs) = and(map(p, xs))$ in the beginning.

Example 5.1.1 First, convert and to folding form and map to build form and then apply fusion law:

$all(p, xs)$

$= and(map(p, xs))$ definition

$= foldr \wedge True\ map(p, xs)$ folding form of and

$= foldr \wedge True\ \textbf{build}\ (\oplus z \mapsto$

       $foldr\ (x\ ys \mapsto p(x) \oplus ys)\ z\ xs)$ build form of map

$= (\oplus z \mapsto foldr\ (x\ ys \mapsto p(x) \oplus ys)\ z\ xs) \wedge True$ fusion law

$= foldr\ (x\ ys \mapsto p(x) \wedge ys)\ True\ xs$ β – reduction

Define a helper function $(first\ f)\ x\ y = f(x)\ y$, which applies function f to the first argument. Then further simplify it to:

$$all\ p = foldr\ (\wedge) \circ (first\ p)\ True$$

Example 5.1.2 A sentence consists of multiple words. We are going to define $join$, a function that concatenates a list of words together interpolated with spaces. To simplify edge case handling, we add additional space to the end of the sentence. Below is a straightforward definition:

$$join(ws) = concat(map(w \mapsto w + ['\ '], ws)) \qquad (5.13)$$

It builds a new list of ws' by appending a space to every word and then concatenates ws' to a result string. However, its performance is poor for appending is an expensive operation. It needs to move from the head to the tail of every word before doing $+$. The intermediate ws' is of the same length as ws and is thrown away finally. We optimize it with the fusion law:

$join(ws)$

   $= concat(map(w \mapsto w + ['\ '], ws))$

   $= build\ (f\ z \mapsto foldr\ (x\ y \mapsto foldr\ f\ y\ x)\ z$

     $map(w \mapsto w + ['\ '], ws))$ by Eq. (5.10), $concat$

   $= build\ (f\ z \mapsto$

$$foldr\ (x\ y \mapsto foldr\ f\ y\ x)\ z\ (\mathbf{build}\ (f'\ z' \mapsto$$
$$foldr\ (w\ b \mapsto f'\ (w \mapsto w \mathbin{+\!\!+} ['\,']) \ b)\ z'\ ws)))$$ by Eq. (5.11), *map*
$$=build\ (f\ z \mapsto$$
$$foldr\ (w\ b \mapsto (x\ y \mapsto foldr\ f\ y\ x)\ (w \mathbin{+\!\!+} ['\,'])\ b)\ z\ ws)$$ fusion law
$$=build\ (f\ z \mapsto$$
$$foldr\ (w\ b \mapsto foldr\ f\ b\ (w \mathbin{+\!\!+} ['\,']))\ z\ ws)$$ β-reduction x, y
$$=build\ (f\ z \mapsto$$
$$foldr\ (w\ b \mapsto$$
$$\mathbf{foldr}\ \mathbf{f}\ \mathbf{b}\ (\mathbf{build}\ (f'\ z' \mapsto$$
$$foldr\ f'\ (foldr\ f'\ z'\ ['\,'])\ w)))\ z\ ws)$$ by Eq. (5.9), $\mathbin{+\!\!+}$
$$=build\ (f\ z \mapsto$$
$$foldr\ (w\ b \mapsto$$
$$foldr\ (foldr\ f\ b\ ['\,'])\ w)\ z\ ws)$$ fusion law
$$=foldr\ (w\ b \mapsto foldr\ (:)\ (\mathbf{foldr}\ (:)\ \mathbf{b}\ ['\,'])\ w)\ [\,]\ ws$$ build with (:) and [ ]
$$=foldr\ (w\ b \mapsto foldr\ (:)\ ('\,' : b)\ w)\ [\,]\ ws$$ evaluate the bold

It finally reduces to:

$$join(ws) = foldr\ (w\ b \mapsto foldr\ (:)\ ('\,' : b)\ w)\ [\,]\ ws$$

or in Curried form:

$$join = foldr\ (w\ b \mapsto foldr\ (:)\ ('\,' : b)\ w)\ [\,]$$ (5.14)

One can further expand *foldr* to obtain a definition with both good readability and performance.

$$join\ [\,]\quad\ = [\,]$$
$$join\ (w{:}ws) = h\ w$$

where

$$h\ [\,]\quad\ ='\,' : join\ ws$$
$$h\ (x{:}xs) = x : h\ xs$$

The original definition of *join* in Eq. (5.13) uses *concat* ∘ *map*(f), which is very common. Many programming environments provide optimal *concatMap* as we deduced:[1]

$$concatMap\ f\ xs = build\ (c\ n \mapsto foldr\ (x\ b \mapsto foldr\ c\ b\ (f\ x))\ n\ xs)$$
(5.15)

This example exposes a problem though it demonstrates the power of fusion law. The reduction process is complex and error prone, with many repeated steps. It is a good place that machine performs better than human. Some programming environments have the fusion law built in the compiler [53]. What we need is to define list computations in the *build* ⋯ *foldr* form; then the rest boring work will be taken by machine. The compiler will help us reduce to optimal program, avoiding throwing away intermediate results or redundant recursions.[2] As time goes on, more and more compilers will support this optimization tool.

5.1.5 Type Constraint

For every abstract tool, we need to be aware of its applicable scope and understand when it will be invalid. Here is an example of contradiction for the fusion law. On one hand:

$$\mathbf{foldr}\ f\ z\ (\mathbf{build}\ (c\ n \mapsto [0])) = (c\ n \mapsto [0])\ f\ z \quad \text{fusion}$$
$$= [0] \quad \text{β-reduction}$$

On the other hand:

$$foldr\ f\ z\ (build\ (c\ n \mapsto [0])) = foldr\ f\ z\ ((c\ n \mapsto [0])\ (:)\ [\ ])\ \textit{build}$$
$$= foldr\ f\ z\ [0] \quad \text{β-reduction}$$
$$= f(0, z)$$

Obviously, $f(0, z)$ is not identical to $[0]$; their types are different.[3] The reason of this contradiction is because the λ expression $(c\ n \mapsto [0])$ is not a function that builds result from c and n. It shows the fusion law $foldr\ f\ z\ (build\ g) = g\ f\ z$ has type constraint for g. It needs to be a function taking two arguments: the first one is a binary operation like f or c; the second one is the starting value like z or n. Formally, the binary operation is polymorphic of type $\forall A.\forall B.\ A \times B \to B$;

---

[1] For example, flatMap in Java and Scala.

[2] Haskell standard library, for example, provides most list functions in *build* ⋯ *foldr* form.

[3] Unless the extreme case that $f = (:), z = [\ ]$.

the starting value hence is of type B, and the result type of g is also B. Written in Curried form, g has the type of below:

$$g : \forall A.(\forall B.(A \to B \to B) \to B \to B)$$

In above example of contradiction, the type of the λ expression $(c\ n \mapsto [0])$ is $\forall A.(\forall B.(A \to B \to B) \to B \to [Int])$ breaks the type constraint. The corresponding type of $build$ is:

$$build : \forall A.(\forall B.(A \to B \to B) \to B \to B) \to \mathbf{List}\ A$$

Because there are two polymorphic types A and B, they are called *rank-2 type polymorphic*.

5.1.6 Fusion Law in Category Theory

This section shows how to deduce the foldr/build fusion law and extends it further in category theory. The foldr/build is one of the fusion laws in category theory. The family of fusion laws are called *shortcut fusion* nowadays. It plays important role in compiler and program optimization [55].

As the initial object, the initial algebra has unique arrow to every F-algebra (see Sect. 4.5.5) as shown in below diagram:

$$
\begin{array}{ccc}
FI & \xrightarrow{\ i\ } & I \\
{\scriptstyle F(\!|a|\!)}\downarrow & & \downarrow{\scriptstyle (\!|a|\!)} \\
FA & \xrightarrow{\ a\ } & A
\end{array}
$$

The arrow from the initial algebra (I, i) to F-algebra (A, a) is a catamorphism of $(\!|a|\!)$. If there are another F-algebra (B, b) and arrow $A \xrightarrow{h} B$, we draw them at the bottom:

$$
\begin{array}{ccc}
FI & \xrightarrow{\ i\ } & I \\
{\scriptstyle F(\!|a|\!)}\downarrow & & \downarrow{\scriptstyle (\!|a|\!)} \\
FA & \xrightarrow{\ a\ } & A \\
{\scriptstyle F(h)}\downarrow & & \downarrow{\scriptstyle h} \\
FB & \xrightarrow{\ b\ } & B
\end{array}
$$

Since (I, i) is the initial algebra, from where there is unique arrow to (B, b). It implies there is an arrow from I to B, which is a catamorphism $(\!|b|\!)$ as shown in below diagram:

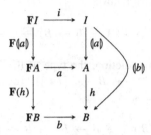

This diagram shows the path from I through A to B, and the path directly from I to B commutes if and only if there is h, such that the square at the bottom commutes. It is called the fusion law of initial algebra, denote as:

$$A \xrightarrow{h} B \Rightarrow h \circ (\!|a|\!) = (\!|b|\!) \iff h \circ a = b \circ \mathbf{F}(h) \tag{5.16}$$

What does this fusion law mean? Catamorphism applies a non-recursive computation to recursive structure (see 193). For list functor **ListF** A, where A is some object, the arrow is $a = f + z$ (coproduct of f and z). The initial arrow for list is $i = (:) + [\,]$. The catamorphism of a is $(\!|a|\!) = foldr(f, z)$. For source object B of $A \xrightarrow{h} B$, let the arrow $b = c + n$. The catamorphism of b is $(\!|b|\!) = foldr(c, n)$. The fusion law says:

$$h \circ foldr(f, z) = foldr(c, n)$$

It simplifies the transformation after folding to one folding. Takano and Meijer in 1995 further abstracted the catamorphism $(\!|a|\!)$ to build some abstract algebra structure $g\,a$ from a, such that the fusion law was extended to [56]:

$$A \xrightarrow{h} B \Rightarrow h \circ g\,a = g\,b \tag{5.17}$$

This extended fusion law is called *acid rain law*.[4] From the initial algebra I, there is unique arrow $I \xrightarrow{(\!|a|\!)} A$; substitute h on the left hand of the acid rain law with $(\!|a|\!)$; substitute a with i; and substitute b with a, we obtain:

$$I \xrightarrow{(\!|a|\!)} A \Rightarrow (\!|a|\!) \circ g\,i = g\,a \tag{5.18}$$

---

[4] The origin fusion law eliminating intermediate results was named as *deforestation* before.

For the list example, the catamorphism $(\![a]\!)$ is $foldr(f, z)$; the initial algebra i for list is $(:) + [\ ]$; define $build(g) = g\ (:)\ [\ ]$, and substitute it into the left side of acid rain law, we obtain the foldr/build fusion law:

$$(\![a]\!) \circ g\ i = g\ a \qquad \text{acid rain law}$$
$$\Rightarrow \quad foldr\ f\ z\ (g\ i) = g\ a \qquad \text{foldr is catamorphism for list}$$
$$\Rightarrow \quad foldr\ f\ z\ (g\ (:)\ [\ ]) = g\ a \qquad \text{the initial algebra } i \text{ for list is } (:) + [\ ]$$
$$\Rightarrow \quad foldr\ f\ z\ (g\ (:)\ [\ ]) = g\ f\ z \qquad \text{algebra } a = f + z$$
$$\Rightarrow \quad \boldsymbol{foldr}\ f\ z\ (\boldsymbol{build}\ g) = g\ f\ z \qquad \text{reverse of } build$$

Hence proves the foldr/build fusion law for list [55].

Exercise 5.1

5.1.1 Verify the definition of *foldl* in *foldr*:

$$foldl\ f\ z\ xs = foldr\ (b\ g\ a \mapsto g\ (f\ a\ b))\ id\ xs\ z$$

5.1.2 Verify below *build* ... *foldr* forms hold:

$$concat\ xss = build\ (f\ z \mapsto foldr\ (xs\ x \mapsto foldr\ f\ x\ xs)\ z\ xss)$$

$$map\ f\ xs = build\ (\oplus\ z \mapsto foldr\ (y\ ys \mapsto (f\ y) \oplus ys)\ z\ xs)$$

$$filter\ p\ xs = build\ \left(\oplus\ z \mapsto foldr\ \left(x\ xs' \mapsto \begin{cases} p(x): & x \oplus xs' \\ \text{otherwise}: & xs' \end{cases}\right)\ z\ xs\right)$$

$$repeat\ x = build\ (\oplus\ z \mapsto let\ xs = x \oplus xs\ in\ xs)$$

5.1.3 Simplify the quick sort algorithm.

$$qsort\ [\ ] \quad = [\ ]$$
$$qsort\ (x{:}xs) = qsort\ [a | a \in xs, a \le x] + [x] + qsort\ [a | a \in xs, x < a]$$

Hint: turn the ZF-expression[5] into *filter*.

---

[5] In set theory, Zermelo-Fraenkel expression $\{f(x) | x \in X, p(x), q(x), \ldots\}$ builds set. We use the square brackets form for list.

5.1.4 Verify the type constraint of fusion law by category theory. Hint: consider the type of the catamorphism.

Answer: page 217

5.2 Make a Century

Knuth gives an exercise ([54, chapter 6], [57, Vol 4]): Write down 1 2 3 4 5 6 7 8 9, only allow to insert + and × symbol in between, parentheses are not allowed. How to make the final calculation result be 100?

For instance:

$$12 + 34 + 5 \times 6 + 7 + 8 + 9 = 100$$

This puzzle for kids is a good programming exercise to find all possible solutions. The straightforward way is exhaustive search with machine brute force. For every position between two digits, there are three possible options: (1) insert nothing; (2) insert +; or (3) insert ×. Since there are eight spaces among nine digits, there are total $3^8 = 6561$ possible options. It's a piece of cake for modern computer to figure out which makes 100 from the 6561 cases.

5.2.1 Exhaustive Search

We need model the expression consists of digits, +, and ×, for example,

$$12 + 34 + 5 \times 6 + 7 + 8 + 9 \tag{5.19}$$

As multiplication is prior to addition, we treat Eq. (5.19) as a list of sub-expressions separated by + and then take sum:

$$sum \ [12, 34, 5 \times 6, 7, 8, 9]$$

In summary, we define an expression as the sum of one or multiple terms (sum-expressions) separated by + as $t_1 + t_2 + \ldots + t_n$, or $expr = sum\ terms$, where $terms = [t_1, t_2, \ldots, t_n]$ is the list of terms.

```
type Expr = [Term]
```

We treat every term (sub-expression) as the product of one or multiple factors separated by × as $f_1 \times f_2 \times \ldots \times f_m$, or $term = product\ factors$, where $factors = [f_1, f_2, \ldots, f_m]$. For example, $5 \times 6 = product\ [5, 6]$. This definition covers single number. For example $34 = product\ [34]$.

```
type Term = [Factor]
```

Every factor consists of digits. For example, 34 has two digits; 5 has 1. In summary $factor = [d_1, d_2, \ldots, d_k]$.

```
type Factor = [Int]
```

Exercise 1.4.2 shows it's a folding process to evaluate the decimal value from a list of digits. For example, $[1, 2, 3] \Rightarrow (((1 \times 10) + 2) \times 10) + 3$. We define it as a function:

$$dec = foldl\ (n\ d \mapsto n \times 10 + d)\ 0$$

The exhaustive search needs to examine every possible expression whether it evaluates to 100 or not. We next define a function to evaluate a expression. This is a top-down recursive process: from expression to terms (sub-expressions), to factors, and finally to decimal values:

$$eval = sum \circ map\ (product \circ (map\ dec))$$

This is a good place for optimization with the fusion law. We leave it as Exercise 5.2.1 and directly give the result here:

$$eval = foldr\ ((+) \circ (foldr\ ((\times) \circ fork\ (dec, id))\ 1\ t))\ 0$$

where $fork(f, g)\ x = (f\ x, g\ x)$. It applies two functions to a variable separately to form a pair. We expand it to below implementation to balance performance and readability:

$$
\begin{aligned}
eval\ [\ ] &= 0 \\
eval(t : ts) &= product\ (map\ dec\ t) + eval(ts)
\end{aligned}
$$

The result is 0 if the expression is empty; otherwise, it takes the first sub-expression, evaluates all its factors, and multiplies them and then adds it to the result of the rest sub-expressions. It repeatedly calls $eval$ for all possible expressions and filters the candidates out unless they evaluate to 100:

$$filter\ (e \mapsto eval(e) == 100)\ es$$

where es is the set of all possible expressions of 1 to 9. To generate es, we start from empty set. Every time, pick a number from 1 to 9 to expand the expressions. The empty set and a number d generate an expression of $[[[d]]]$. The inner most part $[d]$ is a singleton factor that consists of one digit d, i.e., $factor = [d]$. The middle part $[[d]]$ is a term (sub-expression) that consists of this factor $term = [factor] =$

$[[d]]$; the whole $[[[d]]]$ is the final expression $expr = [term] = [[factor]] = [[[d]]]$. Below builds the expression from a single digit:

$$expr(d) = [[[d]]]$$

For the recursive cases, we start from right and repeatedly pick numbers 9, 8, 7, ...to expand the expression. How to expand new expressions from a set of expressions $[e_1, e_2, \ldots, e_n]$ and a digit d? For every expression e_i, try the three options: (1) insert nothing; (2) insert +; and (3) insert ×. Expand e_i as the sum of terms $e_i = t_1 + t_2 + \ldots$, where the first term t_1 is a product of factors $t_1 = f_1 \times f_2 \times \ldots$.

(1) Insert nothing means prepend digit d to the first factor of the first term in e_i. The result $d : f_1$ forms a new factor. For example, let e_i be $8 + 9$, and d be 7. Writing 7 before $8+9$ without inserting anything gives a new expression $78+9$;
(2) Insert × means to use d forms a new factor $[d]$; then prepend it before the first term of e_i. The result $[d] : t_1$ gives a new term. For the same example of $8 + 9$, write 7 before it, and insert a × symbol between 7 and 8. It gives the new expression $7 \times 8 + 9$;
(3) Insert + means to use d forms a new term $[[d]]$, then prepend it to e_i to form a new expression $[[d]] : e_i$. For the same example of $8 + 9$, write 7 before it, and insert + between 7 and 8. It gives the new expression $7 + 8 + 9$.

Summarizes these three options:

```
add d ((ds:fs):ts) = [((d:ds):fs):ts,
                      ([d]:ds:fs):ts,
                      [[d]]:(ds:fs):ts]
```

where e_i is in the form of $(ds{:}fs){:}ts$. The first term of e_i is $ds{:}fs$, and the first factor of the first term is ds. Expand every expression in $[e_1, e_2, \ldots, e_n]$ to three new ones and then concatenate the results together. It is exactly the *concatMap* function in Eq. (5.15):

$$extend\ d\ [\ ] = [expr(d)]$$
$$extend\ d\ es = concatMap\ (add\ d)\ es$$

Below is the complete definition of the exhaustive search:

$$filter\ (e \mapsto eval(e) == 100)\ (foldr\ extend\ [1..9]) \qquad (5.20)$$

5.2.2   Improvement

How can we improve the exhaustive search? Observe the right to left expand process. We write down 9 first, then expand three new expressions with 8. They

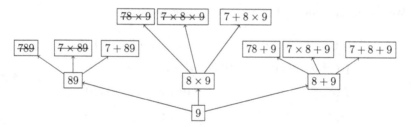

Fig. 5.2 Expanding expressions is like a growing tree

are 89, $8 \times 9 = 72$, and $8 + 9 = 17$. In the next step when we expand with 7, expression 789 exceeds 100 obviously. We may drop it immediately. Any new expressions expanded to the left of 789 can't be 100; hence, it is safe to drop to avoid unnecessary computation. Similarly, 7×89 exceeds 100, need be dropped to avoid further expansion. $7 + 89$ is the only possible expression. Next expand 78×9; it is bigger than 100; hence it is dropped to avoid further expansion.

In summary, we evaluate the expression during the expanding process. Whenever an expression exceeds 100, we immediately drop it to avoid further expansion. The whole process looks like a growing tree. When any expression in a branch exceeds 100, we cut that branch off (Fig. 5.2).

In order to evaluate the expression during expansion easily, we separate the value of the expression into three parts: (1) f is the value of the first factor of the first term; (2) v_{fs} is the product of the rest factors of the first term; and (3) v_{ts} is the sum of the rest terms. The overall value of the expression is calculated out of them as $f \times v_{fs} + v_{ts}$.

When expand a new digit d to the left, the expression and value are updated as the following for each of the three options:

(1) Insert nothing; prepend d before the first factor as its most significant digit. The expression value is $(d \times 10^{n+1} + f) \times v_{fs} + v_{ts}$, where n is the number of digits of f. The three parts of the value update to $f' = d \times 10^{n+1} + f$, v_{fs} and v_{ts} are unchanged;

(2) Insert $\times$; put d as the first factor of the first term. The expression value is $d \times f \times v_{fs} + v_{ts}$. The three parts of the value update to $f' = d$, $v_{fs'} = f \times v_{fs}$, while v_{ts} is unchanged;

(3) Insert +; use d as the new first term. The expression value is $d + f \times v_{fs} + v_{ts}$. The three parts of the value update to $f' = d$, $v_{fs'} = 1$, $v_{ts'} = f \times v_{fs} + v_{ts}$.

To calculate 10^{n+1} in option (1) easily, we record the exponent as the fourth part of the expression value. As such, the value is represented as a 4-tuple (e, f, v_{fs}, v_{ts}). The function calculating the value is as below:

$$value(e, f, v_{fs}, v_{ts}) = f \times v_{fs} + v_{ts} \tag{5.21}$$

We update the 4-tuple when expanding expressions:

$$add\ d\ (((ds:fs):ts), \qquad\qquad (e, f, v_{fs}, v_{ts})) =$$
$$[(((d:ds):fs):ts, \quad (10 \times e, d \times e + f, v_{fs}, v_{ts})),$$
$$(([d]:ds:fs):ts, \qquad (10, d, f \times v_{fs}, v_{ts})),$$
$$([[d]]:(ds:fs):ts, \qquad (10, d, 1, f \times v_{fs} + v_{ts}))] \tag{5.22}$$

For every pair of (expression, 4-tuple), we call *value* function to calculate the value, drop it if exceeds 100 and then concatenate the candidate expressions and the 4-tuples to a list. We redefine the previous *extend* function as below:

$$expand\ d\ [\ ]\ = [(expr(d), (10, d, 1, 0))]$$
$$expand\ d\ evs = concatMap\ ((filter\ ((\leq 100) \circ value \circ snd)) \circ (add\ d))\ evs \tag{5.23}$$

We are ready to fold on *expand* and generate all candidate expressions and 4-tuples that do not exceed 100. Finally, calculate the value from the 4-tuples, and leave only those equal to 100:

$$map\ fst \circ filter\ ((= 100) \circ value \circ snd)\ (foldr\ expand\ [\ ]\ [1, 2, \ldots, 9]) \tag{5.24}$$

The complete program is given in Listings 5.2 and 5.3 in appendix. There are total of seven expressions evaluated to 100:

$$1 : 1 \times 2 \times 3 + 4 + 5 + 6 + 7 + 8 \times 9$$
$$2 : 1 + 2 + 3 + 4 + 5 + 6 + 7 + 8 \times 9$$
$$3 : 1 \times 2 \times 3 \times 4 + 5 + 6 + 7 \times 8 + 9$$
$$4 : 12 + 3 \times 4 + 5 + 6 + 7 \times 8 + 9$$
$$5 : 1 + 2 \times 3 + 4 + 5 + 67 + 8 + 9$$
$$6 : 1 \times 2 + 34 + 5 + 6 \times 7 + 8 + 9$$
$$7 : 12 + 34 + 5 \times 6 + 7 + 8 + 9$$

Exercise 5.2

5.2.1 Use the fusion law to optimize the expression evaluation function:

$$eval = sum \circ map\ (product \circ (map\ dec))$$

5.2.2 How to expand all expressions from left?

5.2.3 The following definition converts expression to string:

$$str = (join \text{ "+"}) \circ (map ((join \text{ " } \times \text{ "}) \circ (map (show \circ dec))))$$

where $show$ converts number to string. Function $join(c, s)$ concatenates multiple strings s with delimiter c. For example,

$$join(\text{"#"}, [\text{"abc"}, \text{"def"}]) = \text{"abc#def"}$$

Use the fusion law to optimize str.

Answer: page 222

5.3 Further Reading

Algorithm deduction, as an application of mathematical deduction, provides a systematic way to optimize program: start from some intuitive, raw, and unoptimized implementation, through a set of formalized rules, and step by step convert to elegant and optimized result. Bird gives many examples in *Pearls of Functional Algorithm Design* [54]. To lay the correctness of programming on mathematical deduction but not human intuition, we need a dedicated theory, that formalizes varies of programs. The foldr/build fusion law is such an example [53]. As the category theory being widely adopted in programming, a series of fusion laws have been developed [55] for algorithm deduction and optimization.

Appendix: Example Source Code

The rank 2 type polymorphic of `build`; the optimized `concatMap`.

Listing 5.1 build and `concatMap`

```
build :: forall a. (forall b. (a → b → b) → b → b) → [a]
build g = g (:) []

concatMap f xs = build (λc n → foldr (λx b → foldr c b (f x)) n xs)
```

Exhaustive search solution for "Making a Century" puzzle:

Listing 5.2 Making a Century

```
type Expr = [Term]        — T₁ + T₂ + ... Tₙ
type Term = [Factor]      — F₁ × F₂ × ... Fₘ
type Factor = [Int]       — d₁d₂ ... dₖ

dec :: Factor → Int
dec = foldl (λn d → n * 10 + d) 0

expr d = [[[d]]]          — single digit expr

eval [] = 0
eval (t:ts) = product (map dec t) + eval ts

extend :: Int → [Expr] → [Expr]
extend d [] = [expr d]
extend d es = concatMap (add d) es where
  add :: Int → Expr → [Expr]
  add d ((ds:fs):ts) = [((d:ds):fs):ts,
                        ([d]:ds:fs):ts,
                        [[d]]:(ds:fs):ts]

sol = filter ((==100) o eval) o foldr extend []
```

The improved exhaustive search solution:

Listing 5.3 Making a Century, Improved solution

```
value (_, f, fs, ts) = f * fs + ts

expand d [] = [(expr d, (10, d, 1, 0))]
expand d evs = concatMap ((filter ((≤ 100) o value o snd)) o (add d)) evs
where
  add d (((ds:fs):ts), (e, f, vfs, vts)) =
    [((((d:ds):fs):ts, (10 * e, d * e + f, vfs, vts)),
     (([d]:ds:fs):ts, (10, d, f * vfs, vts)),
     ([[d]]:(ds:fs):ts, (10, d, 1, f * vfs + vts))]

sol = map fst o filter ((==100) o value o snd) o foldr expand []
```

Chapter 6
Infinity

I see it, but I don't believe it.

Georg Cantor, in a letter to Dedekind in 1877

Escher, Circle limit IV (Heaven and Hell), 1960

Long time ago, our ancestors looked up at starry sky, the vast galaxy, asked how big the world we live in was. As intelligent being, we think about ourselves, our planet, the universe, and infinity. People abstracted numbers from concrete things: from three goats, three apples, and three jars to three of anything. The numbers were not big at first, sufficient for hunting, collecting, and farming. As civilization evolved, people started to trade. A variety of numbering systems were developed to support bigger and bigger numbers. It's nature to ask what the biggest number is. There were two different opinions in ancient time. Some didn't think the question make sense. It's already big enough like thousands or millions in everyday life. Why trouble ourselves with big numbers that would never be used? It's safe to consider, for example, the number of sand grains in the world is infinite. In ancient Greece, people thought 10,000 was a very big number and named it "murias." It later changed to "myriad," meaning infinity [58]. In Buddhism, people also use "the sand in Ganges

© China Machine Press 2024
X. Liu, *Mathematics in Programming*,
https://doi.org/10.1007/978-981-97-2432-1_6

Cover of *The Sand-Reckoner*. Archimedes Thoughtful by Domenico Fetti (1620)

River" for the numbers that are too large to count. In the Mahayana Buddhist classic work *The Diamond Sutra*, it said: "If a virtuous man or woman filled a number of universes, as great as the number of sand-grains in all these rivers, with the seven treasures, and gave them all away in alms (dana), would his or her merit be great?" Others had different opinion. Archimedes believed even the sand grains that filled the whole universe could be represented with a definite number. In his book, *The Sand-Reckoner*, Archimedes said:

> There are some, King Gelon, who think that the number of the sand is infinite in multitude; and I mean by the sand not only that which exists about Syracuse and the rest of Sicily but also that which is found in every region whether inhabited or uninhabited. Again there are some who, without regarding it as infinite, yet think that no number has been named which is great enough to exceed its multitude. And it is clear that they who hold this view, if they imagined a mass made up of sand in other respects as large as the mass of the Earth, including in it all the seas and the hollows of the Earth filled up to a height equal to that of the highest of the mountains, would be many times further still from recognising that any number could be expressed which exceeded the multitude of the sand so taken. But I will try to show you by means of geometrical proofs, which you will be able to follow, that, of the numbers named by me and given in the work which I sent to Zeuxippus, some exceed not only the number of the mass of sand equal in magnitude to the Earth filled up in the way described, but also that of a mass equal in magnitude to the universe.

Archimedes thought it *only* need 10^{63} sand grains to fill the universe. The universe he meant was celestial sphere, which was about 20,000 times the radius of the Earth. The observable universe is about 46.5 billion light-years and consists of around 3×10^{74} atoms.[1] Archimedes was definitely genius in ancient Greek time, he demonstrated how to quantify the "infinite big" things. There are many words in different languages that serve as the unit for big numbers. The following table lists these words in Chinese; they increase for every 10^4 ([59], pp31).

---

[1] Also said to have 10^{80} to 10^{87} elementary particles.

京	10^{16}	载	10^{44}
垓	10^{20}	极	10^{48}
秭	10^{24}	恒河沙	10^{52}
穰	10^{28}	阿僧祇	10^{56}
沟	10^{32}	那由他	10^{60}
涧	10^{36}	不可思议	10^{64}
正	10^{40}	无量大数	10^{68}

Many words come from Buddhism, like "恒河沙," which means the sand grain in Ganges River. It has 52 zeros after 1. Below table lists the unit words in English. Starting from 1, there is a unit for every 1000 magnitude (compare to 10000 magnitude step in Chinese).

thousand	10^3	quattuordecillion	10^{45}	octovigintillion	10^{87}
million	10^6	quindecillion	10^{48}	novemvigintillion	10^{90}
billion	10^9	sexdecillion	10^{51}	trigintillion	10^{93}
trillion	10^{12}	septdecillion	10^{54}	untrigintillion	10^{96}
quadrillion	10^{15}	octodecillion	10^{57}	duotrigintillion	10^{99}
quintillion	10^{18}	novemdecillion	10^{60}	**googol**	10^{100}
sexillion	10^{21}	vigintillion	10^{63}		
septillion	10^{24}	unvigintillion	10^{66}		
octillion	10^{27}	duovigintillion	10^{69}		
noniliion	10^{30}	trevigintillion	10^{72}		
decillion	10^{33}	quattuorvigintillion	10^{75}		
undecillion	10^{36}	quinvigintillion	10^{78}		
duodecillion	10^{39}	sexvigintillion	10^{81}		
tredecillion	10^{42}	seprvigintillion	10^{84}		

The last unit, googol, was coined in 1920 by 9-year-old Milton Sirotta, nephew of US mathematician Edward Kasner. It is digit 1 followed by one hundred 0s. The Internet company Google's name came from this word [60].

6.1 Zeno's Paradoxes

Whether there exists infinity beyond all numbers is not only a mathematical problem but also a philosophical problem. Infinity also leads to the concept of infinitesimal (infinitely small). Zeno of Elea, an ancient Greek philosopher, thought of a set of problems about infinity. Some of them are preserved in Aristotle's *Physics*. Four paradoxes among them are most famous.

The first one is well-known as Achilles and the tortoise paradox. Achilles was the greatest Greek warrior of Homer's *Iliad*, a hero of the Trojan War. In this paradox,

Fig. 6.1 Achilles and the tortoise

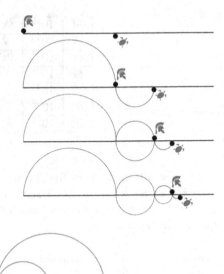

Fig. 6.2 Dichotomy paradox

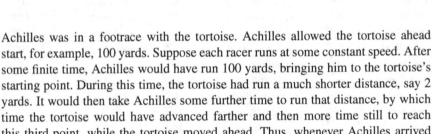

Achilles was in a footrace with the tortoise. Achilles allowed the tortoise ahead start, for example, 100 yards. Suppose each racer runs at some constant speed. After some finite time, Achilles would have run 100 yards, bringing him to the tortoise's starting point. During this time, the tortoise had run a much shorter distance, say 2 yards. It would then take Achilles some further time to run that distance, by which time the tortoise would have advanced farther and then more time still to reach this third point, while the tortoise moved ahead. Thus, whenever Achilles arrived somewhere the tortoise had been, he still had some distance to go before he could even reach the tortoise (Fig. 6.1). Although it contradicts our common sense in real life, the argument is so convincing. This paradox attracted many great thinkers for thousands of years. Lewis Carroll and Douglas Hofstadter even took Achilles and the tortoise as figures in their literary works.

The second is dichotomy paradox. Atalanta, in Greek mythology, was a virgin huntress. Suppose Atalanta was going to walk to the end of a path. Before she could get there, she must travel halfway; before traveling halfway, she must travel a quarter; before traveling a quarter, she must travel one-eighth; and so on (Fig. 6.2). This description requires one to complete infinitely many tasks, which Zeno believed was impossible. The paradoxical conclusion then would be that travel over any finite distance can neither be completed nor begin, and so all motion must be an illusion.

The third is arrow paradox. Zeno stated that for motion to occur, an object must change the position which it occupied. He gave an example of an arrow in flight. In any one (duration-less) instant of time, the arrow is neither moving to where it is, nor to where it is not. It cannot move to where it is not, because no time elapses for it to move there; it cannot move to where it is, because it is already there. In

Fig. 6.3 Arrow paradox

Fig. 6.4 Moving rows paradox

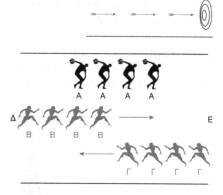

other words, at every instant of time, there is no motion occurring. If everything is motionless at every instant, and time is entirely composed of instants, then motion is impossible. Whereas the first two paradoxes divide space, this paradox starts by dividing time, and not into segments, but into points (Fig. 6.3).

The fourth is called the moving rows paradox, or stadium paradox. It is also about dividing time into atomic points. As in Fig. 6.4, there are three rows in the stadium, each row being composed of an equal number of bodies. At the beginning, they are all aligned. At the smallest time duration, row A stays, and row B moves to the right one space unit, while row Γ moves to the left one space unit. To row B, row Γ actually moves two space units. It means, there should be time duration that Γ moves one space unit relative to B. And it is the half time of the smallest duration. Since the smallest duration is atomic, it involves the conclusion that half a given time is equal to that time.

Zeno's paradoxes are easy to understand. However, the conclusion is surprising. In our common sense, motion and time are so real. Achilles must be able to catch up the tortoise. However, it is hard to solve Zeno's paradox. Aristotle, Archimedes, Russell, Weyl, etc., proposed varies of solutions to Zeno's paradoxes [61].

Zeno was a philosopher in ancient Greece. He was born in Elea, which was a Greek colony located in present-day southern Italy. Little is known for certain about Zeno's life. In the dialogue of Parmenides, Plato describes a visit to Athens by Zeno and Parmenides, at a time when Parmenides was "about 65," Zeno was "nearly 40," and Socrates was "a very young man." Assuming an age for Socrates of around 20 and taking the date of Socrates' birth as 469 BC gives an approximate date of birth for Zeno of 490 BC. Some less reliable details of Zeno's life are given by Diogenes Laërtius in his *Lives and Opinions of Eminent Philosophers*. It said Zeno was the adopted son of Parmenides. He was skilled to argue both sides of any question, the universal critic. And that he was arrested and perhaps killed at the hands of a tyrant of Elea [7].

Zeno was the primary member of the Eleatics, which were a pre-Socratic school of philosophy founded by Parmenides in the early fifth century BC in the ancient town of Elea. It was another important school after Pythagoras. The Eleatics rejected the epistemological validity of sense experience and instead took logical standards of clarity and necessity to be the criteria of truth. Of the members, Parmenides and Melissus built arguments starting from sound premises. Zeno, on the other hand, primarily employed the reduction to absurdity, attempting to destroy the arguments of others by showing that their premises led to contradictions. Zeno is best known for his paradoxes, which Russell described as "immeasurably subtle and profound."

Zeno of Elea (About 490BC–425BC)

Ancient Chinese philosophers developed equivalent paradoxes to some of Zeno's. From the surviving Mohist school of names book of logic which states, "a one-foot stick, every day take away half of it, in a myriad ages it will not be exhausted," Zeno's paradoxes caused deep confusion to the ancient Greeks. The views about time, space, infinity, continuum, and movement are critical to the later philosophers and mathematicians even today. How to understand infinity became a critical problem that must be solved.

6.1.1 Potential Infinity and Actual Infinity

Aristotle studied Zeno's paradoxes deeply. He distinguished the *potential infinite* and the *actual infinite*. This work had fundamental influence up to contemporary times [7]. Potential infinity is a process that never stops. It can be a collection of "things" that continues without terminating, going on, or repeating itself over and over again with no recognizable ending point. The obvious example is natural numbers. No matter where you are while listing or counting out natural numbers, there always exists another number to proceed. For another example, in Euclidean geometry, a line with a starting point could extend on without end, but could still be potentially infinite because one can add on more length to a finite length to allow it to extend.[2] The actual infinite involves never-ending sets or "things" within a space that has a beginning and end; it is a series that is technically "completed" but

---

[2] Euclid avoided to use the term "infinitely extend" in his work. Instead he said a line could be extended any long as needed. This was a common treatment in ancient Greece.

consists of infinitely many members. According to Aristotle, actual infinities cannot exist because they are paradoxical. It is impossible to say that you can always "take another step" or "add another member" in a completed set with a beginning and end, unlike a potential infinite. It was ultimately Aristotle's rejection of the actual infinite that allowed him to refute Zeno's paradox.

Although Aristotle disproved the existence of the actual infinite finally and tended to reject some major concepts in mathematics, the importance of mathematics was never belittled in Aristotle's eyes. Aristotle argued that actual infinity as it was not applicable to geometry and the universal was not relevant to mathematics, making potential infinity all that actually was important. Aristotle's viewpoint to infinity was typical in ancient Greece. Despite these arguments, the ancient Greek mathematicians achieved astonishing result with the concept of potential infinity. For example, Euclid proved there were infinitely many prime numbers. It is considered one of the most beautiful proofs in history.

Proposition 6.1.1 (Euclid's Elements, Book IX, Proposition 20) *Prime numbers are more than any assigned multitude of prime numbers [10].*

Euclid indented to avoid term of "infinitely many" when stated this proposition. Such treatment is very common in *Elements*. It's famous that Euclid used reduction to absurdity in his proof. We show it in modern language.

Proof Assume there are finite prime numbers $p_1, p_2, \ldots, p_n$. Make a new number:

$$p_1 p_2 \cdots p_n + 1$$

Which is the product of the n prime numbers plus one. It is either prime or not.

(1) If it is prime, it definitely does not equal to any of $p_1, \ldots p_n$, is a new prime not in the list;
(2) If it is not prime, then it must be divided by some prime p. However, no one from $p_1 \ldots, p_n$ divides this number; hence p is a new prime beyond the list.

In both cases, we get a new prime number; it contradicts the assumption that there are finite prime numbers. It shows there are infinitely many prime numbers.

$\square$

From today's viewpoint, Euclid applied a kind of indirect "proof of existence," reduction to absurdity. He showed there were infinitely many prime numbers, but did not give a construction to list them. Although quite natural in mathematical proofs nowadays, it led to hot debate about the fundamentals of mathematics in the late nineteenth and early twentieth century.

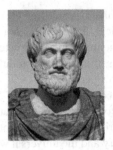

Aristotle (384BC–322BC)

Aristotle was a great philosopher and polymath in ancient Greece. Along with his teacher Plato, he has been known as the "Father of Western Philosophy." Little is known about his life. Aristotle was born in the city of Stagira in Northern Greece in 384BC. His father died when Aristotle was about 10 years old. He was brought up by a guardian, Proxenus of Atarneus. In 367BC, Aristotle joined Plato's Academy in Athens. He remained there for 20 years as Plato's pupil and colleague until Plato died. This period of study in Plato's Academy deeply influenced Aristotle's life.

Shortly after Plato died around 347BC, Aristotle left Athens. The traditional story about his departure records that he was disappointed with the academy's direction after control was passed to Plato's nephew. After that, he traveled around. In 343 BC, Aristotle was invited by Philip II of Macedonia to become the tutor to his 13 years old son Alexander. Aristotle was appointed as the head of the Academy of Macedonia.

In 335BC, Philip II died. Aristotle returned to Athens, establishing his own school there known as the Lyceum. Aristotle conducted courses at the school for the next 12 years. In this period in Athens, between 335 and 323 BC, Aristotle composed many of his works. He studied and made significant contributions to logic, metaphysics, mathematics, physics, biology, botany, ethics, politics, agriculture, medicine, dance, and theatre. There was legend that Aristotle had a habit of walking while lecturing along the walkways covered with colonnades. It was for this reason that his school was named "Peripatetic" (which means "of walking" or "given to walking about"). Aristotle used the language that was much more obscure than Plato's Dialogue. Many of his works are based on lecture notes, and some are even the notes of his students. Aristotle was considered as the first author of textbooks in the Western world.

Following Alexander's death, anti-Macedonian sentiment in Athens was rekindled. In 322 BC, his enemies reportedly denounced Aristotle for impiety, prompting him to flee to his mother's family estate in Euboea. He said "I will not allow the Athenians to sin twice against philosophy"—a reference to Athens's trial and execution of Socrates. He died on Euboea of natural causes later that same year, at the age of 63.

More than 2300 years after his death, Aristotle remains one of the most influential people who ever lived. He contributed to almost every field of human knowledge in ancient time, and he was the founder of many new fields. Among countless achievements, Aristotle was the founder of formal logic, pioneered the study of zoology, and left every future scientist and philosopher in his debt through his contributions to the scientific method.

6.1.2 Method of Exhaustion

Ancient Greek mathematicians developed the method of exhaustion and made great achievements with infinity. The idea of exhaustion originated in the late fifth century BC with Antiphon (About 480BC–410BC). He studied the classic geometric problem, square the circle. To approximate the area of a circle, Antiphon started from an inscribed square and then repeatedly doubled the number of sides to get octagons, hexagons... As the area of the circle gradually "exhausted," the side length of inscribed polygons got smaller and smaller. Antiphon thought the polygon would eventually coincide with the circle. This was the idea of exhaustion. The method was made rigorous a few decades later by Eudoxus of Cnidus, who used it to calculate areas and volumes. The correctness of this method relies on below axiom.

Axiom 6.1.1 (Axiom of Archimedes) Given two magnitudes a and b, there exists some natural number n, such that $a \leq nb$.

This axiom is fundamental. We've seen Euclidean algorithm to compute the greatest common measurement in Sect. 2.2. Axiom of Archimedes shows that Euclid algorithm always terminates for commensurable magnitudes. Eudoxus stated the logic as "Given two different magnitudes, for the larger one, subtract a magnitude larger than its half, then for the remaining, subtract another magnitude larger than its half, repeat this process, there must be some remaining less than the smaller one."

With the method of exhaustion, Eudoxus showed that: the volume of a pyramid is $\frac{1}{3}$ the volume of a prism with the same base and altitude, and the volume of a cone is $\frac{1}{3}$ that of the corresponding cylinder. These result are summarized as propositions in the book 12 of Euclid's *Elements* [7].

Archimedes greatly developed the method of exhaustion and achieved the peak level of result in ancient time. He approximated π; found the formulas to compute circular area, surface area, and volume of sphere and cone; and even found the method to calculate the area under the parabola.

The Fields Medal carries a portrait of Archimedes

Archimedes (287BC–212BC) was a mathematician, physicist, engineer, inventor, and astronomer. He was born in the seaport city of Syracuse, at that time a self-governing colony in Magna Graecia. Archimedes might have studied in Alexandria, Egypt, in his youth. During his lifetime, Archimedes made his work known through correspondence with the mathematicians in Alexandria. Although few details of his life are known, he is considered the greatest mathematician of antiquity and one of the greatest of all time. Various popular stories about him are widely circulated.

The most widely known anecdote about Archimedes tells us of how he uncovered a fraud in the manufacture of a golden crown commissioned by King Hiero II of Syracuse. The king had supplied the pure gold to be used, and Archimedes was asked to determine whether some silver had been substituted by the dishonest goldsmith. Archimedes had to solve the problem without damaging the crown, so he could not melt it down into a regularly shaped body in order to calculate its density. While taking a bath, he noticed that the level of the water in the tub rose as he got in, and realized that this effect could be used to determine the volume of the crown. Archimedes then took to the streets naked, so excited by his discovery that he had forgotten to dress, yelling "Eureka!" (meaning "I have found (it)!" in Greek). The test was conducted successfully, proving that silver had indeed been mixed in. His discovery, "Archimedes' Principle" is known by every student at school. Eureka was later used to describe the moment of inspiration.

In 214BC, the Second Punic War broke out. Legend has it that Archimedes created a giant parabolic mirror to deflect the powerful Mediterranean sun onto the ship's sails, setting fire to them. Archimedes also created a huge crane-operated hook—the Claw of Archimedes—that was used to lift the enemy ships out of the sea before dropping them to their doom. After 2-year-long siege, In 212 BC, the Romans captured Syracuse. The Roman force commander, Marcellus, had ordered that Archimedes, the well-known mathematician, should not be killed. Archimedes, who was now around 78 years of age, continued his studies after the breach by the Romans, and while at home, his work was disturbed by a Roman soldier. The last words attributed to Archimedes are "Do not disturb my circles!" The soldier killed Archimedes despite orders that Archimedes should not be harmed. 137 years after his death, the Roman orator Cicero described visiting the tomb of Archimedes. It

Fig. 6.5 Approximate π by circumscribed and inscribed polygons

was surmounted by a sphere and a cylinder, which Archimedes had requested be placed on his tomb to represent his mathematical discoveries.[3]

Archimedes calculated π with the method of exhaustion around 250BC. Symbol π represents the ratio of a circle's circumference to its diameter; sometimes it's referred to as Archimedes' constant. As in Fig. 6.5, Archimedes drew two regular polygons inscribed and circumscribed a circle with diameter of 1. For a side of the inscribed polygon and the corresponding arc, the length of the arc is greater than the side because the straight line is the shortest between two points. Hence the circumference of the circle is greater than the inscribed polygon. For similar reason, the circumference of the circle is less than the circumscribed polygon. Since the diameter is 1, the circle's circumference equals to π. Below inequality holds:

$$C_i < \pi < C_o$$

where C_i and C_o are the circumferences of the inscribed and circumscribed polygons, respectively. Successively increasing the number of sides approximates the range of π. Archimedes calculated the 96-sided regular polygon and showed $\frac{223}{71} < \pi < \frac{22}{7}$ (i.e., $3.1408 < \pi < 3.1429$). This upper bound of $\frac{22}{7}$ is widely used in practice.[4]

The method of exhaustion, although rigorous, demands specific treatment for different problems; it's complex compare to modern calculus. It's partially because Ancient Greeks attempted to avoid explicit infinity or infinitesimal and rejected irrational numbers and had to make distinction between geometric magnitudes and numbers.

Ptolemy was another ancient Greek mathematician and astronomer who made great achievement with the method of exhaustion. He developed a geocentric model to calculate the celestial motions. It was almost universally accepted until the scientific revolution. His planetary hypotheses presented a physical realization of the

---

[3] A sphere inscribed in a cylinder has 2/3 the volume and surface area of the cylinder including its bases.

[4] Ptolemy improved the approximation to 3.1416 about 400 years later with method of exhaustion. Zu Chongzhi, a Chinese mathematician around 480AD, improved the interval to $3.1415926 < \pi < 3.1415927$ and suggested the approximations $\pi \approx \dfrac{355}{113} = 3.14159292035\ldots$ remained the most accurate approximation of π for the next 800 years.

universe as a set of nested spheres, in which he used the epicycles of his planetary model to compute the dimensions of the universe. He estimated the Sun was at an average distance of 1,210 Earth radii, while the radius of celestial sphere was 20,000 Earth radii.

After Hellenistic period, Greek civilization was destroyed by several forces. The Romans conquered Greece, Egypt, and the Near East. In 47BC, the Romans set fire to the Egyptian ships in the harbor of Alexandria; the fire spread and burned the library—the most expensive ancient libraries. The emperor Theodosius (ruled 379–396) proscribed the pagan religions and in 392 ordered that their temples be destroyed, including the temple of Serapis in Alexandria, which housed the only remaining sizable collection of Greek works. Thousands of Greek books were burned. Many other works written on parchment were expunged. The final blow to the Greek civilization was the conquest of Egypt by the uprising Arab empire in 640. The remaining books were destroyed on the ground that as Omar, the conqueror, put it, "Either the books contain what we also have, in which case we don't have to read them, or they contain the opposite of what we believe, in which case we must not read them." And so for 6 months, the baths of Alexandria were heated by burning rolls of parchment [3].

It always makes people depressed when read about this. The tragedy of burning books happened in South America, in the Qin Empire, and repeated in history. After capture of Egypt, the majority of the scholars migrated to Constantinople, which had become the capital of the Eastern Roman Empire. The Arabs absorbed the Greek works and translated and commented extensively to Greek knowledge. The "House of Wisdom" in Baghdad gradually became the new academy center. After Medieval, Europeans translated the ancient Greek works from Arabic to Latin. Along with the Renaissance in Europe, not only arts and culture but also mathematics and philosophy recovered and were greatly developed.

Johannes Kepler, a German mathematician and astronomer, took the next important step. He inherited valuable observation data from Tycho Brahe and spent 8 years to analyze the observed data and false trails. Kepler's most famous and important results are known today as Kepler's three laws of planetary motion. According to his first law, Kepler broke with the tradition of 2000 years that circle or sphere must be used to describe celestial motions. It states that each planet moves on an ellipse and that the sun is at one (common) focus of each of these elliptical paths. The other radical step Kepler made was he discovered that the planet does not move at a constant velocity. A line segment joining a planet and the Sun sweeps out equal areas during equal intervals of time. This is his second law. It explains that why a planet some times moves fast (close to the Sun), while some times moves slowly (far from the Sun). The third law states that the square of the orbital period of a planet is directly proportional to the cube of the semi-major axis of its orbit. Such complex models required more powerful mathematical tool, the method of exhaustion was not convenient. Kepler then made simplification and used the new method of measuring the volume of wine barrels.

The next major step was made by Descartes and Fermat. Through analytical geometry, numbers and geometry were bridged and finally evolved to calculus by

Newton and Leibniz. Infinitesimal is a central concept in calculus; the integral sum involves infinitely many infinitesimals. As a side word, John Wallis, an important contributor to calculus, introduced ∞ symbol for infinity in 1665.

Although the logic foundation of calculus caused hot debate, this new tool, representing the modern spirit of the Western, broke the waves in its sail in the eighteenth century. The Bernoulli family, Euler, and Lagrange greatly developed calculus and infinite series, solving many hard problems in astronomy, mechanics, and fluid that people never imagined before.

6.2 Programming with Infinity

Computers use limited resources. Numbers are represented in binary forms suitable for electronic logic circuit. There are finitely many binary bits; hence, the numbers represented in computer are also bounded. A binary number of m bits represents numbers at maximum of $2^m - 1$, which is $11\ldots 1$ of length m. For example, the biggest 16 bits number is $2^{16} - 1 = 65535$. For this reason, if the size of a set is represented in binary, then the set can only hold finitely many elements. In early days of programming, arrays were often used to hold multiple elements. To effectively use computer memories, the size of array need be determined before using. For example, below statement in C programming language declares an array that can hold ten integers:

```
int a[10];
```

There are two different types of numbers: ordinal number and cardinal number. In short, ordinal numbers describe a way to arrange a collection of objects in order, one after another, while cardinal numbers measure the size of collection. We'll return to their definitions in Sect. 6.3.6. Both ordinals and cardinals have to be finite in traditional programming. They can't represent infinity directly. It was reasonable in the early days of engineering. The computer devices were very expensive; people never thought to deal with practice problems with infinity. As time goes on, the cost of computation resources keep decreasing. We are not satisfied with the way to "predict" the size of the collection before using it in programming. New tools, like dynamic array, were developed in some programming environments. They were known as containers, allowed to adjust the size on-demand. However, even in dynamic containers, the elements are still finitely many. They cannot exceed the number representation limit. People developed the linked-list (see Sect. 1.5); elements are stored in nodes that are chained together. The last element points to a special empty node indicating the end. Given such a linked-list, one can start from its head, moving to any node without need of knowing its cardinal. As far as the storage allows, a linked-list can be arbitrary long. It brings the possibility to represent potential infinity.

However, there is eventually a gap between linked-list and potential infinity. We consider natural numbers as potential infinity without terminating or some "end point." But when representing natural numbers with linked-list, no matter how long the list is, for instance, n, we have to store all numbers from 0 to n in it. It only represents sequence of $0, 1, \ldots, n$, but not the natural numbers of $0, 1, \ldots, n, \ldots$

To model the potential infinity, we need the concept of lazy evaluation: instead of evaluating an expression or a variable for its value right away, this strategy delays it until the value is needed. For example, any natural number n has a successor $n + 1$ by Peano's axioms, and the first number is 0. Define natural numbers as below:

$$\mathbb{N} = iterate(n \mapsto n + 1, 0) \tag{6.1}$$

where $iterate$ is defined as:

$$iterate(f, x) = x : iterate(f, f(x))$$

Let $succ(n) = n \mapsto n + 1$ denote the successor function; here are the first several steps generating natural numbers:

$$
\begin{aligned}
iterate(succ, 0) &= 0 : iterate(succ, succ(0)) &&\text{definition of } iterate \\
&= 0 : iterate(succ, 1) &&succ(0) = 0 + 1 = 1 \\
&= 0 : 1 : iterate(succ, succ(1)) \\
&= 0 : 1 : iterate(succ, 2) \\
&= 0 : 1 : 2 : iterate(succ, 3) \\
&= \ldots
\end{aligned}
$$

Without lazy evaluation, this process will repeat forever. It cannot solve practical problems. To make the link operation lazy, we convert it to a λ expression.

$$x : xs = cons(x, () \mapsto xs)$$

Define the expression $() \mapsto exp$ as $delay(exp)$. It builds a function without argument; when evaluates (the function), it gives the value of exp.

When link x to xs, the value of xs won't be evaluated, but wrap xs in a λ expression. The evaluation is delayed to future time (Fig. 6.6). With this modification, we change to generate natural numbers as below:

$$
\begin{aligned}
iterate(succ, 0) &= 0 : iterate(succ, succ(0)) &&\text{definition of } iterate \\
&= cons(0, () \mapsto iterate(succ, succ(0))) &&\text{lazy link}
\end{aligned}
$$

Fig. 6.6 Node x points to a λ expression. It will create a new node when force evaluation

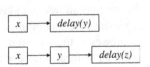

The computation stops here. It won't go on. The result is a list: its first element is 0; the next element is a λ expression. One need force the list evaluation in order to get the successive element:

$$next(cons(x, e)) = e()$$

If apply *next* to $cons(0, () \mapsto iterate(succ, succ(0)))$, it expands to:

$next(cons(0, () \mapsto iterate(succ, succ(0))))$

$=iterate(succ, succ(0))$	definition of *next*
$=iterate(succ, 1)$	definition of *succ*
$=1{:}iterate(succ, succ(1))$	definition of *iterate*
$=cons(1, () \mapsto iterate(succ, succ(1)))$	lazy link

The computation stops again. It generates natural numbers one by one by applying *next* to $\mathbb{N}$ in Eq. (6.1) repeatedly. We call this kind of model *stream*, and use it to represent potential infinity. Below function fetches the first m natural numbers from the potential infinite stream.

$$
\begin{aligned}
take\ 0\ \_ &= [\,] \\
take\ m\ cons(x, e) &= cons(x, take(m - 1, e()))
\end{aligned}
$$

For example, $take\ 8\ N = [0, 1, 2, 3, 4, 5, 6, 7]$. Listing 6.1 and listing 6.2 implement natural numbers with potential infinity in different programming languages.

Exercise 6.1

6.1.1 Equation (1.12) in Chapter 1 implements Fibonacci numbers by folding. How to define Fibonacci numbers as potential infinity with *iterate*?

6.1.2 Define *iterate* by folding.

Answer: page 239

6.2.1   Coalgebra and Infinite Stream ★

Coalgebra in category theory gives the stream model of potential infinity. Readers may skip this section in the following two pages. Revisit coalgebra and F-morphism.

Definition 6.2.1 Let C be a category; $C \xrightarrow{F} C$ is an endo-functor. Object A and arrow (morphism) $A \xrightarrow{\alpha} FA$ forms a pair (A, α), called *F-coalgebra*, where A is the carrier object.

Treat F-coalgebra as object, when the context is clear; denote the object by a pair (A, α). The arrows between F-coalgebra objects are defined as the following:

Definition 6.2.2 A F-morphism is an arrow between F-coalgebra objects:

$$(A, \alpha) \longrightarrow (B, \beta)$$

If the arrow $A \xrightarrow{f} B$ between the carrier objects makes below diagram commutes:

where $\beta \circ f = \mathbf{F}(f) \circ \alpha$

F-coalgebra and F-morphism form F-coalgebra category $\mathbf{CoAlg(F)}$. For F-algebra, we care about the initial algebra; symmetrically, for F-coalgebra, we care about the *final coalgebra*. It is the final object in F-coalgebra category, denoted by (T, μ). For any coalgebra (A, f), there is unique morphism m, such that the below diagram commutes:

By Lambek theorem, the final coalgebra is the fixed point for functor $\mathbf{F}$. There is an isomorphism $T \xrightarrow{\mu} \mathbf{F}T$, such that $\mathbf{F}T$ is isomorphic to T.

We use catamorphism to evaluate the initial algebra. Symmetrically, we use *anamorphism* (ana- means upward) to coevaluate the final coalgebra. For any coalgebra (A, f), the unique arrow to the final coalgebra (T, μ) is represented with the anamorphism as $[\![(f)]\!]$. The brackets are called *lens brackets* because they look like a pair of lenses (analogous to banana brackets for catamorphism).

$$T \xrightarrow{\mu} FT$$

$$m = [\![f]\!], \text{ if and only if } \mu \circ m = \mathbf{F}(m) \circ f$$

We use the final coalgebra and anamorphism to build infinite stream. Anamorphism takes a coalgebra $A \xrightarrow{f} \mathbf{F}A$ of carrier object A; it generates **Fix F**. The fixed point of functor **F** is the final coalgebra. It is in the form of infinite stream:

$$[\![f]\!] = \mathbf{Fix} \circ \mathbf{F}\,[\![f]\!] \circ f \qquad (6.2)$$

Let arrow $(A \to \mathbf{F}A) \xrightarrow{ana} (A \to \mathbf{Fix}\ F)$ be the anamorphism, $fmap$ be the mapping of arrows for functor **F** (see Eq. (4.2)). Substitute in Eq. (6.2):

$$ana\ f = \mathbf{Fix} \circ fmap\ (ana\ f) \circ f$$

Here is a concrete example, functor **F** is defined as below:

```
data StreamF E A = StreamF E A
```

with fixed point:

```
data Stream E = Stream E (Stream E)
```

StreamF E is a normal (none recursive) functor; we intend to name it "stream." The coalgebra of this functor is such a function; it transforms a "seed" a of type A to a pair, containing a, and the next seed.

Coalgebra generates varies of infinite streams. Here are two examples. The first example is Fibonacci numbers. We use $(0, 1)$ as the starting seed. To generate next seed, we take 1 in the pair as the new first value and $0 + 1$ as the second value to form a new seed $(1, 0 + 1)$. Repeat this process, for seed (m, n); generate next seed of $(n, m + n)$. Written in coalgebra gives the following definition:

$$(Int, Int) \xrightarrow{fib} \mathbf{StreamF}\ Int\ (Int, Int)$$
$$fib(m, n) = \mathbf{StreamF}\ m\ (n, m + n)$$

In this definition, carrier object A is a pair of integers. Feed fib into anamorphism to build the infinite stream of Fibonacci numbers. For functor **StreamF** E, the type of the anamorphism is:

$$(A \to \mathbf{StreamF}\ E\ A) \xrightarrow{ana} (A \to \mathbf{Stream}\ E)$$

with implementation $ana\ f = fix \circ f$, where:

$$fix\ (StreamF\ e\ a) = Stream\ e\ (ana\ f\ a)$$

Listing 6.5 gives the example program. Apply the anamorphism to fib and the starting pair $(0, 1)$; it gives infinite stream of Fibonacci numbers:

$$ana\ fib\ (0, 1)$$

Define auxiliary function to take the first n elements from the infinite stream:

```
take 0 _ = []
take n (Stream e s) = e : take (n - 1) s
```

The next example demonstrates how to generate infinite stream of prime numbers with the sieve of Eratosthenes. The start seed is a (infinite) list of natural numbers without 1: $[2, 3, 4, \ldots]$. Next, remove all multiples of 2 from it to obtain the next seed, which is a (infinite) list starting from 3 as $[3, 5, 7, \ldots]$ Next, remove all multiples of 3, and repeat this process. Define below coalgebra accordingly:

$$[Int] \xrightarrow{era} \textbf{StreamF}\ Int\ [Int]$$
$$era(p{:}ns) = \textbf{StreamF}\ p\ \{n \mid p \nmid n, n \in ns\}$$

Then feeding it to anamorphism, we obtain the infinite stream of all prime numbers:

```
primes = ana era [2 ...]
```

For list particularly, anamorphism is called unfold. Anamorphism and catamorphism are mutually inverse. We can turn the infinite stream back to list through catamorphism.

Exercise 6.2

6.2.1 Show **Stream** is the fixed point of **StreamF**. Hint: use the definition of the fixed point in Eq. (4.27).

6.2.2 Define $unfold$.

6.2.3 The fundamental theorem of arithmetic states that, any integer greater than 1 can be uniquely factored as the product of prime numbers. Given a text T, and a string W, does any permutation of W exists in T? Solve this programming puzzle with this theorem and the stream of primes.

Answer: page 242

6.3 Infinity and Set Theory

Aristotle's influence was profound. For more than 2,000 years, mathematicians and philosophers have been thinking about infinity. Most of them accepted the potential infinity. However, there are discordant views about actual infinity. For a long time, people believed the actual infinity was God, or only God mastered actual infinity. Many attempts to reason the actual infinity led to confusion and contradict result. For example, suppose all natural numbers form a set, because natural numbers are separated by even and odd numbers. It's natural to think that even numbers are half of natural numbers. However, doubling every natural number gives an even number; and dividing every even number by 2 gives a natural number and vice versa. There is one-to-one correspondence between natural numbers and even numbers. Do they have same size? If not, which one has more? Natural numbers or even numbers?

Galileo, father of the modern science, made a paradox in his scientific work *Two New Sciences* in 1636. Some numbers are squares, while others are not; therefore, all the numbers, including both squares and non-squares, must be more numerous than just the squares. And yet, for every number there is exactly one square, which forms the sequence 1, 4, 9, 16, 25, . . . ; hence, there cannot be more of one than of the other. This paradox is known as Galileo's paradox.

Not only numbers, people found similar paradox in geometry. As shown in Fig. 6.7, for two circles with the same center, every radius connects two points in each circle, respectively. There is one-to-one correspondence between the points in the big and the small circles. It implies there are same many points in each circle. However, our common sense tells us there must be more points in the big one.

Because of these paradoxes, Aristotle's opinion to avoid actual infinity was widely accepted. Galileo concluded that the ideas of less, equal, and greater couldn't apply to infinite sets like natural numbers. People rejected terms of actual infinity, like "all the natural numbers."

Georg Cantor (1845–1918)

Fig. 6.7 Every point in the
big circle is corresponding to
a point in the small circle

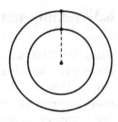

About 200 years after Galileo, Cantor made breakthrough discovery about infinity. Cantor studied those paradoxes. He thought the key problem was in our common sense, the assumption that the whole must be larger than any part. It isn't necessarily right. The development of science demonstrated our common view could be incomplete or wrong. The theory of relativity challenged our understanding to space and time—the space we are living in is not necessarily Euclidean. Quantum theory challenged our common sense of causal—randomness rules the quantum world. When think out of the box, it might open a new world we've never seen, resulting in fundamental revolution. This was exactly what Cantor did. If we accept the counterintuitive assumption that "the part equals to the whole," then the door to the infinity is open. Cantor realized the importance of one-to-one correspondence between two sets and introduced infinite sets.

To compare two sets, Cantor defined if there was one-to-one correspondence between them, such that every element in set M has the unique corresponding element in set N; then the two sets were equinumerous—of the same cardinal (size) or same number of elements (remind "isomorphic" in Sect. 3.2.3), denoted by $M \cong N$. For finite sets, it is obviously true; when it extends to infinite sets, it means there are the same infinitely many even numbers as the natural numbers! The points in small and big circles of the same center are equinumerous... Dedekind, Cantor's friend, gave such a definition to infinite set in 1888: if some proper subset of a set is equinumerous to this set, then it is an infinite set.[5]

6.3.1 Hilbert's Grand Hotel

To explain Cantor's idea, Hilbert told an interesting story in a lecture in 1924 (published in 2013). It was popularized through George Gamow's 1947 book *One Two Three... Infinity*. In this story, there was a Grand hotel with infinitely many rooms. It was fully occupied during the hot season. One evening, a tired driver had passed many "No vacancy" hotels before reaching to this one. He went to see if there might nonetheless be a room for him. The clerk said: "No problem, we can make a room for you." He moved the guest in room 1 to room 2, then moved the

---

[5] Known as Dedekind infinite set.

guest in room 2 to room 3, and moved the guest in room 3 to room 4, ... He moved everyone to the next room. That freed up the first room for this new guest.

The story goes on. On the second day, there came a tour group of infinitely many guests. The clerk said: "No problem, we can make rooms for everyone." He moved the new guest who was in room 1 last night to room 2, then moved the guest in room 2 to room 4, and moved the guest in room 3 to room 6, ... He moved everyone from room i to room $2i$. Since the hotel had infinitely many rooms, after moving, room 2, 4, 6, . . . these even number rooms were occupied with the guests accommodated yesterday, while the room 1, 3, 5, . . . these odd number rooms were freed up. There were infinitely many odd numbers rooms that could accommodate every member in the tour group.

The story does not end. On the third day, there came infinitely many tour groups of infinitely many guests each. Could the magic Hilbert's hotel accommodate them all? Before showing the answer, let us revisit the story on the first 2 days.

As shown in Fig. 6.8, on the first day, the clerk moved every guest to the next room to free up room 1. Essentially, he established a 1-to-1 correspondence for shadowed circles between the two rows as $n \leftrightarrow n + 1$. It reveals a counterintuitive fact that infinity plus one is infinity again. Hilbert's Grand hotel can accommodate not only this one more guest, but also k new guests by repeating this arrangement k times. Something like:

$$\infty + 1 \to \infty$$

$$\infty + k \to \infty$$

Figure 6.9 shows the solution on the second day. The clerk essentially established 1-to-1 correspondence between natural numbers and even numbers, hence freed up infinitely many odd number rooms. Then he established another 1-to-1 correspondence between these empty rooms and infinitely many guests in the tour group, which was exact the 1-to-1 correspondence between odd numbers and natural numbers. The second day story says infinity plus infinity give the infinity back.

$$\infty + \infty \to \infty$$

It's natural to ask whether the infinity on the left hand equals to, less, or greater than the infinity on the right hand. Can we compare infinities? It was exactly this question that led Cantor to study infinity in depth. In Hilbert's Grand hotel story, the clerk established 1-to-1 correspondence between all these infinities; hence they are

Fig. 6.8 First day of
Hilbert's Grand hotel

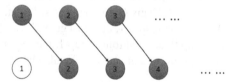

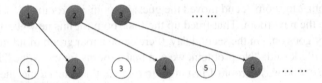

Fig. 6.9 Second day of Hilbert's Grand hotel

Fig. 6.10 One solution to
number the infinity of infinity

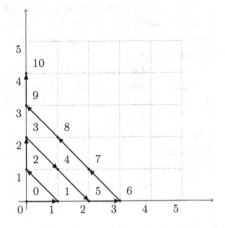

all equinumerous. Such infinities with 1-to-1 correspondence to natural numbers are called *countable* infinities.

To solve the problem on the third day, we need think about how to establish 1-to-1 correspondence between (a) the infinitely many tour groups of infinity many guests and (b) the infinity many rooms. One may come to the idea to arrange the guests in the first tour group to room 1, 2, 3, ... then arrange the guests in the second group to room $\infty + 1, \infty + 2, \ldots$ However, this method does not work. Before arrangement, we don't know which is more numerous, the rooms or the guests. Consider how this process executes: the first guest of the second group will never know when the first group finishes accommodating; this guest has no way to determine which room should move to. Compare to the first day: the new guest could immediately move to room 1 when the original guest moved to room 2; nevertheless the whole infinity accommodating process is endless. Same situation happened on the second day: when the original guest in room 1 moved to room 2, the first guest of the tour group could move to the freed up room 1 immediately. Next, the guest in room 2 moved to room 4, and the guest in room 3 moved to room 6, at this time, the second guest of the tour group could move to room 3...

Figure 6.10 gives a numbering solution. For convenience, we label the first guest 0, the second guest 1, the third guest 2, ...; label the guests already lived in the hotel as group 0, the first tour group 1, the second group 2, ... In this figure, every guest corresponds to a grid point. We also label the hotel rooms from 0.

Fig. 6.11 From cover page
of *Proof without word*

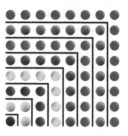

Now we arrange rooms in this order: assign the guest 0 of group 0 to room 0, the guest 1 of group 0 to room 1, the guest 0 of group 1 to room 2, the guest 0 of group 2 to room 3, the guest 1 of group 1 to room 4, ... Along the zigzag path, assign room one by one, without missing any guests. Hence establish a 1-to-1 correspondence between every guest in infinitely many groups and the infinity many rooms. Hilbert's Grand hotel surprisingly holds "two-dimensional" infinity.[6]

Exercise 6.3

6.3.1 In the 1-to-1 correspondence between the rooms and guests with the number-ing scheme shown in Fig. 6.10, which room number should be assigned for guest i of group j? Which guest of which group will live in room k?
6.3.2 For Hilbert's Grand hotel, there are alternative solutions for the problem on the third day. Figure 6.11 shows the cover page of the book *Proof without word*. Can you give a numbering scheme based on this figure?

Answer: page 247

6.3.2 One-to-one Correspondence and Infinite Set

From Hilbert's Grand hotel story, we see the importance of the 1-to-1 correspon-dence in studying infinity. If there exists 1-to-1 correspondence between two sets, then they have the same cardinality. We can further use 1-to-1 correspondence to classify sets. As shown in Chap. 3, for two sets A and B, establish a map $A \xrightarrow{f} B$, such that every element x in A corresponds to element y in B through

---

[6] The traditional solution uses the Euclid's theorem that there are infinitely many prime numbers. Empty the odd numbered rooms by sending the guest in room i to room 2^i; then put the first group's guests in rooms 3^n, the second group's guests in rooms 5^n, ...; put the k-th group's guests in rooms p^n, where p is the k-th prime number. This solution leaves certain rooms empty; specifically, all odd numbers that are not prime powers, such as 15 or 847, will no longer be occupied.

Fig. 6.12 Bijection (1-to-1 correspondence) between two sets

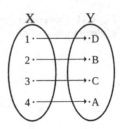

$x \rightsquigarrow y = f(x)$. For sets, we call f function; y is the image of x, and x is preimage. If there is exact one preimage, the function is *injective*; if every element y in B has a preimage, then the function is *surjective* or onto. If both injective and surjective, it is *bijective* or one-to-one correspondence. Figure 6.12 shows a bijection between two sets.

Hilbert's Grand hotel shows the set of natural numbers is equinumerous with its proper subsets like even/odd numbers. And the sets of one-dimensional natural number n and two-dimensional pair (m, n) are also equinumerous. Starting from natural numbers, Cantor extended a series of infinite sets through 1-to-1 correspondence. For example:

1. **Integers**. Establish the following 1-to-1 correspondence:

$$
\begin{array}{ccccccccc}
0 & 1 & -1 & 2 & -2 & \cdots & n & -n & \cdots \\
\updownarrow & \updownarrow & \updownarrow & \updownarrow & \updownarrow & & \updownarrow & \updownarrow & \\
0 & 1 & 2 & 3 & 4 & \cdots & 2n-1 & 2n & \cdots
\end{array}
$$

   Hence extend natural numbers to integers. From another viewpoint, it essentially corresponds odd numbers to positive integers and even numbers to negative integers and zero. It tells that the integers and natural numbers are equinumerous.

2. **Rationals**. Express every rational number as fraction p/q of two integers, where $q \neq 0$, and $q = 1$ for integers. On the third day of Hilbert's Grand hotel, the clerk established 1-to-1 correspondence between a pair (p, q) and a natural number. We adjust it a bit for rational number p/q. Put negative numbers aside for now; we skip whenever the second number q is zero or p/q is reducible (with common divisor). Then we reuse the method for integers and cover negative rational numbers as well. In this way, we construct rational numbers from natural numbers. Below table shows how the first several natural numbers correspond to rational numbers:

$$
\begin{array}{cccccccccc}
0 & 1 & 2 & 3 & 4 & 5 & 6 & 7 & 8 & \cdots \\
\updownarrow & \updownarrow & \updownarrow & \updownarrow & \updownarrow & \updownarrow & \updownarrow & \updownarrow & \updownarrow & \\
0 & 1 & \dfrac{1}{2} & -\dfrac{1}{2} & -1 & -2 & -\dfrac{2}{3} & -\dfrac{1}{3} & \dfrac{1}{3} & \cdots
\end{array}
$$

   It tells natural numbers and rational numbers are equinumerous.

3. **Algebraic numbers**. An algebraic number is some root of a non-zero polynomial in one variable with rational coefficients. Equivalently, by clearing denominators, we can turn the coefficients to integers. For example, $\sqrt{2}$ and $1 \pm \sqrt{3}i$ are algebraic numbers, while π and e are not.

Given an equation

$$a_0 x^n + a_1 x^{n-1} + \ldots + a_n = 0$$

where $a_0, a_1, \ldots a_n$ are integers and $a_0 \neq 0$. All its roots are algebraic numbers. Define a positive integer:

$$h = n + |a_0| + |a_1| + \ldots + |a_n|$$

which is the sum of the degree and coefficients of the equation. We name h as the height of the equation. For any equation, h is a unique natural number. Conversely, given a height h, there could be multiple equations. For example, the height of equations $x - 3 = 0$, $x^3 + 1 = 0$, $x^3 - 1 = 0$, $x^2 + x + 1 = 0$, and $x^2 - x + 1 = 0$ are all 4. Nevertheless, there are finitely many equations for every h. Therefore, we can enumerate all equations: first list all equations of height $h = 1$, then list the equations of height $h = 2, \ldots$, and repeat this process. The equations of the same heights can be listed in arbitrary order. According to the fundamental theorem of algebra, the number of roots equals to the equation degree n. Taking the duplicated roots into consideration, the number of distinct roots is not greater than n. Hence there are finite many roots for equation of height h. We are ready to enumerate all algebraic numbers.

First, enumerate all roots for equations of height $h = 1$ (only one equation $x = 0$), which is 0; then enumerate all roots for equations of height 2. Because different equations may have some same roots, we skip any root if it has been enumerated before. In this way, we establish 1-to-1 correspondence between algebraic numbers and natural numbers. Hence they are equinumerous. In other words, we extend natural numbers to algebraic numbers.

We meet a problem next. Can we extend natural numbers to real numbers? Not only for normal irrationals but also cover transcendental numbers like π and e? Cantor and Dedekind made great breakthrough contribution when they studied this problem.

Georg Cantor was the founder of set theory. He was born in 1845 in the western merchant colony of Saint Petersburg, Russia, and brought up in the city until he was 11. Cantor demonstrated exceptional skill in mathematics in school. But his father wanted Cantor to became "a shining star in the engineering firmament." However, at the age of 17, Cantor had sought his father's permission to study mathematics at university, and he was overjoyed when eventually his father consented [7].

He entered the Polytechnic of Zürich in 1862 and then moved to the University of Berlin in 1863. He attended lectures by Leopold Kronecker, Karl Weierstrass, and Ernst Kummer. He spent the summer of 1866 at the University of Göttingen.

Cantor was a good student, and he received his doctorate degree in 1867 with the dissertation on number theory. Cantor later took up a position at the University of Halle, where he spent his entire career.

At that time, many mathematicians were trying to rebuild the rigorous logical foundation of analysis led by Weierstrass. Cantor was also influenced by this movement. He soon realized the importance to study real numbers as the basis of calculus, which became the beginning of set theory research. In 1872, he met and began a friendship with the young mathematician Richard Dedekind while on holiday in Switzerland. Even in Cantor's honeymoon in Harz mountains, Cantor spent much time in mathematical discussions with Dedekind. They started a long time correspondence between each other.

In 1874, Cantor published his first revolutionary paper about set theory at the age of 29. It marked as the beginning of set theory as a branch of mathematics. With the extraordinary ingenuity, Cantor established set theory in the following 10 years almost alone, leading the revolution of infinity in mathematics. However, he was not well recognized during his most creative period. He desired, but was not able to obtain a professor chair at the University of Berlin. He spent his career at the University of Halle, which was an infamous university with a meager salary. Cantor's theory was originally regarded as so counterintuitive—even shocking. People found paradoxes hidden in infinity sets (see Russell's paradox in the next chapter). Cantor's work encountered resistance from mathematical contemporaries. Among them, some were famous mathematicians, including his teacher, the leading mathematician in Berlin, Leopold Kronecker. He had a famous saying: "God made the integers, all else is the work of man." He objected to Cantor's theory about infinity and transfinite numbers, said it was not mathematics but mysticism. Mathematics was headed for the madhouse under Cantor. Henri Poincaré referred to Cantor's work as a "disease" from which mathematics would eventually be cured. Poincaré said, "There is no actual infinite; the Cantorians have forgotten this, and that is why they have fallen into contradiction." Later Hermann Weyl criticized Cantor's hierarchy of infinities as "fog on fog." Hermann Schwartz was originally a friend of Cantor, but he stopped the correspondence with Cantor as opposition to Cantor's ideas continued to grow.

The tragedy was not the set theory, but Cantor went to the madhouse. Cantor suffered his first known bout of depression in May 1884. During the rest of his life, the depression recurred several times with different intensity, driving him from the community into the mental hospital refuge. Cantor retired in 1913, living in poverty (because of the war conditions). In June 1917, he entered the sanatorium for the last time and continually wrote to his wife asking to be allowed to go home. Cantor died on January 6, 1918, in the sanatorium where he had spent the last year of his life.

Cantor's set theory was publicly acknowledged and praised at the first ICM held in Zurich in 1897. Adolf Hurwitz (1859–1919) openly expressed his great admiration of Cantor and proclaimed him as one by whom the theory of functions has been enriched. Jacques Hadamard expressed his opinion that the notions of the theory of sets were known and indispensable instruments. Over time, people gradually realized the importance of set theory. Hilbert praised Cantor's work as "the finest product of mathematical genius and one of the supreme achievements of purely intellectual human activity." The continuum hypothesis, introduced by Cantor, was presented by Hilbert as the first of his 23 open problems in his address at the 1900 ICM in Paris. When Brouwer, the founder of intuitionism, criticized the paradoxes in set theory, Hilbert defended it by declaring, "No one shall expel us from the paradise that Cantor has created."

Richard Dedekind (1831–1916)

Richard Dedekind was a German mathematician. He was born on October 6, 1831, in Brunswick Germany, which is the hometown of Gauss. When Dedekind was in school, mathematics was not his main interest. He studied science, in particular physics and chemistry. However, they became less than satisfactory to Dedekind with what he considered an imprecise logical structure and his attention turned toward mathematics. He entered the University of Göttingen in the spring of 1850 with a solid grounding in mathematics. He learned number theory from M A Stern and physics from Wilhelm Weber. Gauss was still teaching, although mostly at an elementary level, and Dedekind became his last student. Dedekind did his doctoral work in four semesters under Gauss's supervision and submitted a thesis on the theory of Eulerian integrals. He received his doctorate from Göttingen in 1852.

After that, Dedekind went to Berlin for 2 years of study, where he and Bernhard Riemann were contemporaries; they were both awarded the habilitation degrees in 1854. Dedekind was then qualified as a university teacher, and he began teaching at Göttingen giving courses on probability and geometry. He studied for a while with Dirichlet, and they became good friends. About this time, Dedekind studied the work of Galois, and he was the first to lecture on Galois theory. He also became one of the first people to understand the importance of the notion of groups for algebra and arithmetic.

Dedekind was humble. Many of his achievements were unknown to the people at the time. For example, after Dirichlet's death, Dedekind wrote and published the famous book *Lectures on Number Theory* based on his notes from Dirichlet's lectures. Dedekind was so modest that he published the book under Dirichlet's name, even after adding many additional results of his own in later editions.

Unfortunately, Dedekind's modesty hurt his career; he failed to get tenure at Göttingen and ended up on the faculty of a minor technical university ([9] pp.140), Institute of Technology Brunswick in his hometown.

Dedekind remained in Brunswick for the rest of his life, retiring on April 1, 1894. He lived his life as a professor in Brunswick. "In close association with his brother and sister, ignoring all possibilities of change or attainment of a higher sphere of activity. The small, familiar world in which he lived completely satisfied his demands: ... there he found sufficient leisure and freedom for scientific work in basic mathematical research. He did not feel pressed to have a more marked effect in the outside world: such confirmation of himself was unnecessary." Dedekind died on February 12, 1916.

Dedekind made a number of highly significant contributions to mathematics and his work would change the style of mathematics into what is familiar to us today. While teaching calculus for the first time at Brunswick, Dedekind developed the notion now known as a Dedekind cut, now a standard definition of the real numbers. As well as his analysis of the nature of number, his work on mathematical induction, including the definition of finite and infinite sets, and his work in number theory, particularly in algebraic number fields, are of major importance. He introduced the notion of an ideal which is fundamental to ring theory (later introduced and extended by Hilbert and Emmy Noether).

Even today, there are still different views regarding Cantor's and Dedekind's work. Not to mention the fierce divisions and debates at the beginning of the twentieth century. Most biographies and comments we see today are often too critical to Kronecker, Brouwer, and the intuitionism they represented. They sympathize with Cantor and enthusiastically praise the revolution of infinite sets and transfinite numbers. We recommend the rational readers have your own thoughts, but not be completely influenced by one-sided view or the other. Kronecker had a strong belief in mathematical philosophy. He emphasized that mathematics should deal only with finite numbers and with a finite number of operations. He was the first to doubt the significance of non-constructive existence proofs. We should not think that Kronecker's views of mathematics were totally eccentric. Although it was true that most mathematicians of his day would not agree with those views, and indeed most mathematicians today would not agree with them, they were not put aside. Kronecker's ideas were further developed by Poincaré and Brouwer, who placed particular emphasis upon intuition. Intuitionism stresses that mathematics has priority over logic, the objects of mathematics are constructed and operated upon in the mind by the mathematician, and it is impossible to define the properties of mathematical objects simply by establishing a number of axioms. Poincaré in his popular book *Science and Hypothesis* stated that convention plays an important role in physics. His view came to be known as "conventionalism." He also believed that the geometry of physical space is conventional. His idea inspired Einstein when he developed theory of relativity.

6.3.3 Fibonacci Numbers and Hamming Numbers

Some programming environments support lazy evaluation by default, allowing to perform complex computation directly on infinite streams. Below is alternative definition of natural numbers:

$$\mathbb{N} = 0{:}map(succ, \mathbb{N})$$

where $\mathbb{N}$ is an infinite list of natural numbers. The first number is zero; from the second one, every number is the successor of the previous natural number, as shown in the following table:

	$\mathbb{N}:$	0	1	2	...
$map(succ, \mathbb{N}):$		$0+1$	$1+1$	$2+1$	...
$0{:}map(succ, \mathbb{N}):$		0 1	2	3	...

Based on this idea, below example program defines the infinite list of natural numbers and then takes the first ten numbers:

```
nat = 0 : map (+1) nat

take 10 nat
[0,1,2,3,4,5,6,7,8,9]
```

Similarly, we define Fibonacci numbers as infinite list F. The first number is 0, the second number is 1, and every Fibonacci number after them is the sum of the previous two. We make a table analogous to the one for natural numbers:

	F:	0	1	1	2	3	5	8	...
	F':	1	1	2	3	5	8	13	...
0	1	1	2	3	5	8	13	21	...

where the first row lists all Fibonacci numbers; the second row removes the first number and lists the rest. We can consider it is generated by shifting the first row to left by one cell; in the third row, every cell is the sum of the above two numbers in the same column. It lists all Fibonacci numbers except for the first two: 0 and 1. If prepend them to the left, then the third row gives back all Fibonacci numbers. Below are the corresponding definition and example program:

$$F = 0{:}1{:}[x + y | x \leftarrow F, y \leftarrow F'] \tag{6.3}$$

```
fib = 0 : 1 : zipWith (+) fib (tail fib)
```

For another example, define regular numbers be the numbers whose only prime divisors are 2, 3, or 5. In computer science, they are often called Hamming numbers after Richard Hamming (1915-1998, American mathematician, received ACM Turning award in 1968), who proposed the problem of finding computer algorithms for generating these numbers in ascending order. Here are the first several Hamming numbers:

$$1, 2, 3, 4, 5, 6, 8, 9, 10, 12, 15, 16, 18, 20, 24, 25, 27, 30, 32, 36, 40, 45, 48, 50, 54, 60, \ldots$$

It's not trivial to write a program to generate Hamming numbers. However, there exists an intuitive and efficient method by using infinite stream. Let H be the infinite list of Hamming numbers. The first number is 1. For every number x in H, $2x$ is also a Hamming number. Define $H_2 = [2x|x \leftarrow H]$. It is also true for factor 3 and 5. Define $H_3 = [3x|x \leftarrow H]$ and $H_5 = [5x|x \leftarrow H]$, respectively. If we merge these three lists H_2, H_3, and H_5 together, remove duplicated numbers, and prepend 1, it gives back the list of Hamming numbers:

$$H = 1 : H_2 \cup H_3 \cup H_5$$

$$= 1 : [2x|x \leftarrow H] \cup [3x|x \leftarrow H] \cup [5x|x \leftarrow H]$$

We implement $\cup$ to merge two lists $xs = [x_1, x_2, \ldots]$ and $ys = [y_1, y_2, \ldots]$ in ascending order and drop the duplicated numbers:

$$(x:xs) \cup (y:ys) = \begin{cases} x < y : & x:(xs \cup (y:ys)) \\ x = y : & x:(xs \cup ys) \\ x > y : & y:((x:xs) \cup ys) \end{cases}$$

Here is the corresponding example program that implements this definition; it returns the 1,000,000th Hamming number:

```
ham = 1 : map (*2) ham <> map (*3) ham <> map (*5) ham
  where xxs@(x:xs) <> yys@(y:ys)
    | x == y = x : xs <> ys
    | x < y  = x : xs <> yys
    | x > y  = y : xxs <> ys

ham !! 1000000
519312780448388736089589843750000000000000000000000000000
0000000000000000000000000000000
```

6.3.4 Countable and Uncountable Infinity

From natural numbers, we've seen how to construct integers, rationals, and algebraic numbers (including part of irrational numbers like radicals). There are 1-to-1 correspondences between them and natural numbers; hence they are all equinumerous. A set is called *countable* infinite if it has the same cardinality as natural numbers. Are all infinite sets countable? Are there any larger infinities? On November 29, 1873, Cantor wrote to Dedekind a letter ([1], pp. 198):

> *May I ask you a question, which has a certain theoretical interest for me. but which I can't answer; may be you can answer it and would be so kind as to write to me about it. It goes as follows: take the set of all natural numbers n and denote it $\mathbb{N}$. Further, consider, say, the set of all positive real numbers x and denote it $\mathbb{R}$. Then the question is simply this: can $\mathbb{N}$ be paired with $\mathbb{R}$ in such a way that to every individual of one set corresponds to one and only one individual of the other? At first glance, one says to oneself, "No, this is impossible, for $\mathbb{N}$ consists of discrete parts and $\mathbb{R}$ is a continuum." But nothing is proved by this objection. And much as I too feel that $\mathbb{N}$ and $\mathbb{R}$ do not permit such a pairing, I still cannot find the reason. And it is this reason that bothers me; maybe it is very simple.*

One week after on December 7, Cantor wrote again to Dedekind:

> *Recently I had time to follow up a little more fully the conjecture which I mentioned to you; only today I believe I have finished the matter. Should I have been deceived, I would not find a more lenient judge than you. I thus take the liberty of submitting to your judgement what I have written, in all the incompleteness of a first draft.*

Cantor proved it is impossible to establish 1-to-1 correspondence between natural numbers and real numbers; hence real numbers are uncountable. December 7, 1873, was the day that set theory was born. Cantor had given two proofs. The second one is the popular Cantor's *diagonal argument*.

Cantor used reduction to absurdity in his proof. Suppose the real numbers in open interval $(0, 1)$ are countable. There exists 1-to-1 correspondence to natural numbers. Then one can list all real numbers in this interval as a sequence $a_0, a_1, a_2, \ldots, a_n, \ldots$ in decimals. For any irrational number, its decimal format is endless non-repeating; For rational number, its decimal format can be infinitely repeating finite sequence of digits, for example, $\frac{1}{3} = 0.333\ldots$; for decimals with finite digits, we can append infinitely many zeros, for example, $\frac{1}{2} = 0.5000\ldots$. List all real numbers in interval $(0, 1)$ as below:

$$a_0 = 0.a_{00}a_{01}a_{02}a_{03}\cdots$$

$$a_1 = 0.a_{10}a_{11}a_{12}a_{13}\cdots$$

$$a_2 = 0.a_{20}a_{21}a_{22}a_{23}\cdots$$

$$a_3 = 0.a_{30}a_{31}a_{32}a_{33}\cdots$$

$$\cdots$$

$$a_n = 0.a_{n0}a_{n1}a_{n2}a_{n3}\cdots$$

$$\cdots$$

Note $a_0, a_1, a_2, \ldots$ are not necessarily in ascending order. Next construct a number $b = 0.b_0 b_1 b_2 b_3 \cdots b_n \cdots$, where the n-th digit $b_n \neq a_{nn}$. To achieve this, we simply make a rule: if $a_{nn} \neq 5$, then let $b_n = 5$, else let $b_n = 6$:

$$b_n = \begin{cases} 5 & : a_{nn} \neq 5 \\ 6 & : a_{nn} = 5 \end{cases}$$

The constructed number b must not equal to any number in the above list. This is because at least the n-th digit is different. That the diagonal digits are different. Highlight them in bold font in below table.

$$a_0 = 0.\mathbf{a_{00}}a_{01}a_{02}a_{03} \cdots$$

$$a_1 = 0.a_{10}\mathbf{a_{11}}a_{12}a_{13} \cdots$$

$$a_2 = 0.a_{20}a_{21}\mathbf{a_{22}}a_{23} \cdots$$

$$a_3 = 0.a_{30}a_{31}a_{32}\mathbf{a_{33}} \cdots$$

$$\cdots$$

$$a_n = 0.a_{n0}a_{n1}a_{n2}a_{n3} \cdots \mathbf{a_{nn}} \cdots$$

$$\cdots$$

Because we assume all real numbers in interval $(0, 1)$ are enumerated without one missing, b should be in this interval, but it does not equal to any a_i. The 1-to-1 correspondence misses b, hence leading to contradiction. It shows impossible to establish 1-to-1 correspondence between real numbers and natural numbers. This is the Cantor's diagonal argument.

One may argue why we can't add b to the list of $a_0, a_1, a_2, \ldots$. Suppose after adding b to the list, its position is the m-th number, we can construct another new number c, where its m-th digit does not equal to b_m. Hence we get another number not being included.

This proof shows a surprising result: the set of real numbers in interval $(0, 1)$ is uncountable! It is the first infinite set that people found more numerous than natural numbers.[7] As the next step, establish 1-to-1 correspondence: $y = \pi x - \frac{\pi}{2}$. It sends every real number in $(0, 1)$ to interval $(-\frac{\pi}{2}, \frac{\pi}{2})$ without any missing. We immediately conclude that the real numbers in this new interval are uncountable. As

---

[7] Courant and Robbins give another intuitive geometric proof in book *What is mathematics*. Suppose the points in the unit line segment $(0, 1)$ can be enumerated as $a_1, a_2, a_3, \ldots$ Cover point a_1 by an interval of length $1/10$, cover point a_2 by an interval of length $1/100$, ... cover point a_n by an interval of length $1/10^n$, and so on. Then the unit line segment $(0, 1)$ is completely covered (there can be overlaps) by these subintervals of length $1/10, 1/100, 1/1000, \ldots$. However, the total length of these subintervals, which is the sum of a geometric series, is $1/10 + 1/100 + 1/1000 + \ldots = 1/9 < 1$. It is impossible to cover the line segment of length 1 by an interval of total length $1/9$; hence, the assumption cannot hold, and the points in the line segment are uncountable [63].

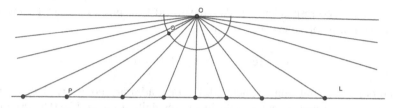

Fig. 6.13 Map a semicircle of length 1 to all real numbers

the final attack, establish another 1-to-1 correspondence: $y = tan(x)$. It sends every real number between $-\frac{\pi}{2}$ and $\frac{\pi}{2}$ to the infinite set of all real numbers without any missing. There is another geometric method to establish the 1-to-1 correspondence between the unit line segment to all real numbers. Bend the line segment to a semicircle of length 1; then draw an infinite line L outside the circle. From any point P in L, connect it with the centre of the circle; it intersects with the arc at a point Q, as in Fig. 6.13.

Cantor therefore found the set of real numbers was not countable. It is a higher level of infinity than countable sets. Cantor named it uncountable set, denoted by C.

For points and line, in Euclid's *Elements*, a point is defined as "A point is that which has no part," and line consists of points. According to Hippasus's finding, there are irrational numbers in the line. In other words, rationals can't fully cover a line, while real numbers can. The above proof further shows the points in the unit line segment, the points in a line segment of any length, and the points in infinitely long line, which is the number axis, are all equinumerous. They are all uncountable sets. So as the points in circles of the same centre.

Compare rationals and irrationals. Given any two rational numbers, there are infinitely many irrational numbers between them; given any two irrational numbers, there are also infinitely many rational numbers between them. It seems they are equinumerous, but according to Cantor's finding, rational numbers are countable, but irrational numbers are uncountable. Irrationals are more numerous than rationals. Further, we've seen that algebraic numbers are countable, but the transcendental numbers like π and e are uncountable; they are more numerous than algebraic numbers.

On the third day of Hilbert's Grand hotel, the clerk established 1-to-1 correspondence between one-dimensional countable infinite numbers and two-dimensional infinite grids, thus demonstrating rational numbers were countable. Consider the points of real numbers in one-dimensional line segment, and points in two-dimension plane, which are more numerous? or equinumerous? Cantor raised this question in a letter to Dedekind in January 1874. He seemed sure the latter, the two-dimension square, had more points than one-dimensional line segment, but was not able to prove it. Four years later, Cantor surprised to find he was wrong. He managed to figure out an interesting 1-to-1 correspondence. He sent the proof to Dedekind, asking for review in June 1877. In that letter, Cantor said "I see it, but I don't believe it!", quoted at the header of this chapter.

This 1-to-1 correspondence makes the world in a piece of sand. Facing two infinite sets of points, one is unit square:

$$E = \{(x, y)|0 < x < 1, 0 < y < 1\}$$

the other is a unit ling segment $(0, 1)$. Take arbitrary point (x, y) in the unit square; represent both x and y in decimals (for finite decimal, like 0.5, write it in $0.4999\cdots$, see Exercise 6.4.1). Then group the fractional part after the decimal point; every group ends at the first nonzero digits, for example,

$$x = 0.3 \quad 02 \quad 4 \quad 005 \quad 6 \quad \cdots$$
$$y = 0.01 \quad 7 \quad 06 \quad 8 \quad 04 \quad \cdots$$

Next, construct a number $z = 0.3\,01\,02\,7\,4\,06\,005\,8\,6\,04 \cdots$ by taking the group of digits from x and y in turns. For this example, first write down 0 and decimal point; then take the first group from x, which is 3; then take the first group from y, which is 01; next take the second group from x, which is 02; next take the second group from y, which is 7, ... number z definitely belongs to the unit line segment. For every two different points in the square, their decimals of x and y must have different digits. Hence the corresponding z are different. It means $(x, y) \rightsquigarrow z$ is an injection. Conversely, for any point z in the unit line segment, group the fractional part, then append all the odd groups after 0 and decimal point to form x, and use all the even groups to form y. Pair (x, y) is a point in the unit square. It implies $(x, y) \rightsquigarrow z$ is also a surjection, hence a bijection (1-to-1 correspondence). It shows that the points in the unit line and unit square have the same cardinality; both are uncountable. Similarly, we can further show that a line segment has equinumerous points as three-dimensional space and even n-dimensional space.

Before Cantor, there were only finite sets and "the infinite," which was considered a topic for philosophical, rather than mathematical, discussion. Cantor was the first distinguished infinite sets of different sizes. He did not stop with the distinction between countable and uncountable infinities, but went on seeking if there were more numerous infinities. Could there be an end point of the path along "infinity of infinity"?

Exercise 6.4

6.4.1 Let $x = 0.9999\cdots$, then $10x = 9.9999\cdots$. Subtract them gives $10x - x = 9$. Solving this equation gives $x = 1$. Hence $1 = 0.9999\cdots$. Is this reasoning correct?

Answer: page 258

6.3.5 Dedekind Cut

In order to make the foundation of calculus rigorous, mathematicians in the nineteenth century went back to inspect the concepts like infinitesimal and infinite series. Through the work of Cauchy, Weierstrass, and so on, the standard of rigor, including limit and convergence, was set up. However, there was still a critical problem remaining, the concept of real numbers. Calculus was built on top of the continuity of real numbers, while it lacked a satisfied definition of real number. People thought rationals could present line, but later found there were "gaps" between rational numbers. It was not completeness or continuous, while the line was demanded to be complete, continuous, and without any gaps.

When Dedekind was thinking how to teach differential and integral calculus, the idea of cut came to him on November 24, 1858. He developed this idea and published the result in 1872. Dedekind found although rationals were dense—for any two rational numbers, no matter how they are close to each other, there were other rational numbers in between—they were not continuous. Consider a continuous number line; use an infinitely thin knife, the knife of thought, to cut the line into two parts ([7] pp. 196).

Because the line is continuous without any gaps, no matter how thin the knife, it must cut at a point, but not pass through a gap. (For the line of rationals on the contrary, the knife may cut at a point or through a gap between two rational numbers, for example, cutting at position $\sqrt{2}$.) Suppose cutting at point A, then A is either on the left or on the right. It cannot be on both sides, nor on neither side. This point cannot be divided or disappeared. In other words, since the line is continuous, wherever it is cut into two parts, one must have an end point, while the other not.

Dedekind defined a cut (A_1, A_2), where A_1 was called "closed downward," and A_2 was "closed upward," where all numbers in downward A_1 were less than every number in upwards A_2, such that A_1 represented the left half line of the cut and A_2 represented the right half. For any such a cut, either A_1 has a greatest number, or A_2 has a smallest number. There must be one and only one case.

When apply Dedekind cut to all rational numbers, it shows that rationals are not continuous. For example, A_1 contains all numbers ≤ 2, and A_2 contains all numbers > 2; then this cut defines rational number 2. However for counterexample, the downward A_1 contains all rational numbers with their square less than or equal to 2. The upward A_2 contains the remaining. In this cut, there is no greatest rational number in the downward, while there is no smallest rational number in the upward too. It shows there is a gap in rational numbers. When cut at this point, the knife will pass through. This cut actually defines a new number $\sqrt{2}$, and it is not a rational number, as shown in Fig. 6.14.

Every real number r divides the rational numbers into two subsets, namely, those greater than r and those less than r. Dedekind's brilliant idea was to represent the real numbers by such divisions of the rationals. Every cut of rationals defines a real number. The cut through a gap (no greatest in A_1, and no smallest in A_2) is an irrational number; the cut at a point (either A_1 has the greatest or A_2 has

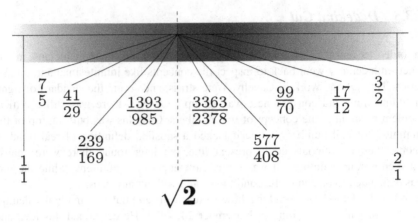

Fig. 6.14 Define $\sqrt{2}$ with Dedekind cut

the smallest) is a rational number. And real numbers contain both rationals and irrationals. Hence Dedekind cut defines real numbers. Every point in the number line is a real number. It gives the foundation of the continuity of real numbers.

From Hippasus finding the irrational number to Dedekind defining real numbers, it takes 2,000 years.[8] Dedekind cut always divides numbers into two *finalized* infinite parts (sets). It is no longer limited by potential infinities.

6.3.6 *Transfinite Numbers and Continuum Hypothesis*

Cantor considered power set on the way to find more numerous infinities. For set A, its power set is the set of all possible subsets of A. For example $A = \{a, b\}$, its power set contains $\{\varnothing, \{a\}, \{b\}, \{a, b\}\}$, total four elements. For a set of three elements, its power set contains eight subsets. Generally, for a set of n elements, because every element can be select or skip for a subset, the size of its power set is 2^n. For any finite set, it's obvious that power set has greater cardinality than the original set.

Cantor proved in 1891 that even for infinite set, the power set had a strict greater cardinality than the original set. This result is called *Cantor's theorem* nowadays (see Appendix "Appendix: Proof of Cantor's Theorem" for the proof). This theorem opens the door to infinite of infinities. Cantor introduced notation $\aleph_0$ for the cardinality of countable infinite set, like natural numbers ($\aleph$ is the first Hebrew letter). The cardinality of the power set for countable infinite set is $2^{\aleph_0}$.

---

[8] In the same year of 1872, Weierstrass defined irrational numbers as limits of convergent series; Cantor defined irrational numbers as convergent sequences of rational numbers. The theory of real numbers hence were established through these different paths.

According to Cantor's theorem, $\aleph_0 < 2^{\aleph_0}$, on top of that, Cantor repeatedly took power set to generate greater and greater infinities:

$$\aleph_0, 2^{\aleph_0}, 2^{2^{\aleph_0}}, \ldots \tag{6.4}$$

Cantor named the series of leveled infinite cardinal numbers as *transfinite cardinal numbers*. The story of Hilbert's Grand hotel actually demonstrates the arithmetic rules of transfinite numbers like $\aleph_0 + 1 = \aleph_0$, $\aleph_0 + k = \aleph_0$, $\aleph_0 + \aleph_0 = \aleph_0$, ...

Besides power set, Cantor found another method to generate greater infinities. Compare to cardinal, there is concept of ordinal. It is defined recursively as below:

(1) 0 is an ordinal number;
(2) If a is an ordinal number, then $a \cup \{a\}$ is also an ordinal number, denoted $a + 1$. It is called the successor of a;
(3) If S is a set of ordinals (all elements are ordinal numbers), then $\cup S$ is an ordinal number;
(4) Any ordinal number is obtained from above three steps.

Below lists first several ordinal numbers from 0:

$$0$$
$$1 = 0 \cup \{0\}$$
$$2 = 1 \cup \{1\} = 0 \cup \{0\} \cup \{0 \cup \{0\}\}$$
$$3 = 2 \cup \{2\} = 1 \cup \{1\} \cup \{1 \cup \{1\}\} = \ldots$$
$$\ldots$$

$\cup S$ is the union of all elements of S, called *infinitary union*. By (1) and (2) in the definition of ordinals, natural numbers $0, 1, 2, 3, \ldots, n, \ldots$ are all ordinals. Let ω be the set of natural numbers. Because all natural numbers are ordinals, ω is a set of ordinals. Take its infinitary union:

$$\cup \omega = \{0, 1, 2, \ldots\} = \omega \tag{6.5}$$

By (3) in the ordinal definition, ω is an ordinal. It is a *limit ordinal*,[9] and is the smallest infinite ordinal. Append it to the end of natural numbers to form a new series:

$$0, 1, 2, \ldots, \omega$$

---

[9] A nonzero ordinal that is not a successor is called a limit ordinal.

Start from ω, repeatedly applying (2) in ordinal definition; it gives a new ordinal series:

$$\omega + 1, \omega + 2, \omega + 3, \ldots, \omega + n, \ldots$$

Combine the above two series into one set, denoted by $\omega \cdot 2$. Because the infinitary union $\cup \omega \cdot 2 = \omega \cdot 2$, it is an ordinal, and it is a limit ordinal. From $\omega \cdot 2$, repeat above process; it generates an infinite of infinite ordinal series:

$$0, 1, 2, \ldots, n, \ldots$$
$$\omega, \omega + 1, \omega + 2, \ldots, \omega + n, \ldots$$
$$\omega \cdot 2, \omega \cdot 2 + 1, \omega \cdot 2 + 2, \ldots, \omega \cdot 2 + n, \ldots$$
$$\cdots$$
$$\omega \cdot k, \omega \cdot k + 1, \omega \cdot k + 2, \ldots, \omega \cdot k + n, \ldots$$
$$\cdots \tag{6.6}$$
$$\omega^2, \omega^2 + 1, \omega^2 + 2, \ldots, \omega^2 + n, \ldots$$
$$\cdots$$
$$\omega^3, \omega^3 + 1, \omega^3 + 2, \ldots, \omega^3 + n, \ldots$$
$$\cdots$$
$$\omega^\omega, \omega^\omega + 1, \omega^\omega + 2, \ldots, \omega^\omega + n, \ldots$$
$$\cdots$$

The first row lists natural numbers, all others rows are infinite ordinals, and the first of every row is the limit ordinal. The ordinals obtained by this way are far from what people could imagined before. It extends natural numbers to a world of infinite ordinals. What's surprising, these ordinals are all countable! As a set, it has 1-to-1 correspondence with natural numbers. We'll soon see, there exist uncountable ordinals; further, there exist greater and greater infinite ordinal series one by one (Fig. 6.15).

Among these infinite ordinals, which one is the best as the cardinal number for countable infinite set? It's natural to select the smallest limit ordinal ω. Here is the definition for cardinal number.

Definition 6.3.1 An ordinal a is a cardinal if there is no ordinal $b < a$ with $a \cong b$.

An ordinal a is a cardinal, if for any ordinal $b < a$, the cardinality of b is less than of a. This definition asserts every natural number n is a cardinal, and ω is a cardinal. When using the ordinal ω as cardinal, it's written as $\aleph_0$, that is, $\aleph_0 = \omega$. We've shown how to use $\aleph_0$ as the cardinal for all infinite countable sets.

Except ω, although all rest ordinals in series Eq. (6.6) are greater than ω, their cardinals are the same as ω (all equal to countable infinity). Hence they are not cardinals. In order to obtain greater cardinals, we form a new set containing all ordinals in series Eq. (6.6); denote it by ω_1.

$$\omega_1 = \{a | a \text{ is ordinal, and} |a| \leq \aleph_0\}$$

Fig. 6.15 Infinite ordinals

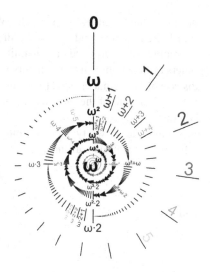

where $|a|$ is the cardinal of a. We skip the proof that ω_1 is a cardinal, and it is the first uncountable cardinal. Then, repeat this method, and expand a new infinite ordinal series from ω_1:

$$\omega_1, \omega_1 + 1, \ldots, \omega_1 \cdot 2, \ldots, \omega_1^2, \ldots, \omega_1^\omega, \ldots$$

All elements in this series have the same cardinality. The smallest one is ω_1, which satisfies the cardinal definition. It is the second infinite cardinal $\aleph_1 = \omega_1$. Using similar process to construct ω_1, we form another set:

$$\omega_2 = \{a | a \text{ is ordinal, and } |a| \le \aleph_1\}$$

It gives the third infinite cardinal $\aleph_2 = \omega_2$. Repeating this, we obtain a series of infinite cardinals. In summary, for any ordinal a, define the infinite cardinal $\aleph_a$; then form a set:

$$\omega_{a+1} = \{b | b \text{ is ordinal, and } |b| \le \aleph_a\}$$

It gives a cardinal greater than $\aleph_a$ as $\aleph_{a+1} = \omega_{a+1}$. For any ordinal a, there is an infinite cardinal $\aleph_a$. All these cardinals also form a series:

$$\aleph_0, \aleph_1, \aleph_2, \ldots, \aleph_n, \ldots, \aleph_\omega, \ldots \tag{6.7}$$

From left to right, these cardinals become greater and greater, and there are not any other infinite cardinals between any two consecutive $\aleph$s. The infinite ordinals and infinite cardinals are also called transfinite ordinals and transfinite cardinals or transfinite numbers as a whole. Where will these more and more numerous transfinite numbers lead to? Cantor thought it would be God.

People were surprised at transfinite numbers. Some praised Cantor's innovation; others criticized transfinite numbers were "fog on fog," and "Cantor was building a disease of mathematics need to be cured." Although there was hot debate, transfinite number was one of the most astonishing achievements of thought in the nineteenth century.

Cantor found two types of infinite cardinal series: one was power sets, and the other was transfinite cardinals:

$$\aleph_0, 2^{\aleph_0}, 2^{2^{\aleph_0}}, \ldots$$

and

$$\aleph_0, \aleph_1, \aleph_2, \ldots$$

$\aleph_0$ is the cardinal of the infinite countable set; $\aleph_1$ is the next transfinite cardinals next to $\aleph_0$. By Cantor's theorem of power set, $2^{\aleph_0}$ is more numerous than $\aleph_0$, but whether is it more or less numerous than $\aleph_1$? Cantor conjectured $2^{\aleph_0} = \aleph_1$, that there wasn't any other transfinite cardinals between $\aleph_0$ and $2^{\aleph_0}$; hence $2^{\aleph_0}$ was the first transfinite cardinal more numerous than infinite countable set. In 1874, Cantor proved $2^{\aleph_0} = C$ and showed all subsets of natural numbers have the same cardinality as real numbers. Therefore, Cantor's conjecture essentially states that, there exists no set whose cardinality is strictly greater than that of natural numbers $\aleph_0$ and less than that of real numbers C. Because real numbers are often called continuum, this conjecture is called *continuum hypothesis*, abbreviated as CH. Extend continuum hypothesis: for any ordinal a, whether $2^{\aleph_a} = \aleph_{a+1}$ holds. This conjecture is called generalized continuum hypothesis, abbreviated as GCH.

Cantor raised continuum hypothesis in a paper in 1878. He believed it to be true and tried for many years to prove it. Sometimes he thought he had proved it false; then the next day found his mistake. Again he thought he had proved it true only again to quickly find his error. His inability to prove the continuum hypothesis caused him considerable anxiety till his death in 1918. The problem, whether we can prove continuum hypothesis true or false became the first on Hilbert's list of 23 important open questions that was presented at ICM in 1900 in Paris. The continuum hypothesis came from the practical problems from geometry, mechanics, and physics. Hilbert expressed this view:

> *But even this creative activity of pure thought is going on, the external world once again reasserts its validity, and by thrusting new questions upon us through the phenomena that occur, it opens up new domains of mathematical knowledge.*

Because the continuum hypothesis is the most central open problem at the foundation of mathematical logic and axiomatic set theory, it has been studied by many mathematicians for almost a century. Although many significant progresses were made, it is not completely resolved. Gödel proved in 1938 that the negation of the CH, the existence of a set with intermediate cardinality, could not be proved in standard set theory, also known as ZFC set theory (stands for **Z**ermelo-**F**raenkel axioms for set theory together with the **A**xiom of **C**hoice (or AC). Informally, AC says that given any collection of nonempty bins (even infinite), it is possible to make a selection of exactly one object from each bin). The second half of the independence of the CH, unprovability of the nonexistence of an intermediate-sized set, was proved in 1963 by Paul Cohen with a new powerful technique called *forcing*. There was an interesting story said that Cohen, the young US mathematician, was not sure about his proof ([7], pp. 280). He came to Princeton and knocked on Gödel's house. Gödel was struggling with paranoia at the time. He slightly open the door, such that Cohen could pass his proof through. Then Cohen was shut out. Two days later, Gödel invited Cohen came in to drink tea. Cohen was awarded Fields Medal in 1966.

Gödel and Cohen showed CH was undecidable from ZFC set theory; CH was independent to the axioms in ZF system. Similar things also happened to AC. Gödel, and Cohen's result also shows that AC is undecidable from ZF system. Accepting AC gives the consistent mathematical system, while rejecting AC gives another consistent system. With AC, accepting or rejecting CH both gives consistent mathematics, respectively. CH and AC are independent to ZF set theory [62].

6.4 Infinity in Art and Music

Infinity-inspired art and music when facing the vast galaxy, the sea, and the mystery nature. *Starry Night* by Vincent van Gogh is impressive among countless works of art about sky (Fig. 6.16). Van Gogh's night sky is a field of roiling energy. Below the

Fig. 6.16 Vincent van Gogh, *The Starry Night*, Museum of Modern Art, New York

exploding stars, the village is a place of quiet order. Connecting Earth and sky is the flame like cypress, a tree traditionally associated with graveyards and mourning. Van Gogh said "Looking at the stars always makes me dream." He painted in June 1889, in the mental hospital at Saint-Rémy-de-Provence in France, just before sunrise, with the addition of an idealized village. Van Gogh stayed there for 108 days. During this period, he pictured about 150 oil canvas and hundreds of sketches.

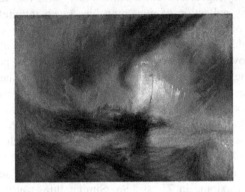

Joseph Mallord William Turner, *Snow Storm*, 1842, Tate Modern, London UK

Snow Storm: Steam-Boat off a Harbour's Mouth is a painting by Turner. The picture may recall a particularly bad storm in January 1842. In order to feel the power of sea, Turner got the sailors to tie him to the mast to observe. "I was lashed for four hours. and I did not expect to escape, but I felt bound to record it if I did." However, the critical response to the painting was largely negative at the time, with one critic calling it "soapsuds and whitewash." Turner said "I did not paint it to be understood, but I wished to show what such a scene was like." As time going, people finally realized, as John Ruskin, the leading English art critic of the Victorian era said "It is one of the very grandest statements of sea-motion, mist and light, that has ever been put on canvas."

Infinity has been a critical problem in philosophy and theology from the concrete things in nature. In Ptolemy's model, the universe is realized as a set of nested spheres, with planets moving along them. The outmost boundary is celestial sphere. By the medieval, the church had largely accepted this model developed by Aristotle and Ptolemy. People believed the Earth, created by God, was the center of the universe, and the celestial sphere was bounded.

Un missionnaire du moyen âge raconte qu'il avait trouvé le point
où le ciel et la Terre se touchent...

Camille Flammarion, L'Atmosphere: Météorologie Populaire (Paris, 1888), pp. 163

A woodcut published in Paris in 1888 reflected how people understood the bounded universe at that time. The original caption bellow the picture translated to "A medieval missionary (Bruno) tells that he has found the point where heaven and Earth meet..." If someone stands at the boundary of stellar sphere, can he raise his stick out to across the boundary of universe? It's hard to prevent him from doing that; but if he can, what is the space outside the physical world? This was the paradox of the universe boundary. To solve it, scholars in medieval refactored Aristotle's theory and proposed an idea of progressive boundary. Others believed, if you throw a spear outside the world boundary, it would enlarge the universe. The world of matter is bounded, but the boundary is surrounded by endless void.

During the Renaissance, artists adopted mathematics and science into their works. Leonardo da Vinci had wide interest including architecture, anatomy, mathematics, and engineering. He intended to use perspective disciplines and experiment different aesthetic proportions in his works. Albrecht Dürer studied human proportions and the use of transformations to a coordinate grid to demonstrate facial variation. His book *Four Books on Measurement* introduced both painting theory and research on geometric and perspective principles. Kepler and Desargues independently developed the concept of "point at infinity" in projective geometry.

Desargues developed an alternative way of constructing perspective drawings by generalizing the use of vanishing points to include the case when these are infinitely far away. He made Euclidean geometry, where parallel lines are truly parallel, into a special case of an all-encompassing geometric system.

Non-Euclidean geometry brought a new view to artists about infinity. Euclidean geometry has been a perfect example of rigor reasoning for 2,000 years. However, mathematicians' addiction to perfection continued questioning Euclid's fifth postulate. The first four are intuitive and obvious, for example, to draw a line from any point to any point; all right angles are equal to one another, while the fifth is disparate complex. In Euclid's original formulation "If a straight line falls on two straight lines in such a manner that the interior angles on the same side are together less than two right angles, then the straight lines, if produced indefinitely, meet on that side on which are the angles less than the two right angles." It is also known as "parallel postulate," because it is equivalent to the following statement: for any given line and a point not on it, there is exactly one line through the point that does not intersect the line. People doubted if this postulate could be deduced from the first four. Actually, the fifth postulate hasn't been used in a large portion in *Euclid's Elements*. Many mathematicians attempted to prove the fifth postulate but all failed. Saccheri tried to prove by absurdity. He assumed the fifth postulate was false, but obtained a series of obscure results. Saccheri believed they were too strange, and Euclid's fifth postulate must be true.

In nineteenth century, Gauss, János Bolyai, and Nikolai Ivanovich Lobachevsky independently realized it's impossible to establish such proof; parallel postulate was independent to other postulates. Gauss did not publish his result; people found it after Gauss died in 1885. Bolyai and Lobachevsky independently published treatises on non-Euclidean geometry around 1830. In this new geometry (known as hyperbolic geometry nowadays), Lobachevsky replaced Euclid's postulate: in a plane, given a line and a point not on it, there were multiple lines through a point that did not intersect the line. Then he developed a whole set of consistent results. Many of them are different from Euclidean geometry. For example, in the new geometry, the inner angle of a triangle is less than two right angles.

Bernhard Riemann, in 1854, constructed another geometry that was different from both Euclid and Lobachevsky. He founded the field of Riemannian geometry, and the simplest one is called elliptic geometry. It's non-Euclidean due to its lack of parallel lines, meaning every two lines must intersect in a plane. In elliptic geometry, the inner angle of a triangle is greater than two right angles. Eugenio Beltrami, in 1868, showed that Euclidean geometry and hyperbolic geometry were equiconsistent so that hyperbolic geometry was logically consistent if and only if Euclidean geometry was. In 1871, Felix Klein defined a metric method to describe the non-Euclidean geometries. Klein's influence has led to the current usage of the term "non-Euclidean geometry" to mean either "hyperbolic" or "elliptic" geometry.

Non-Euclidean geometry brings such possibility that infinite space can be bounded. Poincaré, in his popular science book *Since and Hypothesis*, described an interesting world [64]. The world is encapsulated in an infinite bounded sphere. The temperature at the center is very high. Along the distance away from the center,

Fig. 6.17 M.C. Escher.
Circle limit III, 1959

the temperature decreases in proportion. When it arrived at the sphere surface, it decreases to absolute zero K. Let the radius of the world sphere be R, for a point with the distance to the center as r; the temperature is proportion to $R^2 - r^2$. In this special world, due to thermal expansion, object size is proportional to the temperature; the closer to the sphere boundary, the smaller. If a citizen in this world is moving toward the sphere, as the temperature decreases lower and lower, he also becomes smaller and smaller. The pace keeps slowing down. He will never reach the boundary, although this world is bounded. What Poincaré described was exactly a world ruled by a kind of hyperbolic geometries. Inspired by Poincaré, Escher painted a series of woodcuts, named circle limit, to illustrate the bounded (Figs. 6.1 and 6.17), infinite world. In his painting, angles, devils, and fishes become smaller and smaller when closing to the circle; they can never walk out of this infinite but bounded world.

Besides art, there are music about infinity. In May 1747, Johann Sebastian Bach visited Sanssouci Palace in Postdam. King Frederick II invited Bach to try out his new Silbermann pianos. After his return to Leipzig, he composed a subject, which he had received from the King, under the title of "Musical Offering." He sent a copy to the King on July 7. This collection of music is catalogued as BWV1079, in which there is a special piece with title "Canon per Tonos," known as endlessly rising canon. It pits a variant of the king's theme against a two-voice canon at the fifth. However, it modulates and finishes one whole tone higher than it started out at. It thus has no final cadence (Fig. 6.18). Douglas Hofstadter wrote:

What makes this canon different from any other, however, is that when it concludes-or, rather, seems to conclude-it is no longer in the key of C minor, but now is in D minor. Somehow Bach has contrived to modulate (change keys) right under the listener's nose. And it is so constructed that this "ending" ties smoothly onto the beginning again; thus one can repeat the process and return in the key of E, only to join again to the beginning. These successive modulations lead the ear to increasingly remote provinces of tonality, so that after several of them, one would expect to be hopelessly far away from the starting key. And yet magically, after exactly six such modulations, the original key of C minor has been restored! All the voices are exactly one octave higher than they were at the beginning, and here the piece may be broken off in a musically agreeable way. Such, one imagines, was

Fig. 6.18 Part of the endlessly rising canon

Bach's intention; but Bach indubitably also relished the implication that this process could go on ad infinitum, which is perhaps why he wrote in the margin "As the modulation rises, so may the King's Glory." To emphasize its potentially infinite aspect [4].

Exercise 6.5

6.5.1 Light a candle between two mirrors, what image can you see? Is it potential or actual infinity?

Answer: page 270

Appendix: Example Programs

Define natural numbers with stream; take the first 15 numbers.

Listing 6.1 Stream of natural numbers (Java 1.8)

```
IntStream.iterate(1, i -> i + 1);

IntStream.iterate(1, i -> i + 1)
        .limit(15).forEach(System.out::println);
```

Listing 6.2 Stream of natural numbers (Python 3)

```
def naturals():
    yield 0
    for n in naturals():
        yield n + 1
```

Define the infinite set of natural numbers recursively in Haskell.

Listing 6.3 Stream of natural numbers (Haskell)

```
nat = 1 : (map (+1) nat)

take 15 nat
```

Define the infinite set of Fibonacci numbers recursively in Haskell; then fetch the 1500th Fibonacci number.

Listing 6.4 Fibonacci numbers

```
fib = 0 : 1 : zipWith (+) fib (tail fib)

take 15 fib
[0,1,1,2,3,5,8,13,21,34,55,89,144,233,377]

fib !! 1500
13551125668563101951636936867148408377786010712418497242133543153221487310
87352875061225935403571726530037377881434732025769925708235655004534991410
29242495959974839822286992875272419318113250950996424476212422002092544399
20196960465321438498305345893378932585393381539093549479296194800838145996
187122583354898000
```

Define the infinite stream of prime numbers with coalgebra in Haskell.

Listing 6.5 Stream of primes

```
data StreamF e a = StreamF e a
data Stream e = Stream e (Stream e)

ana :: (a → StreamF e a) → (a → Stream e)
ana f = fix ∘ f where
  fix (StreamF e a) = Stream e (ana f a)

takeStream 0 _ = []
takeStream n (Stream e s) = e : takeStream (n - 1) s

era (p:ns) = StreamF p (filter (p `notdiv`) ns)
  where notdiv p n = n `mod` p ≠ 0

primes = ana era [2..]

takeStream 15 primes
[2,3,5,7,11,13,17,19,23,29,31,37,41,43,47]
```

Appendix: Proof of Cantor's Theorem

Theorem 6.4.1 (Cantor's Theorem) *For every set, $|S| < |2^S|$, where $|S|$ is the cardinality of S; 2^S is the power set of S, which contains all subsets of S.*

Proof The proof has two steps. The first step is to prove $|S| \leq |2^S|$. For any x, let $f(x) = \{x\}$, which is the singleton set of x. It's obvious that for different elements $x_1 \neq x_2$, the corresponding singleton sets are not identical $\{x_1\} \neq \{x_2\}$, which means $f(x_1) \neq f(x_2)$. The map $S \xrightarrow{f} 2^S$ is injective, hence:

$$|S| \leq |2^S|$$

The second step is to prove $|S| \neq |2^S|$. Using reduction to absurdity, suppose they are equal. Then there exists 1-to-1 correspondence $S \xrightarrow{\phi} 2^S$, such that for every $x \in S$, its image $\phi(x) \in 2^S$. It means $\phi(x)$ is some subset of S, i.e., $\phi(x) \subseteq S$. Now we ask whether x belongs to $\phi(x)$. Either $x \in \phi(x)$, or $x \notin \phi(x)$. We put all x that not belonging to $\phi(x)$ to form a new set S_0:

$$S_0 = \{x \mid x \in S, \text{and} x \notin \phi(x)\} \tag{6.8}$$

Obviously, S_0 is a subset of S, which means $S_0 \subseteq S$. Hence $S_0 \in 2^S$. Because ϕ is bijection, there must exist some x_0, such that $\phi(x_0) = S_0$. According to the logical law of excluded middle, either $x_0 \in S_0$, or $x_0 \notin S_0$. Both must be one and only one is true:

(1) If $x_0 \in S_0$, according to the definition of S_0 in Eq. (6.8), $x_0 \notin \phi(x_0)$. But since $\phi(x_0) = S_0$, hence $x_0 \notin S_0$.
(2) If $x_0 \notin S_0$, because $S_0 = \phi(x_0)$, we have $x_0 \notin \phi(x_0)$. But according to the definition of S_0 in Eq. (6.8), $x_0 \in S_0$ should hold.

No matter x_0 belongs to S_0 or not, it leads to contradiction. There cannot be 1-to-1 correspondence between S and 2^S, therefore, $|S| \neq |2^S|$.

Summarize the result: $|S| \leq |2^S|$ and $|S| \neq |2^S|$; it proves Cantor's theorem:

$$|S| < |2^S|$$

□

The second step in the proof reminds the popular Russell's paradox: let S be a set containing all sets that each not belong to itself, then does S belong to itself? We'll show Russell's paradox and Gödel's incompleteness theorems in the next chapter.

Chapter 7
Paradox

I know that I know nothing

Socrates

Escher, Waterfall, 1961

In 1996, the 26th Summer Olympic game was hold in Atlanta, United States. More than 10,000 athletes from 197 nations challenged human limit of speed, strength, and team work in 26 sports. At the same time, there was another interesting competition ongoing. Deep Blue, a computer made by IBM, was challenging the world chess champion Garry Kasparov in a six-game match. Deep Blue won the first game, but Kasparov won three and drew two, defeating Deep Blue by a score of 4:2. The next year, heavily upgraded Deep Blue challenged again to Kasparov. On May 11, computer defeted human: two wins for Deep Blue, one for Kasparov, and three draws. Deep Blue was a super computer of 1270 kilogram weight, with 32 processors. It could explorer 200 million possible moves in a second. The design

© China Machine Press 2024
X. Liu, *Mathematics in Programming*,
https://doi.org/10.1007/978-981-97-2432-1_7

team input 2 million grandmaster games in the past 100 years as the knowledge base for Deep Blue. The machine created by human intelligence, defected human at the first time in the field of intelligence. This result led to attention, fear, and hot debate among mass media.

Most people believed this was a significant progress in artificial intelligence at that time. Although computer could defect human for chess, there was still a big gap in board game Go. There are 8 rows and 8 columns in chess board and 32 pieces. Computer needs to search among a big game tree containing about 10^{123} possible moves. Even Deep Blue could explore 2 million moves per second; it would take about 10^{107} years to exhaust the tree. The design team optimized the program to narrow down the search space, such that Deep Blue only need explore 12 moves ahead from current game, while human grandmasters can only evaluate about 10 moves ahead. However for Go game, there are 19 rows and 19 columns, two players can put black and white pieces in 361 grids. The scale of the game tree is about 10^{360}, which is far bigger than chess. For a long time after Deep Blue, people did not believe computer could defect human in Go.

Deep Blue v.s. Kasparov. from *Scientific American*

20 years later in 2016, a computer program "AlphaGo" challenged top human Go master. Korean professional 9-dan Go player, Lee Sedol, lost the game in a 1:4 series matches. One year later, the successor program "AlphaGo master" beat Chinese professional Ke Jie, the world number one ranked player, in a three-game match. Go had previously been regarded as a hard problem in artificial intelligence that was expected to be out of reach for the technology of the time. It was considered the end of an era. Facing the emotionless machine, Ke Jie was unwilling and burst into tears. As human beings, our feelings were mixed. Even the programmer community, doing intellectual work, was feeling the pressure from machine: will machine replace us eventually?

Traditionally, we thought areas with culture background, inner emotions, and human characters, like art, literature, and music, could not be dominated by machine. In 2015, three researchers Gatys, Ecker, and Bethge from University of Tübingen, a small town 30 km south of Stuttgart, Germany, applied machine learning to art style. By using deep convolutional neural network, they transformed a landscape photo of Tübingen into art painting of different styles [65]. No matter the exaggerated emotion of the post-impressionist Van Gogh, or Turner's romantic turbid light and shadow effect, all vivid imitated by machine, as if the artists painted by themselves (Fig. 7.1).

Fig. 7.1 Artworks in different styles generated by machine learning: (**a**), Landscape photo of Tübingen; (**b**), *The Shipwreck of the Minotaur* by J.M.W. Turner, 1805; (**c**), *The Starry Night* by Vincent van Gogh, 1889; (**d**), *Der Schrei* by Edvard Munch, 1893; (**e**), *Femme nue assise* by Pablo Picasso, 1910; (**f**), *Composition VII* by Wassily Kandinsky, 1913

In the following years, artificial intelligence and machine learning conquered a variety of areas in accelerated speed. Machines generated different styles of music and played them with moods and rhythms of tension, relaxing, and so on. It was not the monotonous electronic sound anymore. Machine batch translated news and academic papers, which was comparable to human professional translators. Machine processed X-ray photos, CT, and MRI medical images to diagonal diseases, and the accuracy exceeded human doctors. Self-driven cars, powered by artificial intelligence traveled on streets, successfully overtake other vehicles and avoid pedestrians. Automated groceries suddenly appeared on the street; people could pick the products and walk out without being checked out by a cashier....
As humans we can't stop asking: Are we eliminating jobs faster than creating? Will human be replaced by machine completely? Will machine rule people in the future?

All these lead to a critical question: does there exist boundary of computation? If yes, where is it?

7.1   Boundary of Computation

Gu Sen described the hesitated feelings when facing a long-running program in his popular book *Fun of Thinking*: Will this program finish? Shall I wait or kill it? Is there a compiler that could tell if my program will run endlessly ([66] pp.228)?

Why not possible? It seems more realistic than time machine. We may see such a scene in a scientific film: a programmer typed something in the dark screen, then hit enter. A highlighted, bold warning popped up immediately "Warn: the program with the given input will run forever. Continue? (Y/N)". If this became true one day, what fantastic cool things will you do? Do you believe that I can make big money with it? I'll firstly use it to prove the Goldbach's conjecture. I can write a program, enumerate all even numbers one by one, examine if it is the sum of two primes. If yes, then check the next even number; otherwise output the counter example and quit. The next thing is to compile my program. Can't the compiler determine my program terminates or not? If the magic compiler warns me that my program will run endlessly, haven't I proved Goldbach's conjecture? Or if the compiler tells me the program will terminate, doesn't it mean Goldbach's conjecture is falsehood? Either case, I'll be the first one that solve the Goldbach's conjecture, and leave my name in history. What's the next? I will modify that program to explore the twin primes, then compile it to see if there are really infinitely many twin primes. And next, are there infinitely many Mersenne primes? This is also an open question in number theory for a long time. I can easily solve it in this way. The $3x + 1$ conjecture? It's a piece of cake to write a "proof program" in a few minutes, then win the 500 dollars prize offered by Paul Erdős. There are enough mathematical open questions, I'll never worry about nothing to do. Martin LaBar in 1984 asked if a 3×3 magic square can be constructed with nine distinct square numbers. The award has accumulated to 100 dollars, 100 euros, and a bottle of champagne. Search "Unsolved problems in mathematics" from internet, filter in those about discrete things and with award, then write a few programs to solve them all...

In 1936, Alan Turing, the pioneer of computer science and artificial intelligence, proved that a general algorithm to determine an arbitrary computer program would finish running, or continue to run forever, could not exist. A key part of the proof was a mathematical definition of a computer and program, which became known as a Turing machine. This problem is called *halting problem* nowadays.

We use reduction to absurdity method to prove Turing's halting problem. Suppose there exists a algorithm $halts(p)$ that can determine arbitrary program p terminates or not. First, construct a never halting program:

$$forever() = forever()$$

This is an infinitely recursive call. Then define a special program G as below:[1]

$$G() = \begin{cases} halts(G) = \text{True}: & forever() \\ \text{otherwise}: & \text{halt} \end{cases}$$

In program G, it utilizes $halts(G)$ to examine whether G itself will halt or not. If it halts, then calls $forever()$ to let it run forever. It exactly means G will not halt in this case; hence $halts(G)$ should be false. However, according to the second clause, it will halt. Therefore $halts(G)$ should be true. whether $halts(G)$ is true or false, we get conflicted result. Hence our assumption can't hold. There does not exist a general algorithm to solve the halting problem for all possible program input.

There is another equivalent two-step proof ([67] pp.268). Most are the same except that G accepts another argument p; it applies p to itself and then passes to $halts$:

```
G(p) = if halts(p(p)) then forever() else 'Halted'
```

Pass G to itself as $G(G)$. If $halts(G(G))$ returns true, then it calls $forever()$, such that $G(G)$ never finishes. Thus it exactly means $halts(G(G))$ should return false; hence the program enters the **else** branch and halts. But again, it means $halts(G(G))$ should return true. Whether halts or not, it leads to absurdity.

The halting problem clearly provides an incomputable problem and breaks the bubbles of all above magic ideas. It reminds of the proof of Cantor's theorem in appendix 6.4, where we applied quite similar method to show that for all sets, including infinite sets, the cardinals are strict less than their power sets. Actually, halting problem is closely related to many interesting logic paradoxes.

---

[1] We intend to use G for special meaning. It's Gödel's initial letter, exactly the same name for nondeterministic proposition in Gödel's incompleteness theorem.

7.2 Russell's Paradox

The history of paradox came back to ancient Greece. We've seen Zeno's paradoxes about infinity and continuum. Logic paradox is often an interesting problem, with strict reasoning but deduced to conflicting result. About the fourth century BC, Eubulides of Miletus, a philosopher in ancient Greek, raised a proposition: "I am lying." How to determine whether it is true or false? If the proposition is false, then what it states (lying) should be true, it conflicts; however if the proposition is true, since it states I am laying, it should be false and lead to conflict again. Whether Eubulides' saying is true or not, all falls into contradiction. This confusing problem is called "liar paradox."

There is variance of liar paradox, appeared as two separated statements:

Achilles: The tortoise is a liar, he always lies. Do not trust him.
Tortoise: Dear Achilles, you are honest, you always speaks truth.

Is it true or false for what the tortoise said? If the tortoise tells truth, then what Achilles states is true. However, Achilles claims the tortoise is lying; it leads to contradiction. On the contrary, if the tortoise lies, then what Achilles says is wrong; hence the tortoise should be true. We end up with a wired loop: whether the tortoise speaks truth or not; it leads to absurdity.

This two-segment liar paradox sometimes appears as a joke. You receive a piece of note: "The other side is nonsense." While when you flip to the other side, it writes: "The other side is true." Which side is true? In a similar way, it reduces to contradiction.

Such paradox also appears in children's story. A lion caught a rabbit. He was so happy, that he promised the rabbit: If you can guess what I am going to do, I'll let you go; otherwise I'll eat you. The clever rabbit then answered: I guess you are going to eat me.

If the lion eats the rabbit, then the rabbit guesses correct. The lion should keep his promise to let the rabbit go. However, if he let the rabbit go, it means the rabbit guesses wrong. Hence the lion should eat the rabbit. The lion falls into the dilemma; he should neither eat the rabbit nor let the rabbit go. We may imagine the rabbit silently runs away when the lion keeps deep thinking.

According to the legend in ancient time, after Greek army defected Persian, the king decided to do something "kind" to the captives—let them choose the way to be killed. According to what the captive said, if it is true, then cut their head off, otherwise hang. A clever captive said: "I think you are going to hang me." If the king hanged him, then what the captive said was true. Hence his head should be cut according to the rule. But if cut his head off, then it did not follow what the captive said. Hence he spoke falsehood and should be hanged. Whether to cut the head or hang, the king's rule would not be conducted correctly. Facing such struggled situation, the king let this clever man go and released all captives.

E. O. Plauen *Father and Son*, 1930s

In Cervantes's novel *Don Quixote*, there is a similar paradox in Part II, Chapter 51:

A deep river divides a certain lord's estate into two parts... over this river is a bridge, and at one end a gallows and a sort of courthouse, in which four judges sit to administer the law imposed by the owner of the river, the bridge and the estate. It runs like this: "Before anyone crosses this bridge, he must first state on oath where he is going and for what purpose. If he swears truly, he may be allowed to pass; but if he tells a lie, he shall suffer death by hanging on the gallows there displayed, without any hope of mercy." ... Now it happened that they once put a man on his oath, and he swore that he was going to die on the gallows there – and that was all. After due deliberation the judges pronounced as follows: "If we let this man pass freely he will have sworn a false oath and, according to the law, he must die; but he swore that he was going to die on the gallows, and if we hang him that will be the truth, so by the same law he should go free."

Besides the liar paradox, the barber paradox is another popular puzzle. It was told by Russell in 1919. In a small village, the barber sets up a rule for himself: "He only shaves all those, and those only, who do not shave themselves." Then the question is, does the barber shave himself? If he shaves himself, then according to his rule, he should not shave himself; but if he does not, then he should serve and shave himself. The barber falls into his own trap.

Russell discovered the paradox in set theory early in 1901. He collected and summarized a series of paradoxes and formalized them as a fundamental problem in set theory. People called this kind of paradoxes as *Russell's paradox*. In Cantor's naive set theory, Russell considered the problem about if any set belonged to itself. Some sets do, while others not. For example, the set of all spoons is obviously not another spoon, while the set of anything that is not a spoon is definitely not a spoon. Russell considered the latter and extended it to all such cases. He constructed a set R, which contained all sets that were not members of themselves:

$$R = \{x \mid x \notin x\}$$

Russell next asked was R a member of R? According to logical law of excluded middle, an element either belongs to a set or does not. For a given set, it makes sense to ask whether the set belongs itself. But this well-defined, reasonable question falls into contradiction.

If R is a member of R, then according to its definition, R only contains the sets that are not members of themselves; hence R should not belong to R; on the contrary, if R is not a member of R, again, from its definition, any set does not belong to itself should be contained, hence R is a member of R. Whether it is a member or not gives contradiction. Formalized as:

$$R \in R \iff R \notin R$$

Russell explicitly gave the paradox in Cantor's set theory.

Bertrand Russell (1872–1970)

Bertrand Russell was born in 1872 in Monmouthshire into a family of the British aristocracy. Both his parents died before he was 3, and his grandfather died in 1878. His grandmother, the countess was the dominant family figure for the rest of Russell's childhood and youth. Her favorite Bible verse, "Thou shalt not follow a multitude to do evil," became his motto.

Russell was educated at home by a series of tutors. When Russell was 11 years old, his brother Frank introduced him to the work of Euclid, which he described in his autobiography as "one of the great events of my life, as dazzling as first love." In 1890, Russell won a scholarship to read for the Mathematical Tripos at Trinity College, Cambridge, where he became acquainted with Alfred North Whitehead. He quickly distinguished himself in mathematics and philosophy, graduating as seventh Wrangler in the former in 1893 and becoming a fellow in the latter in 1895.

Russell started an intensive study of the foundations of mathematics. He discovered Russell's paradox. In 1903 he published *The Principles of Mathematics*, a work on foundations of mathematics. It advanced a thesis of logicism that mathematics and logic are one and the same. The three-volume *Principia Mathematica*, written with Whitehead, was published between 1910 and 1913. This, along with the earlier *The Principles of Mathematics*, soon made Russell world-famous in his field.

After the 1950s, Russell turned from mathematics and philosophy to international politics. He opposed nuclear war. The Russell-Einstein Manifesto was a document calling for nuclear disarmament and was signed by 11 of the most prominent nuclear physicists and intellectuals of the time. Russell was arrested and imprisoned twice. The second time he was in jail was at the age of 89, for "breach of peace" after taking part in an antinuclear demonstration in London. The magistrate offered to exempt him from jail if he pledged himself to "good behavior," to which Russell replied: "No, I won't." In 1950 Russell won the Nobel Prize for Literature. The committee described him as "in recognition of his varied and significant writings in which he champions humanitarian ideals and freedom of thought."

Russell died of influenza on February 2, 1970, at his home in Penrhyndeudraeth. In accordance with his will, there was no religious ceremony; his ashes were scattered over the Welsh mountains later that year.

Russell was sad after discovering the paradox in the central of set theory. "What makes it vital, what makes it fruitful, is the absolute unbridled Titanic passion that I have put into it. It is passion that has made my intellect clear, ...it is passion that enabled me to sit for years before a blank page, thinking the whole time about one probably trivial point which I could not get right ..." ([7] pp.231). Russell wrote to mathematician and logician Gottlob Frege about his paradox. Frege was about to build the foundation of arithmetic. The second volume of his *Basic Laws of Arithmetic* was about to go to press. Frege was surprised, he wrote: "Hardly anything more unwelcome can befall a scientific writer than that one of the foundations of his edifice be shaken after the work is finished. This is the position into which I was put by a letter from Mr Betrand Russell as the printing of this volume was nearing completion...." Russell's paradox in set theory was disastrous. Further, since set theory was seen as the basis for axiomatic development of all other branches of mathematics, Russell's paradox threatened the foundation. This motivated a great deal of research around the turn of the twentieth century to develop a consistent (contradiction-free) mathematics.

Exercise 7.1

7.1.1 We can define numbers in natural language. For example "the maximum of two digits number" defines 99. Define a set containing all numbers that cannot be described within 20 words. Consider such an element: The minimum number that cannot be described within 20 words. Is it a member of this set?

7.1.2 "The only constant is change" said by Heraclitus. Is this Russell's paradox?

7.1.3 Is the quote saying by Socrates (the beginning of this chapter) Russell's paradox?

Answer: page 364

7.3 Philosophy of Mathematics

To solve Russell's paradox at the foundation of mathematics and logic, many mathematicians continued discussing, debating, and proposed varies of solutions at the beginning of the twentieth century. Over 2,000 years, mathematics had long been regarded as the truth with non-doubtful absoluteness and uniqueness in rational thinking. In this hot discussion, people finally realized that different mathematics can coexist under different philosophical views.

7.3.1 Logicism

Gottlob Frege, 1848–1925

Gottlob Frege was a German philosopher, logician, and mathematician. Frege was the early representative of logicism. His goal was to show that mathematics grew out of logic, and in so doing, he devised techniques that took him far beyond the traditional logic. Frege treated the naive set theory as a part of logic; he defined natural numbers with logic. For example, 3 as a number represents three persons, three eggs, three jars, and so on; all these collections are a *class*[2] containing three elements. Which one should be used to represent natural number 3? Frege's idea was "all," all such classes that 1-to-1 corresponded. This is an infinite, abstract class that defines natural number 3. Although a bit complex, it's a great definition that free from culture limitation. No matter what language or what symbol is using, there won't be any ambiguities to understand number 3 through Frege's way, because no symbol is needed in this definition. As such, Frege managed to define number— which was the class of all classes. On top of this definition and logical laws, Frege developed his theory of natural numbers and established logical arithmetic. As the next step, he was going to develop all mathematics except geometry from logic. This is what Frege wanted to achieved in his book *Basic Laws of Arithmetic*.

---

[2] Frege's work was prior to Cantor's; he used the term "class," while Cantor later used "set."

Frege believed logical axioms were reliable and widely accepted. Once his work completed, mathematics would be "fixed on an eternal foundation."

Just during the preparation of press, Russell's letter arrived "in time." Frege fell into confusions about Russell's paradox. His corner stone—using logic to define the concept of numbers—is exactly about class of all classes. Such definition could directly lead to logical paradox.

Russell took over the torch. He tried to develop mathematics from logic in another way. He believed all mathematics is symbolic logic. Russell's logicism was largely influenced by Peano. In 1900, when attended the International Congress of Philosophy in Paris, he wrote: "The Congress was the turning point of my intellectual life, because there I met Peano... It became clear to me that his (Peano's) notation afforded an instrument of logical analysis such as I had been seeking for years...." After he came back, Russell and Whitehead discussed the basic concepts of mathematics every day. After their hard work, they finally wrote *Principia Mathematica*.[3] The three volumes of classic work about mathematical logic were published from 1910 to 1913. To solve the paradox, Russell pointed that: "An analysis of the paradoxes to be avoided shows that they all result from a kind of *vicious circle*. The vicious circles in question arise from supposing that a collection of objects may contain members which can only be defined by means of the collection as a whole." He suggested "Whatever involves all of a collection must not be one of the collection" and call this the "vicious-circle principle." To carry out this restriction, Russell and Whitehead introduced "theory of types."

The theory of types classified sets into levels. Individual elements, such as a person, a number, or a particular book, are of type 0; the sets of elements in type 0 are of type 1; the set of elements in type 1, which are sets of sets are of type 2... Every set is of a well-defined type. The objects in a proposition must belong to its type. Thus if one says a belongs to b, then b must be of higher type than a. Also one cannot speak of a set belonging to itself. Although this approach can avoid paradox, it is exceedingly complex in practice. It took 363 pages till the definition of number 1 in *Principia Mathematica*. Poincaré remarked: "eminently suitable to give an idea of the number 1 to people who have never heard it spoken of before." The theory of types requires all works at their proper type levels; propositions about integers have to be at the level of integers; propositions about rationals have to be at the level of rationals. $n/1$ and n are at different levels and hence should not be handled in one proposition. And the common statements like "all the real numbers..." are not valid any more, as multiple types of sets are involved.

The most questionable part is about *axiom of reducibility*, *axiom of choice*, and *axiom of infinity*. In order to handle natural numbers, real numbers, and transfinite numbers, Russell and Whitehead accepted the axiom of infinity to support the concept of infinite classes. They also accepted one can chose elements from nonempty set or even infinite set to form new set. Such two arguable axioms exist

---

[3] Russell and Whitehead gave this Latin name in honor of Isaac Newton's *Philosophiæ Naturalis Principia Mathematica*.

in set theory too. Many people opposed to the axiom of reducibility particularly. To support mathematical induction, this axiom says any proposition at a higher level is coextensive with a proposition at type 0 level. Poincaré pointed out it was disguised form of mathematical induction. But mathematical induction is part of mathematics and is needed to establish mathematics; hence we cannot prove consistency.

Later Russell himself became more concerned: "Viewed from this strictly logical point of view, I do not see any reason to believe that the axiom of reducibility is logically necessary, which is what would be meant by saying that it is true in all possible worlds. The admission of this axiom into a system of logic is therefore a defect, even if the axiom is empirically true." [3]

7.3.2 Intuitionism

L. E. J. Brouwer (1881–1966)

Some mathematicians took opposite approach to build the foundation of mathematics called intuitionism. They thought mathematics was purely the result of the constructive mental activity of humans rather than the discovery of fundamental principles claimed to exist in an objective reality. Intuitionism can be backtracked to Pascal. Kronecker was the pioneer mathematician hold intuitionism philosophy. Many mathematicians, including Bolyai, Lebesgue, Poincaré, and Weyl supported intuitionism. The founder was Luitzen Egbertus Jan Brouwer. Brouwer was born in 1881 in Overschie near Rotterdam, Netherlands. He entered University of Amsterdam in 1897 and soon demonstrated good mathematics capability. While still an undergraduate, Brouwer proved original results on continuous motions in four-dimensional space and published his result in the Royal Academy of Science in Amsterdam in 1904. Other topics which interested Brouwer were topology and the foundations of mathematics.

Influenced by Hilbert's list of problems proposed at ICM in Paris 1900, Brouwer put a very large effort to study typology from 1907 to 1913. The best known is Brouwer's fixed point theorem. It states that in the plan every continuous function from a closed disk to itself has at least one fixed point. He also extended this theorem to arbitrary finite dimension. Specially, every continuous function from a closed ball of Euclidean space into itself has a fixed point. In 1910, Brouwer proved topological invariance of degree and then gave the rigor definition of topological dimension. Because of the outstanding contribution to topology, he was elected a member of the Royal Netherlands Academy of Arts and Sciences.

When Brouwer was a post graduate student, he was interested in the ongoing debate between Russell and Poincaré on the logical foundations of mathematics.[4] His doctoral thesis in 1907 criticized the logical foundations of mathematics and marked the beginning of the intuitionist school. His views had more in common with those of Poincaré, and if one asked which side of the debate he came down on, then it would have with the latter. Brouwer died in 1966 at the age of 85 in a car accident.

Brouwer's intuitionism came from his philosophy: mathematics is an intellectual human activity. It does not exist outside our mind. Therefore, it is independent from the real physical world. The mind recognizes basic and clear intuitions. These intuitions are not perceptual or empirical, but directly admit certain mathematical concepts, like integers. Brouwer believed that mathematical thinking was a process of intellectual construction. It built its own world that was independent of experience and was limited only by the basic mathematical intuition. The basic intuitive concepts should not be understood as undefined in axiomatic theory, but should be conceived as something; as long as they were indeed useful in mathematical thinking, they could be used to understand various undefined concepts in a mathematics.

In his 1908 paper, Brouwer rejected in mathematical proofs the principle of the excluded middle. It states that any mathematical statement is either true or false; no other possibility is allowed. Brouwer denied that this dichotomy applied to infinite sets. In 1918 he published a set theory, the following year a theory of measure, and by 1923 a theory of functions, all developed without using the principle of the excluded middle.

Brouwer's constructive theories were not easy to set up since the notion of a set could not be taken as a basic concept but had to be built up using more basic notions. Because of this, intuitionism rejected non-constructive existence proofs. For example, Euclid's proof about the existence of infinitely many primes was not acceptable according to Brouwer because it did not give a way to construct the prime number.

---

[4] Poincaré distinguished three kinds of intuition: an appeal to sense and to imagination, generalization by induction, and intuition of pure number—whence comes the axiom of induction in mathematics. The first two kinds cannot give us certainty, but, he says, "who would seriously doubt the third, who would doubt arithmetic?" [68].

In general, intuitionism was more critical than construction in the first decades of the twentieth century. Intuitionism denied a large number of mathematical achievements, including irrational numbers, function theory, and Cantor's transfinite numbers. Many reasoning methods, like the principle of the excluded middle, were rejected. It caused strongly objection by some mathematicians. Hilbert said: "For, compared with the immense expense of modern mathematics, what would wretched remnants mean, the few isolated results incomplete and unrelated, that the intuitionists have obtained."

7.3.3   Formalism

David Hilbert (1862–1943)

The third mathematical school of thought is the formalism led by David Hilbert. Hilbert was one of the most influential and universal mathematicians of the nineteenth and early twentieth centuries. He was born in 1862 in Königsberg, Eastern Prussia. He didn't shine at school at first, but later received the top grade for mathematics. In 1880, Hilbert entered the University of Königsberg, where he met and developed lifelong friendship with Minkowski and Hurwitz. In 1895, invited by Felix Klein, he obtained the position of Professor of Mathematics at the University of Göttingen. He remained there 48 years for the rest of his life. During the Klein and Hilbert years, Göttingen became the preeminent institution in the mathematical world.

Hilbert contributed to many branches of mathematics. There are so many terms, theorems named after him, that Hilbert himself did not know. He once asked other colleagues in Göttingen what "Hilbert space" was. "If you think of mathematics as a world. which it is, then Hilbert was a world conqueror." When he died, *Nature* remarked that there was scarcely a mathematician in the world whose work did not derive from that of Hilbert [69].

Hilbert put forth a most influential list of 23 unsolved problems at ICM in Paris 1900. This is generally reckoned as the most successful and deeply considered compilation of open problems ever to be produced by an individual mathematician. These were the problems which he considered most significant in mathematics at that time: not isolated questions but problems of such a general character that their solution was bound to have an enormous influence on the shape of future mathematics.

From 1933, the Nazis purged many of the prominent faculty members. The new Nazi Minister of Education, Bernhard Rust, asked whether "the Mathematical Institute really suffered so much because of the departure of the Jews." Hilbert replied, "Suffered? It doesn't exist any longer, does it!" Hilbert died in 1943 at age of 81. The epitaph on his tombstone in Göttingen consists of the famous lines he spoke: "We must know. We will know."

Hilbert's *Foundations of Geometry* published in 1899 proposes a formal set, called Hilbert's axioms, substituting for the traditional axioms of Euclid. It is the representative work of axiomatization. Hilbert's approach signaled the shift to the modern axiomatic method. From 1904, Hilbert started studying the foundation of mathematics. In 1920 he proposed explicitly a research project that became known as Hilbert's program. He wanted mathematics to be formulated on a solid and complete logical foundation. It opened the way for the development of the formalist school. According to the formalist, mathematics is manipulation of symbols according to agreed-upon formal rules. It is therefore an autonomous activity of thought. The main goal of Hilbert's program was to provide secure foundations for all mathematics. In particular this should include:

(1) *A formulation of all mathematics.* All mathematical statements should be written in a precise formal language and manipulated according to well-defined rules.
(2) *Completeness*: A proof that all true mathematical statements can be proved in the formalism.
(3) *Consistency*: A proof that no contradiction can be obtained in the formalism of mathematics. This consistency proof should preferably use only "finitistic" reasoning about finite mathematical objects.
(4) *Conservation*: A proof that any result about "real objects" obtained using reasoning about "ideal objects" (such as uncountable sets) can be proved without using ideal objects.
(5) *Decidability*: There should be algorithm for deciding true or false of any mathematical statement.

To execute his program, Hilbert initiated *metamethematics*, to study the mathematics itself using mathematical methods. This study produces metatheories, which are mathematical theories about other mathematical theories. Such approach actually differentiates three different mathematical systems:

(1) Mathematics that is not formalized G: It's normal mathematics that allows classic logical reasoning. For example, applying principle of excluded middle on infinite set.
(2) Formalized mathematics H: All symbols, formulas, axioms, and propositions are formalized. They are undefined concepts without any concrete meanings before explanation. Once explained, they are the concepts in G. In other words, G is the model of H, and H is formalized G. As Hilbert described: "One

must be able to say at all times—instead of points, straight lines, and planes—tables, chairs, and beer mugs." With this approach, the specific meanings and background in Euclid geometry are put aside; we only focus on the relations between the undefined concepts, which are reflected through a collection of axioms.

(3) Metamathematics K: This is the metatheories to study H. All reasoning in K should be admitted intuitively. For example, without applying principle of excluded middle on infinite set.

While Hilbert was working on the project, a young mathematician Gödel proved incompleteness theorems, which showed that most of the goals of Hilbert's program were impossible to achieve.

7.3.4 Axiomatic Set Theory

Different from the mathematical schools of logicism, intuitionism, and formalism, the members of the set-theoretic school did not formulate their distinct philosophy at the beginning, but they gradually gained adherents and a program. The set-theoretic school can be traced back to Cantor and Dedekind's work. Although both were primarily concerned with infinite sets, they found, by establishing the concept of natural numbers on basis of set, all of mathematics could then be derived. When Russell's paradox was found in set theory, some mathematicians believed that the paradox was due to the informal introduction of sets. Cantor's set theory is often described nowadays as "naive set theory." The set theoretic thought that a carefully selected axiomatic foundation would remove the paradoxes, just as the axiomatization of geometry and of number system had resolved logical problems in those areas. Ernst Zermelo was the first who took the axiomatization approach in set theory in 1908.

Zermelo believed the paradoxes arose because Cantor had not restricted the concept of a set. He therefore stressed with clear and explicit axioms to clarify what was meant by a set and what properties sets should have. In particular, he wanted to limit the size of possible sets. He had no philosophical basis but sought only to avoid the contradictions. His axiom system contained the undefined fundamental concepts of set and the relation of one set being included in another. These and the defined concepts were to satisfy the statements in the axioms. No properties of sets were to be used unless granted by the axioms. In his system, the existence of infinite sets, the operations as the union of sets, and the formation of subsets were provided as axioms. Zermelo also used the axiom of choice [3].

(a) Ernst Zermelo (1871-1953)   (b) Abraham Fraenkel (1891-1965)

Zermelo's system of axioms was improved by Abraham Fraenkel in 1922. Zermelo did not distinguished set property and the set itself; they were used as synonymous. The distinction was made by Fraenkel. The system of axioms used mostly common by set theorists is known as Zermelo-Fraenkel system, abbreviated as ZF system. They both saw the possibility of refined and sharper mathematical logic available in their time, but did not specify the logical principles, which they thought were outside of mathematics and could be confidently applied as before [3].

Zermelo provided seven axioms in his 1908 paper. Then in 1930, Fraenkel, Skolem, and von Neumann suggested to add another two axioms. These axioms are as below:

(1) **Axiom of extensionality**: Two sets are equal if they have the same elements. For set A and B, if $A \subseteq B$ and $B \subseteq A$, then $A = B$.
(2) **Empty set**: The empty set exists.
(3) **Axiom schema of separation**: Also known as axiom schema of specification. Any property that can be formalized in the language of the theory can be used to define a set. For set S, if proposition $p(x)$ is defined, then there exists set $T = \{x | x \in S, p(x)\}$.
(4) **Axiom of power set**: One can form the power set (the collection of all subsets of a given set) of any set. This process can be repeated infinitely.
(5) **Axiom of union**: The union over the elements of a set exists.
(6) **Axiom of choice**: abbreviated as AC, see page 6.3.6 in Sect. 6.3.6.
(7) **Axiom of infinity**: There exists a set Z, containing empty set. For any $a \in Z$, then $\{a\} \in Z$. This axiom asserts infinite set exits.
(8) **Axiom schema of replacement**: This axiom is introduced by Fraenkel in 1922. For any function $f(x)$ and set T, if $x \in T$, and $f(x)$ is defined, there exits a set S, that for all $x \in T$, there is a $y \in S$, such that $y = f(x)$. It says that the image of a set under any definable function will also fall inside a set.
(9) **Axiom of regularity**: Also known as axiom of foundation. It was introduced by Von Neumann in 1925. x does not belong to x.

Fig. 7.2 Banach-Tarski paradox: a solid ball can be decomposed and put back together into two copies of the original ball

As such, set theory was abstracted to an axiomatic system. Set turned to be an undefined concept that satisfies these axioms. They do not permit "all inclusive set"; hence, they avoid the paradox and fixed the defects in naive set theory. However, there were still debate about which axioms were acceptable; particularly axiom of choice was arguable.

In 1924, Stefan Banach and Alfred Tarski proved a theorem called Banach-Tarski paradox.[5] This theorem states that, if you accept axiom of choice, then for a solid ball in three-dimensional space, there exists a decomposition of the ball into a finite number of disjoint subsets, which can then be put back together in a different way to yield two identical copies of the original ball (Fig. 7.2). Indeed, the reassembly process involves only moving the pieces around and rotating them without changing their shape. Banach and Tarski were going to reject axiom of choice through this theorem. However, their proof looked so natural that mathematicians tend to consider it only reflects the counterintuitive fact about axiom of choice. Some set theorists insist not to including axiom of choice; such axiomatic set theory is called ZF system, while the one included axiom of choice is called ZFC system (see Sect. 6.3.6).

7.4   Gödel's Incompleteness Theorems

By 1930, there had been four separated, distinct, and more or less conflicting approaches about mathematics foundation. Their supporters adhere to their own mathematical schools. One could not say a theorem is correctly proven, because by 1930, he had to add by whose standard it was proven correct. The consistency of mathematics, which motivated these new approaches, was not settled at all except if one argue that it's the human intuition guarantees consistency [3]. Hilbert was still planning his project optimistically to prove the completeness and consistency of mathematics. All these were ended up by a young mathematician and logician, Gödel.

---

[5] Also known as Hausdorff-Banach-Tarski theorem or "doubling sphere paradox"

Kurt Gödel, 1906–1978

Kurt Gödel was born in 1906 in Brünn, Austria-Hungary Empire (now Brno, Czech Republic), into a German family. In his family, young Gödel was known as "Mr. Why" because of his insatiable curiosity. In 1924, Gödel entered the University of Vienna. He hadn't decided whether to study mathematics or theoretic physics until he learned number theory. He decided to take mathematics as his main subject in 1926. Gödel was also interested in philosophy and took part in seminars about mathematical logic. The exploration of philosophy and mathematics set Gödel's life course.

In 1929, at the age of 23, he completed his doctoral dissertation. He was going to further study Hilbert's program to prove the completeness and consistency of mathematics in finite steps. However, he soon developed an excepted result. In 1930 Gödel attended the Second Conference on the Epistemology of the Exact Sciences, held in Königsberg, where he delivered his first incompleteness theorem and, soon, proved the second incompleteness theorem.

Gödel worked at University of Vienna from 1932. In 1933, the Nazis rose in influence in Austria academy over the following years. In 1936, Moritz Schlick, whose seminar had aroused Gödel's interest in logic, was assassinated by one of his former students. This triggered "a severe nervous crisis" in Gödel. He developed paranoid symptoms, including a fear of being poisoned. After World War II broken out, Gödel accepted the invitation from the Institute for Advanced Study in Princeton and moved to the United States. To avoid the difficulty of an Atlantic crossing, the Gödels took the Trans-Siberian Railway to the Pacific, sailed from Japan to San Francisco, and then crossed the United States by train to Princeton. He met Albert Einstein in Princeton, who became a good friend. They were known to take long walks together to and from the Institute for Advanced Study. Einstein's death in 1955 impacted him a lot. In his later life, logician and mathematician Wang Hao was his close friend and commentator.

Gödel married Adele Nimbursky, whom he had known for over 10 years, on September 20, 1938. Their relationship had been opposed by his parents on the grounds that she was a divorced dancer, 6 years older than Gödel. Later in his life, Gödel suffered periods of mental instability and illness. He had an obsessive fear of being poisoned; he would eat only food that Adele prepared for him. Late in 1977, she was hospitalized for 6 months and could subsequently no longer prepare her husband's food. In her absence, he refused to eat, eventually starving to death. He weighed 29 kilograms (65 lb) when he died. His death certificate reported that he died of "malnutrition and inanition caused by personality disturbance" in 1978. Because of the outstanding contributions in logic, he was regarded as the greatest logician since Aristotle.

In 1931, Gödel published his paper titled *On Formally Undecidable Propositions of Principia Mathematica and Related Systems*. Where "Principia Mathematica" is the work of Russell and Whitehead. In that article, he proved for any computable axiomatic system that is powerful enough to describe the arithmetic of the natural numbers:

(1) If a system is consistent, it cannot be complete.
(2) The consistency of axioms cannot be proved within their own system.

For any consistent formal system, Gödel gave an undecidable statement G that could neither be proved nor disproved. This theorem is called Gödel's first incompleteness theorem. It shows that consistent formalized system is incomplete. As far as the system is powerful enough to contain arithmetic of natural numbers, there will be problems that exceed it. One may ask, since G is undecidable, what if you accept G or negation of G as additional axiom, to obtain a more powerful system? However, Gödel soon proved the second incompleteness. It shows that if a formal system containing elementary arithmetic, then the consistency cannot be proved within its own system. Whether to accept or reject G, the new system is still incomplete. There always exists undecidable statement in the higher level.

In Euclidean geometry, for example, we can exclude the fifth postulate, to obtain the axiomatic system with the first four postulates. However, we cannot prove the fifth postulate true or false. Whether to accept or reject the fifth postulate gives consistent geometry—Euclidean geometry or varies of non-Euclidean geometries. In axiomatic set theory ZF system, we cannot prove axiom of choice to be true or false. Accepting it gives the consistent ZFC system, while rejecting it gives another consistent system. After adding axiom of choice to establish ZFC system, we cannot prove the continuum hypothesis to be true or false in ZFC. Accepting continuum hypothesis gives a consistent system, while rejecting continuum hypothesis gives another consistent system.

Gödel's first and second incompleteness theorems ended attempts in half-century, beginning with the work of Frege and culminating in Principia Mathematica and Hilbert's formalism, to find a set of axioms sufficient for all mathematics. Even the elementary arithmetic system is consistent; such consistency cannot be proved within itself. As André Weil said: "God exists since mathematics is consistent, and the Devil exists since we cannot prove it" [7] (Fig. 7.3).

Fig. 7.3 Escher, Angels and
Devils, 1941

7.5   Proof Sketch for the First Incompleteness Theorem

According to Hilbert's project, there are two steps: (1) formalize all the mathematics
into a system. It needs to represent every mathematical branch as a formal system,
which only contains finitely many axioms; (2) study with metamathematics, and
prove it is complete and consistent. Among them, the most fundamental system is
the arithmetic of natural numbers. Many mathematical branches are isomorphic to
it. We've seen how to extend natural numbers to integers, rationals in Sect. 6.3.2, and
define real numbers through Dedekind cut of rationals; by establishing the bijective
correspondence between points and real numbers, we can further include geometry
as arithmetic to coordinates.

7.5.1   Formalize Arithmetic

Here we use the method and terms in *Gödel, Escher, Bach: An Eternal Golden Braid*
by Hofstadter to introduce the proof sketch. Gödel's proof started from modeling a
formal system. We call this system "Typographical Number Theory," abbreviated
as TNT. It happens to be the same abbreviation of Trinitrotoluene, a powerful
explosive. Hofstadter intended to give this name to hint it's powerful enough to
destroy the building itself. TNT formalizes the number theory in natural language
into a series of typographical strings. Although it sounds complex, we realize it step
by step on top of Peano's Axioms in Sect. 7.7. The first step is to define numbers.

By Peano's axioms, zero is natural number; every natural number has its successor. Define typographical string for numbers as below:

zero	0
one	S0
two	SS0
three	SSS0
...	...

where "S" means successor; two letters of "SS" is the successor of a successor. A hundred letters of S followed by 0 are the 100 times successor of zero, which is the natural number 100. Although the string is very long, the rule is simple enough. With natural numbers being defined, the second step defines variables. To make the system as simple as possible, we only use five typographical letters a, b, c, d, e. When we need more, we simply add primes like $a', a'', a''', \ldots$ Next use "+" for addition, "·" for multiplication, and the parenthesis to control the arithmetic orders. To formalize the proposition, we add "=" and "¬" for negation and → for implication. Here are some examples of formal propositions (no matter they are true or false):

(a) one plus two equals four: $(S0 + SS0) = SSSS0$
(b) two plus two is not equal to five: $\neg(SS0 + SS0) = SSSSS0$
(c) if one equals to zero, then zero equals one: $(S0 = 0) \rightarrow (0 = S0)$

A proposition can have free variables, for example:

$$(a + SS0) = SSS0$$

It states a plus 2 equals 3. Obviously the value of a determines whether this proposition is true or false. We add universal quantifier ∀, existential quantifier ∃, and colon ":" to indicate quantifier scope. The following proposition:

$$\exists a : (a + SS0) = SSS0$$

means there exists a, such that a plus 2 equals 3. For another example:

$$\forall a : \forall b : (a + b) = (b + a)$$

This is commutative law of addition for natural numbers. When remove the quantifier for a, it changes to:

$$\forall b : (a + b) = (b + a)$$

This is an open formula since a is free. It expresses an unspecified number a commutes with all numbers b. It may or may not be true. In order to compose propositions, we need logical conjunction (and) $\wedge$, logical disjunction (or) $\vee$. Although there are few symbols, TNT is very expressive. Here are some examples:

(a) 2 is not the square of any natural numbers: $\neg\exists a : (a \cdot a) = SS0$
(b) Fermat's last theorem when n equals 3: $\neg\exists a : \exists b : \exists c : ((a \cdot a) \cdot a) + ((b \cdot b) \cdot b) = ((c \cdot c) \cdot c)$

Here are typographical symbols to express propositions so far. To construct TNT system, the next step is to define axioms and reasoning rules.

7.5.1.1 Axioms and Reasoning Rules

Following Peano's axioms, define five axioms for TNT system:

(1) $\forall a : \neg Sa = 0$, zero is not the successor of any number;
(2) $\forall a : (a + 0) = 0$, any number plus 0 equals itself;
(3) $\forall a : \forall b : (a + Sb) = S(a + b)$, addition for natural numbers;
(4) $\forall a : (a \cdot 0) = 0$, any number multiplies zero equals zero;
(5) $\forall a : \forall b : (a \cdot Sb) = ((a \cdot b) + a)$, multiplication for natural numbers.

Next establish reasoning rules. For example, from axiom (1), 0 is not the successor of any number, to deduce a special case: 0 is not the successor of 1. In order to do this, we introduce the rule of specification:

Rule of Specification Suppose u is a variable occurs in string x. If $\forall u : x$ is a theorem, then x is also a theorem, and any replacement of u in x wherever it occurs also gives a theorem.

There is a restriction; the term that replaces u must not contain any variables that is quantified in x. And the replacement should be consistent. The opposite rule to specification is the rule of generalization. It allows us to add the universal quantifier before a theorem.

Rule of Generalization Suppose x is a theorem, u is a free variable in it. Then $\forall u : x$ is a theorem.

For example, $\neg S(c + S0) = 0$ means there is no such a number, that plus 1, then take the successor gives 0; generalized as: $\forall c : \neg S(c + S0) = 0$.

The next rule is about how to convert universal and existential quantifiers.

Rule of Interchange Suppose u is a variable, then the string $\forall u : \neg$ and $\neg\exists u :$ are interchangeable.

When applying this rule to axiom (1), for example, it transforms to $\neg\exists a : Sa = 0$. The next rule allows to put an existential quantifier before a string.

Rule of Existence Suppose a term appears once or multiple times in a theorem, then it can be replaced with a variable, and add a corresponding existential quantifier in front.

Use axiom (1), for example, $\forall a : \neg Sa = 0$, we can replace 0 with a variable b and add the corresponding existential quantifier to give $\exists b : \forall a : \neg Sa = b$. It states that, there exists a number, such that any natural number is not its successor.

Next define rules of symmetry and transitivity for equality. Let r, s, t all stand for arbitrary terms.

Rules of Quality

(a) *Symmetry*: if $r = s$ is a theorem, then $s = r$ is also theorem;
(b) *Transitivity*: if $r = s$ and $s = t$ are theorems, then $r = t$ is also theorem.

define below rules to add or remove the successorship S:

Rules of Successorship

(a) *Add*: If $r = t$ is theorem, then $Sr = St$ is a theorem;
(b) *Drop*: If $Sr = St$ is theorem, then $r = t$ is a theorem.

So far, the TNT system is powerful to support constructing complex theorems.

Exercise 7.2

7.2.1 Translate Fermat's last theorem into a TNT string.
7.2.2 Prove the associative law of addition with TNT reasoning rules.

Answer: page 365

7.5.2 ω Incomplete

With axioms and reasoning rules in TNT system, we can prove a series of theorems:

$$(0 + 0) \quad\;\; = 0$$
$$(0 + S0) \quad = S0$$
$$(0 + SS0) \;\; = SS0$$
$$(0 + SSS0) = SSS0$$
$$\cdots \qquad\quad \cdots$$

By axiom (2), we deduce the first theorem by replacing a with 0; on top of this theorem and by axiom (3), we obtain the second theorem; every theorem can be deduced from the previous one. Observe this pattern, it follows with the question: why can't summarize them to a theorem?

$$\forall a : (0 + a) = a$$

Note it is different from axiom (2). Unfortunately, we can't deduce this theorem with all the rules in TNT so far. We may want to add an additional rule: if a series of strings are all theorems, then the universally quantified string which summarizes them is a theorem. However, only human outside TNT has this insight. It's not a valid rule for the formal system.

Lack of capability to do summarize shows TNT is incomplete. Accurately speaking, a system with this kind of "defect" is called ω incomplete, where ω is the cardinal of countable infinite set (Eq. 6.5 in Sect. 6.3.6). A system is ω incomplete if all the strings in a series are theorems, but the universal quantified summarizing string is not a theorem. It happens that the negation of the summarizing string:

$$\neg \forall a : (0 + a) = a$$

is not a theorem of TNT too. It means the string is undecidable within TNT system. The capability of TNT is insufficient to determine this string is theorem or not. It just likes the same situation that with only the first four postulates in Euclidean geometry, the fifth postulation is undecidable. We can either accept to add the fifth postulation to obtain Euclidean geometry or reject to add its negation to obtain non-Euclidean geometry. Similarly, we can either add this string or its negation to TNT to construct different formal systems.

It looks a bit counterintuitive if we chose the negation as theorem: zero plus any number does not equal to this number anymore. It's quite different from the familiar elementary arithmetic. It exactly shows the concrete meaning of a formal system is undefined. We give it the meaning of addition for natural numbers only for the purpose of easy understanding (remind Hilbert's quotation "One must be able to say at all times–instead of points, straight lines, and planes–tables, chairs, and beer mugs").

The ω incompleteness shows TNT is missing an important rule—you may have already thought of—Peano's fifth axiom that corresponding to mathematical induction. We add this last piece of tile to the puzzle.

Rule of Induction Suppose u is a variable in string X, denoted as Xu. If it is a theorem when replace u with 0, and $\forall u : Xu \rightarrow XSu$. It means if X is a theorem for u, so as it is when replace u to Su. Then $\forall u : Xu$ is also a theorem.

With mathematical induction supported, TNT system now has the same capability as Peano's arithmetic.

Exercise 7.3

7.3.1 Show $\forall a : (0 + a) = a$ with the induction rule.

Answer: page 366

Table 7.1 A Gödel
numbering to TNT

Symbol	Number	Symbol	Number
0	666	S	123
=	111	+	112
.	236	(	362
)	323	a	262
'	163	∧	161
∨	616	→	633
¬	223	∃	333
∀	626	:	636

7.5.3 Gödel Numbering

Gödel took a critical step called Gödel numbering. TNT system is powerful enough
to mirror other formal systems, is it possible to mirror TNT by itself? What Gödel
thought was to "arithmetize" the reasoning rules. To do this, he assigned all symbols
with a number (Table 7.1).

Axiom (1) translates to below numerals:

$$\forall \quad a \quad : \quad \neg \quad S \quad a \quad = \quad 0$$
$$626 \quad 262 \quad 636 \quad 223 \quad 123 \quad 262 \quad 111 \quad 666$$

The numbering scheme is not unique. It does not matter if one assigns different
numbers. With Gödel numbering, every TNT string is represented as a number
(although it may be a very big number). The problem is in converse direction: given
any number, can we determine if it represents a TNT theorem? We know the first five
TNT numbers, which represent the five axioms. With TNT reasoning rules, one can
construct infinitely many TNT numbers from these five numbers. As such, below is
a valid proposition in number theory:

a is a TNT number.

For example, 626,262,636,223,123,262,111,666 is a TNT number (we add
commas to make it easy to read); it represents axiom (1). Its negation is:

¬a is a TNT number.

For example, 123,666,111,666 is not a TNT number. It means we can replace a
by a string of 123666111666 letters of S followed by 0. This huge string actually
means $S0 = 0$ is not a TNT theorem. TNT system can really speak about itself.
It is not an accidental feature, but because of the fact that all formal systems are
isomorphic to number theory N. Hence we formed a circle: A TNT string has its
interpretation in number theory N, while the statement in N can have a second
meaning, which is the metalanguage interpretation about TNT (Fig. 7.4).

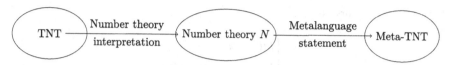

Fig. 7.4 TNT → number theory N → Meta-TNT

7.5.4 Construct the Self-reference

Gödel's last step was to construct a self-reference. It's such a TNT string, called G, which is about itself:

G is not a theorem of TNT.

Detonate TNT by asking whether is G a theorem of TNT or not. If G is a theorem, then it states the truth that "G is not a theorem." Here we see the power of self-reference. As a theorem, G can't be falsehood. Because TNT is not assumed to treat falsehood as theorem, we have to draw the conclusion that G is not a theorem. While known that G is not a theorem, we should admit that G is truth. TNT does not meet our expectation—we find a string, it states the truth, but is not a theorem. Further, considering the fact that G has its number theory interpretation, which is a statement of arithmetic property about natural numbers. From the reasoning outside TNT, we confirm this statement is true, and the string is not a theorem of TNT. However, when asked TNT whether this string is true, TNT could never say "Yes" or "No."

G is that undecidable proposition. This is the sketch of the proof of Gödel's first incompleteness theorem.

7.6 Universal Program and Diagonal Argument

What does Gödel's incompleteness theorems mean to programming? There is an exactly isomorphic problem of programming. We start from a formal computer programming language. This language supports primitive recursive function. It is a kind of functions in number theory, maps from natural numbers to natural numbers, and follows below five axioms:

(1) Constant function: The 0-ary constant function 0 is primitive recursive;
(2) Successor function: The 1-ary successor function $S(k) = k + 1$ is primitive recursive;
(3) Projection function: The n-ary function, which accepts n arguments, and returns its i-th argument P_i^n is primitive recursive;
(4) Composition: The result of finitely many times composition of primitive functions is still primitive recursive;

(5) Primitive recursion:

$$\begin{cases} h(0) = k \\ h(n+1) = g(n, h(n)) \end{cases}$$

We say h is computed from g through primitive recursion. It can be extended to the case of multiple arguments.

The programming language that supports basic arithmetic like addition, subtraction, if-then branches, equal, less than prediction, and bounded loop is called primitive recursive programming language. The bounded loop means the number of loops are determined beforehand. It can be loop without go-to statement or for-loop that does not allow to alter the loop variable inside it. However, it cannot be while loop or repeat-until loop. Because of these limitations, all primitive recursive programs must halt.

An important property of primitive recursive function is that, all primitive recursive programs are recursively enumerable. Suppose we can list all primitive recursive functions that, with one input and one output, store them in an infinitely big library. We can number each of them,[6] from 0, 1, 2, 3, ... And denote these programs as $B[0]$, $B[1]$, $B[2]$, ... For the i-th program, when input n, it gives result $B[i](n)$.

Now, construct a special function $f(n)$, when input n, its output is what the n-th program's output for n plus 1, as below:

$$f(n) = B[n](n) + 1$$

Such f is definitely computable. Now we ask whether f is a primitive recursive program stored in the library. If yes, suppose its number in the library is m. According to our numbering method, when input m to the m-th program, the result should be $B[m](m)$. However, from the definition of f, its output should be $f(m) = B[m](m) + 1$. These two results are not equal obviously. The contradiction proves there exists computable function that is not primitive recursive.

This proof uses the same method as Cantor's diagonal argument in Sect. 6.3.4. We have to relax the bounded loop limitation to make the programming language more powerful. To do that, we allow go-to statement in the loop to jump out; the loop variable can be altered inside; we introduced while loop, repeat-until loop, and general recursive functions. As such, we extend from primitive recursive function to *total recursive function*. This kind of programming language is called Turing-complete language. Most computer programming languages are Turing-complete, which are isomorphic to the formal systems like arithmetic of natural numbers. However, there is "defect" in Turning-complete language, as one can

---

[6] One numbering method is to concatenate all ASCII codes of the program to form a number, then sort them in ascending order. Since all programs are different, their ASCII codes are distinct.

construct the primitive recursive function of Turing halting problem. It shows there exits incomputable problem. Is it possible to relax the limitation further, empower Turning-complete language, to design a universal program? The answer is no. Turning-complete is at the highest level that reaches to the limitation of formal system. There is no other limitation can be relaxed any more. Gödel's incompleteness theorems assert, once the formal system is powerful enough to include arithmetic of natural numbers, there must be undecidable problem in it.

7.7 Epilogue

As human being, our rational thinking is great. It leads us back to thousands of years and talk to ancient sages; it sends us across the universe and step onto the unreachable planet; it foresees elementary parcels that are invisible to eyes; it breaks through intuition and reaches to high-dimensional magic world. Looking up the sky, through the clouds, dust, and stars, we feel the insignificance of ourselves. We are just passing passengers in the long river of time, just like a drop in the vast sea.

The problem we've seen in this chapter is essentially about ourselves as human. Does there exist the boundary of our rational thinking? Are we swallowing ourselves along a strange circle? These are questions everybody will ask at the era of artificially intelligence. People are making machine isomorphic to people, implementing huge machinery computing isomorphic to brain and rational thinking. Just as Escher illustrated in *Dragon* (Fig. 7.5), it tried best to break free from the two-dimensional picture. It has found the two slits in the paper. Its head and neck pokes through one slit, and the tail through the other, with the head biting the tail, it wants to pull itself out to the three-dimensional world. As the observer outside, we clearly know all still happen on the two-dimensional paper; the dragon's hard work

Fig. 7.5 Escher, *Dragon*, 1952

is in vain. All these are "like a dream, an illusion, a bubble and a shadow, like dew and lightning."

Even it was about a hundred years ago, the hot debate about mathematical foundation, the genius proof by Gödel still have their practical significance today. As human beings, we are humbly in awe of the nature, the universe, our ancestors, and ourselves.

Appendix A
Answers

Answer of Exercise 1

1.1 It is a classic exercise to program a tick-tack-toe game. It is trivial to test if three numbers sum up to 15. Use this point to implement a simplified tick-tack-toe program that never loses.

We use Lo Shu Square to model the tick-tack-toe game as they are isomorphic (equivalent). We set up two sets X, O of the cells that each player has occupied. For the game in the preface, it starts with $X = \varnothing$, $O = \varnothing$, and ends with $X = \{2, 5, 7, 1\}$, $O = \{4, 3, 8, 6\}$. To determine if either player wins, we write a program to check if there are any three elements add up to 15.

There are two methods to do this checking. One is to list all the rows, columns, and diagonals, total eight tuples: $\{\{4, 9, 2\}, \{3, 5, 7\}, \ldots, \{2, 5, 8\}\}$. Then check if anyone is the subset of the cells occupied by a player. The other method is interesting. Suppose a player covers cells $X = \{x_1, x_2, \ldots, x_n\}$, sorted in ascending order as they appear in the Lo Shu Square. First pick x_1, and then let two pointers l, r point to the next element and the last element. If the sum of three numbers $s = x_1 + x_l + x_r$ equals to 15, it means the player lines up and wins. If it is less than 15, because the elements are in ascending order, we can increase the left pointer l by 1 and test again. Otherwise, if it is greater than 15, then we decrease the right pointer r by 1. When the left and right pointers meet, then no tuple adds up to 15 when fixing x_1. We next pick x_2 and do the similar thing. In the worst case, after $(n - 2) + (n - 3) + \ldots + 1$ checks, we know whether a player wins or not.

```
def win(s):
    n = len(s)
    if n < 3:
        return False
    s = sorted(s)
    for i in range(n - 2):
```

© China Machine Press 2024
X. Liu, *Mathematics in Programming*,
https://doi.org/10.1007/978-981-97-2432-1

```
            l = i + 1
            r = n - 1
            while l < r:
                total = s[i] + s[l] + s[r]
                if total == 15:
                    return True
                elif total < 15:
                    l = l + 1
                else:
                    r = r - 1
        return False
```

Given X and O, we test if the game is in end state—either a player wins or draws with all nine cells being occupied. Next we use the classic **min-max** method in AI to realize the game. For each state, we evaluate its score. One player tries to maximize the score, while the opponent tries to minimize it. Define the score as 0 for any draw state, 10 if player X wins, and -10 if player O wins. These numbers are arbitrary. They can be any other values without impacting the correctness.

```
WIN = 10
INF = 1000

# Lo Shu magic square
MAGIC_SQUARE = [4, 9, 2,
                3, 5, 7,
                8, 1, 6]

def eval(x, o):
    if win(x):
        return WIN
    if win(o):
        return -WIN
    return 0

def finished(x, o):
    return len(x) + len(o) >= 9
```

For any game state, let the program explore ahead till an end state (one side wins, or draw). The explore method is to cover all unoccupied cells exhaustively, then switch to the opponent player, and consider what is the best move to beat the other. For all candidate moves, select the highest score for player X, or the lowest score for player O.

```
def findbest(x, o, maximize):
    best = -INF if maximize else INF
    move = 0
    for i in MAGIC_SQUARE:
        if (i not in x) and (i not in o):
            if maximize:
                val = minmax([i] + x, o, 0, not maximize)
```

```
                    if val > best:
                        best = val
                        move = i
                else:
                    val = minmax(x, [i] + o, 0, not maximize)
                    if val < best:
                        best = val
                        move = i
        return move
```

The **min-max** search is recursive. To speed up, we take the number of steps into account. For payer X, subtract the recursion depth from the score, while for player O, add the depth to the score.

```
def minmax(x, o, depth, maximize):
    score = eval(x, o)
    if score == WIN:
        return score - depth
    if score == -WIN:
        return score + depth
    if finished(x, o):
        return 0  # draw
    best = -INF if maximize else INF
    for i in MAGIC_SQUARE:
        if (i not in x) and (i not in o):
            if maximize:
                best = max(best, minmax([i] + x, o, depth + 1,
                    not maximize))
            else:
                best = min(best, minmax(x, [i] + o, depth + 1,
                    not maximize))
    return best
```

We obtain a program that never loses to human player. It uses the Lo Shu Square behind essentially.

```
def board(x, o):
    for r in range(3):
        print "-----------"
        for c in range(3):
            p = MAGIC_SQUARE[r*3 + c]
            if p in x:
                print "|X",
            elif p in o:
                print "|O",
            else:
                print "| ",
        print "|"
    print "-----------"

def play():
```

```
x = []
o = []
while not (win(x) or win(o) or finished(x, o)):
    board(x, o)
    while True:
        i = int(input("[1..9]==>"))
        if i not in MAGIC_SQUARE or MAGIC_SQUARE[i-1] in x
            or \
            MAGIC_SQUARE[i-1] in o:
            print "invalid move"
        else:
            x = [MAGIC_SQUARE[i-1]] + x
            break
    o = [findbest(x, o, False)] + o
board(x, o)
```

Answer of Exercise 1.2

1.2.1 Define 1 as the successor of 0, and prove $a \cdot 1 = a$ holds for all natural numbers.

Proof We reuse the result: $0 + a = a$ proven in Appendix B:

$$a' \cdot 1 = a' \cdot 0' \qquad\qquad \text{1 as the successor of 0}$$
$$= a' \cdot 0 + a' \qquad\qquad \text{2nd rule of multiplication}$$
$$= 0 + a' \qquad\qquad \text{1st rule of multiplication}$$
$$= a'$$

$\square$

1.2.2 Prove the distribution of multiplication over addition.

Proof We only prove the left side distribution $c(a+b) = ca + cb$. By induction on b, when $b = 0$:

$$c(a + 0) = ca \qquad\qquad \text{1st rule of +}$$
$$= ca + 0 \qquad\qquad \text{1st rule of + (converse)}$$
$$= ca + c0 \qquad\qquad \text{1st rule of multiplication (converse)}$$

Next assume $c(a + b) = ca + cb$ holds, and we are going to show $c(a + b') = ca + cb'$

$$
\begin{aligned}
c(a + b') &= c(a + b)' && \text{2nd rule of +} \\
&= c(a + b) + c && \text{2nd rule of multiplication} \\
&= ca + cb + c && \text{induction assumption} \\
&= ca + (cb + c) && \text{associativity of +} \\
&= ca + cb' && \text{2nd rule of multiplication (converse)}
\end{aligned}
$$

□

1.2.3 Prove the associativity and commutativity of multiplication.
We only prove the associativity of $(ab)c = a(bc)$. For the commutativity, we provide a proof outline.

Proof Induction on c, when $c = 0$:

$$
\begin{aligned}
(ab)0 &= 0 && \text{1st rule of multiplication} \\
&= a0 && \text{1st rule of multiplication (converse)} \\
&= a(b0) && \text{1st rule of multiplication (converse)}
\end{aligned}
$$

Next assume $(ab)c = a(bc)$, and we are going to show $(ab)c' = a(bc')$:

$$
\begin{aligned}
(ab)c' &= (ab)c + ab && \text{2nd rule of multiplication} \\
&= a(bc) + ab && \text{induction assumption} \\
&= a(bc + b) && \text{distribution proven above} \\
&= a(bc') && \text{2nd rule of multiplication (converse)}
\end{aligned}
$$

□

To prove the commutativity of multiplication, we take three steps, all with mathematical induction. First we prove $1a = a$, then prove the right side distribution $(a + b)c = ac + bc$, and finally, prove the commutativity.

1.2.4 How to verify $3 + 147 = 150$ with Peano axioms?
Here is the classic proof of $2 + 2 = 4$:

$$
\begin{aligned}
2 + 2 &= 0'' + 0'' && \text{2 is the successor of successor of 0} \\
&= (0'' + 0')' && \text{2nd rule of +} \\
&= ((0'' + 0)')' && \text{2nd rule of +}
\end{aligned}
$$

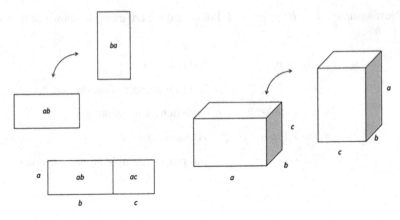

Fig. A.1 Geometric explanation for arithmetic laws

$$= ((0'')')' \qquad \text{1st rule of } +$$

$$= 0'''' = 4 \qquad \text{4 times successor of 0}$$

It will be too long to prove $3 + 147 = 150$ in this way. One solution is to apply the commutative law of addition and then prove $147 + 3 = 150$; another solution is to prove $3 + a = a'''$ with mathematical induction.

1.2.5 Give the geometric explanation for associativity, commutativity of multiplication, and distribution.

See Fig. A.1.

Answer of Exercise 1.3

1.3.1 Define square of natural number $()^2$ with *foldn*.

We define square from the iterative relation $(n + 1)^2 = n^2 + 2n + 1$

$$()^2 = 2nd \circ foldn \ h \ (0, 0)$$

where h accepts a pair (i, s), containing number i and its square s. It increases i by 1 and then uses the iterative relation to calculate the next square.

$$h \ (i, s) = (i + 1, s + 2i + 1)$$

1.3.2 Define $()^m$ with *foldn*, which gives the m-th power of a natural number.

One may tend to reuse the definition of $m^0 = foldn((\times m), 1)$ in Eq. (1.9):

$$()^m = 2nd \circ foldn \ h \ (0, 0)$$

where

$$h\,(i, b) = (i + 1, (i + 1)^m)$$

It looks tricky, as all the intermediate results are dropped. Alternatively, leverage Newton's binomial theorem:

$$(n + 1)^m = n^m + \binom{m}{1} n^{m-1} + \ldots + \binom{m}{m-1} n + 1$$

We can establish the iterative relation:

$$(n)^m = 2nd(foldn\ h\ (1, 1)\ (n - 1))$$

where

$$h(i, x) = (i + 1, C \cdot X)$$

$C \cdot X = \sum c_j x_j$ is the dot product between binomial coefficients and the powers. The powers can be calculated by repeatedly dividing x by i, and the binomial coefficients can be iterated from Pascal triangle. Below is an example program that puts them together:

```
exp m n = snd $ foldn h (1, 1) (n - 1) where
  cs = foldn pascal [1] m
  h (i, x) = (i + 1, sum $ zipWith (*) cs xs) where
    xs = take (m + 1) $ iterate ('div' i) x

pascal = gen [1] where
  gen cs (x:y:xs) = gen ((x + y) : cs) (y:xs)
  gen cs _ = 1 : cs
```

1.3.3 Define sum of odd numbers with *foldn*, what sequence does it produce? $1 + 3 + 5 + \ldots$ can be defined with *foldn* as $2nd \circ foldn\ h\ (1, 0)$, where:

$$h\,(i, s) = (i + 2, s + i)$$

As shown in Fig. 1.6, the sum of odd numbers is always a square number.

1.3.4 There is a line of holes in the forest. A fox hides in a hole. It moves to the next hole every day. If we can only check one hole a day, is there a way to catch the fox? Prove this method works. What if the fox moves more than one hole a day?

No matter which hole the fox hides in, we only examine the odd ones: 1, 3, 5, ... To show that we can always catch the fox, observe the below table:

Day	1	2	3	...	m
Check	1	3	5	...	$2m - 1$
Fox	m	$m + 1$	$m + 2$	...	$2m - 1$

Suppose the fox hides in the m-th hole on day one, solving equation $m + k = 2k + 1$ gives that, after $k = m - 1$ days, we will examine the $(2m - 1)$-th hole, while the fox moves exactly to it. Below *foldn* program demonstrates this process:

$$fox \ m = foldn \ h \ (1, m) \ (m - 1)$$
$$\text{where: } h \ (c, f) = (c + 2, f + 1) \tag{A.1}$$

If the fox hides in the p-th hole and moves to q holes every day, denote such pair as (p, q), then map them to the natural numbers with the method we introduced in Fig. 6.10. With this method, we enumerate all (p, q) combinations and catch the fox.

Answer of Exercise 1.4

1.4.1 What does the expression *foldr(cons, nil)* define?
It defines the list itself.
1.4.2 Read in a sequence of digits (string of digit numbers), and convert it into decimal with *foldr*. How to handle hexadecimal digit and number? How to handle the decimal point?
Assume the lowest digit is on the left, and the highest on the right of the input list, and convert it as below:

$$foldr \ (c \ d \mapsto 10d + c) \ 0$$

However, if the lowest digit is on the right, and the elements in the list are characters but not digit, then we need to adjust it to:

$$1st \circ foldr \ (c, (d, e) \mapsto ((toInt \ c)e + d, 10e)) \ (0, 1)$$

We make it process the hexadecimal numbers by replacing 10 to 16. When meeting the decimal point, divide d, the result so far, by e to calculate the fractional part.

$$1st \circ foldr \ h \ (0, 1)$$

where

$$h\,(c, (d, e)) = \begin{cases} c = `.' & (d/e, 1) \\ \text{otherwise} & ((toFloat\ c)e + d, 10e) \end{cases}$$

1.4.3 Jon Bentley gives the maximum sum of sub-vector puzzle in *Programming Pearls*. For integer list $[x_1, x_2, \cdots, x_n]$, find the range i, j that maximizes the sum of $x_i + x_{i+1} + \cdots + x_j$. Solve it with *foldr*.

If all numbers are positive, then the maximum sum is the sum of the whole list. This is because addition is monotone increasing upon positive numbers. If all numbers are negative, then the maximum sum should be zero, which is the sum of the empty list. For any sub-list, the sum increases when adding a positive number, while it decreases when adding a negative number. We maintain two things during folding: One is the maximum sum found so far S_m, and the other is the sum of the sub-lists till the current number being examined S. When adding the next number, if S exceeds S_m, it means we find a larger sub-list sum. Hence replace S_m with S; if S becomes negative, it means we complete the previous sub-list and should start a new one.

$$max_s = 1st \circ foldr\ f\ (0, 0)$$

where: $f\ x\ (S_m, S) = (S'_m = \max(S_m, S'), S' = \max(0, x + S))$. Here is the example program that implements this solution.

```
maxSum :: (Ord a, Num a) ⇒ [a] → a
maxSum = fst ∘ foldr f (0, 0) where
  f x (m, mSofar) = (m', mSofar') where
    mSofar' = max 0 (mSofar + x)
    m' = max mSofar' m
```

If want to return the sub-list together with the maximum sum, we need to maintain two pairs P_m and P during folding, and each pair contains the sum and the sub-list as (S, L).

$$max_s = 1st \circ foldr\ f\ ((0, [\,]), (0, [\,]))$$

where: $f\ x\ (P_m, (S, L)) = (P'_m = \max(P_m, P'), P' = \max((0, [\,]), (x + S, x{:}L)))$.

1.4.4 The longest sub-string without repeated characters. Given a string, find the longest sub-string without any repeated characters in it. For example, the answer for string "abcabcbb" is "abc." Solve it with *foldr*.

We give two methods. One solution is to maintain the longest sub-string of distinct characters during folding and record and check if the character c has met before and its last appeared position. If c never occurred, or it appears before the current sub-string we are examining, then append c to the current sub-string, and compare with the longest one we have found so far. Otherwise, it means the current sub-string contains a repeated character, and we need to go back to its last occurred position, move ahead one, and then restart searching.

$$longest(S) = fst2 \circ foldr\ f\ (0, |S|, |S|, \varnothing)\ zip(\{1, 2, \ldots\}, S)$$

where folding starts from a tuple of four elements. It contains the length of the longest sub-string we have found so far, the right boundary of the longest sub-string, the right boundary of the current sub-string, and a map recorded the last occurred position for different characters. Function $fst2$ extracts the first two elements from the tuple. To obtain the position of each character easily during folding, we zip the string S and natural number sequence together. The most critical function f is defined as below:

$$f\ (i, c)\ (n_{max}, e_{max}, e, Idx) = (n'_{max}, e'_{max}, e', Idx[c] = i)$$

where:

$$n'_{max} = \max(n_{max}, e' - i + 1)$$

$$e' = \begin{cases} c \notin Idx: & e \\ Idx[c] = j: & \min(e, j - 1) \end{cases}$$

$$e'_{max} = \begin{cases} e' - i + 1 > n_{max}: & e' \\ \text{otherwise}: & e_{max} \end{cases}$$

Here is an example program that implements this solution. It returns the maximum length and the *end* position of the sub-string.

```
longest :: String → (Int, Int)
longest xs = fst2 $ foldr f (0, n, n, Map.empty::(Map
                    Char Int)) (zip [1..] xs) where
  fst2 (len, end, _, _) = (len, end)
  n = length xs
  f (i, x) (maxlen, maxend, end, pos) =
      (maxlen', maxend', end', Map.insert x i pos) where
    maxlen' = max maxlen (end' - i + 1)
    end' = case Map.lookup x pos of
      Nothing → end
      Just j → min end (j - 1)
    maxend' = if end' - i + 1 > maxlen then end'
    else maxend
```

We record ending position because *foldr* starts from right. While the traditional way starts from the left, for example:

```
1: function LONGEST(S)
2:      Idx ← ∅
3:      n_max ← 0, s_max ← 0, s ← 0
4:      for i ∈ {0, 1, ... |S|} do
5:          if S[i] ∈ Idx then
6:              j ← Idx[S[i]]
7:              s = max(s, j + 1)
8:          if i − s + 1 > n_max then
9:              s_max ← s
10:         n_max ← max(n_max, i − s + 1)
11:         Idx[S[i]] = i
12:     return S[s_max ... s_max + n_max]
```

The second method is a number theory solution by leveraging prime numbers. We map each unique character c to a prime number p_c. For any string S, calculate the product of prime numbers mapped from its characters as below.

$$P = \prod_{c \in S} p_c$$

For any new character c', we check whether the corresponding prime number p' divides P or not to know if c' appears in S. Based on this, we design an algorithm. It maintains the product of primes during folding. When there is a character, that its corresponding prime number divides the product, then we find a repeated character. We need to drop off the part containing the repeated character and go on folding. During this process, we also need to update the longest sub-string.

$$longest = fst \circ foldr\ f\ ((0, [\,]), (0, [\,]), 1)$$

where the folding starts from a tuple of three elements. The first two are pairs, representing the longest sub-string (length and content) and the current substring. The last one in the tuple is the product of primes, starting from 1. Function f is defined as the following:

$$f\ c\ (m, (n, C), P) = \begin{cases} p_c | P : & update(m, (n + 1, c{:}C), p_c \times P) \\ otherwise : & update(m, (|C'|, C'), \prod_{x \in C'} p_x) \end{cases}$$

where:

$$update(a, b, P) = (\max(a, b), b, P)$$
$$C' = c : takeWhile \,(\neq c)\, C$$

Answer of Exercise 1.5

1.5.1 Starting from $(0, 1)^T$, Fibonacci numbers are isomorphic to natural numbers under the matrix multiplication:

$$\begin{pmatrix} F_n \\ F_{n+1} \end{pmatrix} = \begin{pmatrix} 0 & 1 \\ 1 & 1 \end{pmatrix}^n \begin{pmatrix} 0 \\ 1 \end{pmatrix}$$

Write a program to compute the power of 2×2 square matrix, and use it to give the n-th Fibonacci number.

First we need to define multiplication for 2×2 square matrix and the multiplication between matrix and vector:

$$\begin{pmatrix} a_{11} & a_{12} \\ a_{21} & a_{22} \end{pmatrix} \times \begin{pmatrix} b_{11} & b_{12} \\ b_{21} & b_{22} \end{pmatrix} = \begin{pmatrix} a_{11}b_{11} + a_{12}b_{21} & a_{11}b_{12} + a_{12}b_{22} \\ a_{21}b_{11} + a_{22}b_{21} & a_{21}b_{12} + a_{22}b_{22} \end{pmatrix}$$

and

$$\begin{pmatrix} a_{11} & a_{12} \\ a_{21} & a_{22} \end{pmatrix} \times \begin{pmatrix} b_1 \\ b_2 \end{pmatrix} = \begin{pmatrix} a_{11}b_1 + a_{12}b_2 \\ a_{21}b_1 + a_{22}b_2 \end{pmatrix}$$

When calculating the n-th power of M^n, we need not repeat multiplication n times. If $n = 4$, we can first calculate M^2 and then multiply the result to itself to obtain $(M^2)^2$. There are total two times of multiplication; if $n = 5$, we only need to compute $M^4 \times M$; hence there are three times of multiplication. We can recursively compute the exponential fast based on n's parity.

$$M^n = pow(M, n, I)$$

where $I = \begin{pmatrix} 1 & 0 \\ 0 & 1 \end{pmatrix}$ is the identity matrix, and the function *pow* is defined as below:

$$pow(M, n, A) = \begin{cases} n = 0: & A \\ n \text{ is even}: & pow(M \times M, \frac{n}{2}, A) \\ \text{otherwise}: & power(M \times M, \lfloor \frac{n}{2} \rfloor, M \times A) \end{cases}$$

In fact, we can represent n in binary format and then perform folding on these 0, 1 bit to fast compute M^n.

Answer of Exercise 2.1

2.1.1 Euclidean algorithm in Eq. (2.2) is recursive. Eliminate recursion, and implement it and the extended Euclidean algorithm only with loops.

Euclidean algorithm is tail recursive, and it can be converted into loop as below:

function GCM(a, b)
    while $b \neq 0$ **do**
        $a, b \leftarrow b, a \bmod b$
    return a

However, it is not obvious to convert the extended Euclidean algorithm. Observe the three sequences of r, s, t, where r is the remainder and $as + bt$ is the linear combination of Bézout identity.

$$r_0 = a, r_1 = b$$
$$s_0 = 1, s_1 = 0$$
$$t_0 = 0, t_1 = 1$$
$$\dots$$
$$r_{i+1} = r_{i-1} - q_i r_i, \quad \text{where}: q_i = \lfloor r_i / r_{i-1} \rfloor$$
$$s_{i+1} = s_{i-1} - q_i s_i$$
$$t_{i+1} = t_{i-1} - q_i t_i$$
$$\dots$$

Obviously, the sequences terminate when $r_{k+1} = 0$. From Euclidean algorithm:

$$gcm(a, b) = gcm(r_{k-1}, r_k) = gcm(r_k, 0) = r_k$$

Moreover, we show Bézout's identity holds:

$$\gcm(a, b) = r_k = as_k + bt_k$$

Proof By mathematical induction on i of r_i, for 0 and 1:

$$r_0 = a \qquad\qquad as_0 + bt_0 = a \cdot 1 + b \cdot 0 = a$$

$$r_1 = b \qquad\qquad as_1 + bt_1 = a \cdot 0 + b \cdot 1 = b$$

Assume $r_{i-1} = as_{i-1} + bt_{i-1}$ and $r_i = as_i + bt_i$ hold, for $i + 1$:

$$\begin{aligned}
r_{i+1} &= r_{i-1} - q_i r_i & \text{sequence definition} \\
&= (as_{i-1} + bt_{i-1}) - q_i(as_i + bt_i) & \text{induction assumption} \\
&= a(s_{i-1} - q_i s_i) + b(t_{i-1} - q_i t_i) & \text{rearrange} \\
&= as_{i+1} + bt_{i+1} & \text{sequence definition}
\end{aligned}$$

Hence the sequences always satisfy Bézout's identity.   □

Based on this, we define the non-recursive extended Euclidean algorithm as below.

> **function** Ext-Gcm(a, b)
> $s', s \leftarrow 0, 1$
> $t', t \leftarrow 1, 0$
>   **while** $b \neq 0$ **do**
> $q, r \leftarrow \lfloor a/b \rfloor, a \bmod b$
> $s', s \leftarrow s - qs', s'$
> $t', t \leftarrow t - qt', t'$
> $a, b \leftarrow b, r$
>   **return** (a, s, t)

2.1.2 Most programming environments require integers for modulo operation. However, the length of segment is not necessarily integer. Implement modulo operation that manipulates segments. What is about its efficiency? Analogue to the compass and edge construction, intercept segments with compass to get the remainder.

> **function** Mod(a, b)
>   **while** $b < a$ **do**
> $a \leftarrow a - b$
>   **return** a

The efficiency is linear. Below lemma provides a tool for optimization.

Lemma 2.2.3 (Recursive Remainder) *If* $r = (a \bmod 2b)$, *then:*

$$a \bmod b = \begin{cases} r \leq b : & r \\ r > b : & r - b \end{cases}$$

This lemma allows to speed up the modulo from linear to logarithmic time:

$$a \bmod b = \begin{cases} a \leq b : & a \\ a - b \leq b : & a - b \\ \text{otherwise} : & \begin{cases} a' \leq b : & a' \quad \text{where } a' = a \bmod (b + b) \\ a' > b : & a' - b \end{cases} \end{cases}$$

Inspired by Fibonacci numbers, Robot Floyd and Donald Knuth managed to eliminate the recursion and developed a purely iterative modulo operation:

function MOD(a, b)
    if $a < b$ **then**
        return a
    $c \leftarrow b$
    while $c \leq a$ **do**
        $c, b \leftarrow (b + c, c)$ ▷ Increase c like Fibonacci numbers
    while $b \neq c$ **do**
        $c, b \leftarrow (b, c - b)$ ▷ Decrease c back
        if $c <= a$ **then**
            $a \leftarrow a - c$
    return a

2.1.3 In the proof of Euclidean algorithm, we mentioned $b > r_0 > r_1 > r_2 > \cdots \geq 0$. As the remainder can not be negative and b is finite, we must reach to $r_{n-1} = r_n q_{n+1}$ within finite steps. Can r_n infinitely approximate zero, but not be zero? Does the algorithm always terminate? What does the precondition that a and b are commensurable ensure?

For the commensurable magnitudes, the *well-ordering principle* assumes Euclidean algorithm terminates. This principle says every nonempty set of natural numbers has the minimum number. It can be extended to the set of integers, rationals, or even to the finite, nonempty subset of real numbers. From the definition of commensurable, we know the remainders form a finite set.

2.1.4 For the linear Diophantine equation $ax + by = c$, let x_1, y_1 and x_2, y_2 be two pairs of integral solution. Proof that the minimum of $|x_1 - x_2|$ is $b/gcm(a, b)$, and the minimum of $|y_1 - y_2|$ is $a/gcm(a, b)$.

Let the greatest common divisor be $g = gcm(a, b)$. If x_0, y_0 is a pair of solution to $ax + by = c$, then the following x, y also forms a pair of solution:

$$\begin{cases} x = x_0 - k\dfrac{b}{g} \\ y = y_0 + k\dfrac{a}{g} \end{cases}$$

It is easy to verify:

$$ax + by = a(x_0 - k\frac{b}{g}) + b(y_0 - k\frac{a}{g})$$

$$= ax_0 + by_0 - ak\frac{b}{g} + bk\frac{a}{g}$$

$$= c - 0 = c$$

We next show that every solution is in this form. Let x, y be an arbitrary pair of solution. Since $ax + by = c$ and $ax_0 + by_0 = c$ both hold, therefore:

$$a(x - x_0) + b(y - y_0) = c - c = 0$$

Divide both sides with the greatest common divisor of a and b:

$$\frac{a}{g}(x - x_0) + \frac{b}{g}(y - y_0) = 0$$

$$\frac{b}{g}(y - y_0) = -\frac{a}{g}(x - x_0)$$

Because $\dfrac{b}{g}$ divides the left side, it divides the right side too. But $(\dfrac{a}{g}, \dfrac{b}{g}) = 1$, i.e., they are coprime; hence $\dfrac{b}{g}$ must divide $(x - x_0)$. Let $x - x_0 = k\dfrac{b}{g}$ for some $k \in \mathbb{Z}$, and change it to:

$$x = x_0 + k\frac{b}{g}$$

Substitute it back to above equation to get:

$$y = y_0 - k\frac{a}{g}$$

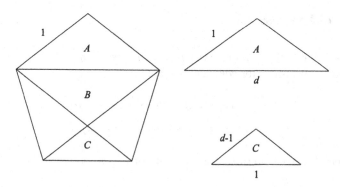

Fig. A.2 The unit regular pentagon

Proves every pair of solution is in this form. Obviously, for any two such pairs, the minimized difference is obtained when $k = 1$. Hence the minimum of $|x_1 - x_2|$ is $b/gcm(a, b)$, and the minimum of $|y_1 - y_2|$ is $a/gcm(a, b)$.   □

2.1.5 For the regular pentagon with side of 1, how long is the diagonal? Show that in the pentagram in Fig. 2.7a, segment AC and AG are incommensurable. What is their ratio in real number?

The solution given here is from Lockhart [70]. As shown in Fig. A.2, we divide the regular pentagon into three triangles. It is easy to show that triangles A and B are congruent, and they are similar to triangle C (can you prove it?). The length of the side is 1, let the diagonal length be d, then the base of triangle C is 1, and its two other edges both are $d - 1$. From the similar triangles:

$$1/d = (d - 1)/a$$

Solving this quadratic equation gives $d = \frac{\sqrt{5}+1}{2}$. We drop the other $d = \frac{\sqrt{5}-1}{2}$ for it is shorter than the side (in fact, it is the length of the hypotenuse of triangle C).

For the segments AC and AG in Fig. 2.7a, map to the three triangles we divided, they are actually the side and the diagonal of the pentagon. Assume they are commensurable, then the base and hypotenuse of triangle C are commensurable. In the recursive inner pentagram, the side and diagonal are also commensurable. We could repeat this infinitely. Therefore, our assumption cannot hold. The side and diagonal are incommensurable. The ratio is about $0.6180339887498949\ldots$ in decimals.

Answer of Exercise 2.2

2.2.1 Use λ conversion rules to verify $tail \ (cons \ p \ q) = q$.
The λ expressions for $cons$ and $tail$ are

$$cons = a \mapsto b \mapsto f \mapsto f \ a \ b$$

$$tail = c \mapsto c \ (a \mapsto b \mapsto b)$$

From the two definitions, we can verify $tail \ (cons \ p \ q) = q$ holds.

$$tail \ (cons \ p \ q) = (c \mapsto c \ (a \mapsto b \mapsto b)) \ (cons \ p \ q)$$

$$\xrightarrow{\beta} (cons \ p \ q) \ (a \mapsto b \mapsto b)$$

$$= ((a \mapsto b \mapsto f \mapsto f \ a \ b) \ p \ q) \ (a \mapsto b \mapsto b)$$

$$\xrightarrow{\beta} ((b \mapsto f \mapsto f \ p \ b) \ q) \ (a \mapsto b \mapsto b)$$

$$\xrightarrow{\beta} (f \mapsto f \ p \ q) \ (a \mapsto b \mapsto b)$$

$$\xrightarrow{\beta} (a \mapsto b \mapsto b) \ p \ q$$

$$\xrightarrow{\beta} (b \mapsto b) \ q$$

$$\xrightarrow{\beta} q$$

2.2.2 We can define numbers with λ calculus. The following are called Church numbers:

0	$\lambda f. \lambda x. x$
1	$\lambda f. \lambda x. f \ x$
2	$\lambda f. \lambda x. f \ (f \ x)$
3	$\lambda f. \lambda x. f \ (f \ (f \ x))$
...	...

Define $+$ and $\times$ for Church numbers.
As a Church number, n means to apply function f to x by n times. First define successor as:

$$succ = \lambda n. \lambda f. \lambda x. f \ (n \ f \ x)$$

It means $f^{n+1}(x) = f(f^n(x))$. Then define addition as:

$$plus = \lambda m. \lambda n. \lambda f. \lambda x. m \ f \ (n \ f \ x)$$

It means $f^{m+n}(x) = f^m(f^n(x))$. While the multiplication is defined as:

$$mul = \lambda m.\lambda n.\lambda f.\lambda x.m \ (n \ f) \ x$$

it means $f^{mn} = (f^n)^m(x)$.

2.2.3 Below are Church Boolean values and the relative logic operators:

true	$\lambda x.\lambda y.x$
false	$\lambda x.\lambda y.y$
and	$\lambda p.\lambda q.p \ q \ p$
or	$\lambda p.\lambda q.p \ p \ q$
not	$\lambda p.p$ **false true**

where **false** is defined as same as Church number 0. Use the λ conversion rules to (1) prove: **and true false = false**; (2) define "if ... then ... else ..." expression.

$$\textbf{and true false} = (\lambda p.\lambda q.p \ q \ p) \ \textbf{true false}$$

$$\xrightarrow{\beta} \textbf{true false true}$$

$$= (\lambda x.\lambda y.x) \ \textbf{false true}$$

$$\xrightarrow{\beta} \textbf{false} \qquad\qquad \Box.$$

Define "if ... then ... else ..." as: $\lambda p.\lambda a.\lambda b.p \ a \ b$.

Answer of Exercise 2.3

2.3.1 Define *depth*, which counts for the maximum depth of a binary tree.

$$depth = foldt(one, x, d, y \mapsto d + max(x, y), 0)$$

where $one(x) = 1$; alternatively:

$$depth = foldt(id, (x, k, y \mapsto 1 + max(x, y)), 0)$$

where $id \ x = x$ is the identity map.

2.3.2 Define *toList*, which converts a binary tree into list. It is also known as the "flatten" operation.

If the tree is empty, then the list is [ ]; otherwise, recursively flatten the left and right, and then put the key x in the middle:

$$toList\ nil\quad = [\,]$$
$$toList\ (l, x, r) = toList\ l \mathbin{+\!\!\!+} [x] \mathbin{+\!\!\!+} toList\ r$$

Alternatively, use *foldt* to define *toList*:

$$toList = foldt\ id\ (as\ b\ bs \mapsto as \mathbin{+\!\!\!+} b{:}bs)\ [\,]$$

2.3.3 Someone thought the abstract fold operation for binary tree, *foldt*, should be defined as the following:

$$foldt(f, g, c, nil)\qquad\qquad = c$$
$$foldt(f, g, c, node(l, x, r)) = foldt(f, g, g(foldt(f, g, c, l), f(x)), r)$$

That is to say $g : (B \times B) \to B$ is a binary operation like add. Can we use this *foldt* to define *mapt*?

This *foldt* cannot define tree mapping. A tree should be mapped to another tree with the same structure. Each value in the tree is sent to another value. Note the type of f is $f : A \to B$, and it sends the element of type A in the tree to type B. While the type of g is $g : (B \times B) \to B$, it only maps values of B but cannot preserve the tree structure.

2.3.4 The binary search tree (BST) is a special tree that the type A is comparable. For any nonempty $node(l, k, r)$, all elements in the left sub-tree l are less than k, and all elements in the right sub-tree r are greater than k. Define function $insert(x, t) : (A \times Tree\ A) \to Tree\ A$ that inserts an element into the tree.

$$insert(x, nil)\qquad\quad = node(nil, x, nil)$$
$$insert(x, node(l, k, r)) = \begin{cases} x < k : & node(insert(x, l), k, r) \\ x > k : & node(l, k, insert(x, r)) \end{cases}$$

We ignore the case $x = k$, as it inserts a duplicated key.

2.3.5 Can we define mapping for multi-trees with folding? If not, how should we modify *foldm*?

Similar to Exercise 2.3.3, we need to modify *foldm* to preserve the tree structure:

$$foldm(f, g, c, nil)\qquad\quad = c$$
$$foldm(f, g, c, node(x, ts)) = g(f(x), map(foldm(f, g, c), ts))$$

where *map* applies to list. With this tree folding tool, we can define multi-tree map:

$$mapm(f) = foldm(f, node, nil)$$

Answer of Exercise 2.4

2.4.1 What are the area and circumference of the Koch snowflake?

Let the side of the equilateral triangle be a, and then the area is $s = \frac{\sqrt{3}}{4}a^2$. In every recursion, each side is transformed into four shorter segments (of the 1/3 length). The below table lists the number of recursions, sides, and incremental areas:

Recursion	0	1	2	...	n	...
Number of sides	3	12	48	...	3×4^n	...
Side length	a	$\frac{a}{3}$	$\frac{a}{9}$	...	$\frac{a}{3^n}$ ...	
Incremental area		$\frac{s}{3}$	$\frac{4s}{27}$	...	$3 \times 4^{n-1}\frac{s}{9^n}$	...

The total area is s plus all the incremental areas:

$$s + \frac{s}{3} + \frac{4s}{27} + \cdots + \frac{3s}{4}(\frac{4}{9})^n + \ldots$$

$$= s + \frac{3s}{4}\frac{4}{9} + \frac{3s}{4}(\frac{4}{9})^2 + \cdots + \frac{3s}{4}(\frac{4}{9})^n + \ldots \quad \text{a geometric sub-series of } \frac{4}{9}$$

$$= \frac{s}{4} + \frac{3s}{4}\frac{1}{1 - \frac{4}{9}} \qquad\qquad\qquad \text{change } s \text{ to } \frac{s}{4} + \frac{3s}{4}$$

$$= \frac{8s}{5}$$

While the circumference equals to the product of side length and the number of sides, which is $3a(\frac{4}{3})^n$. It diverges. Koch snowflake demonstrates an infinite curve encloses finite area.

Answer of Exercise 3.1

3.1.1 Write a program to test whether a binary tree is reflexive symmetric.

$$reflexive\ \varnothing \qquad = True$$
$$reflexive\ (l, k, r) = eq\ l\ r$$

where:

$$eq\ \varnothing\ \varnothing \qquad\qquad\quad = True$$
$$eq\ \varnothing\ (l, k, r) \qquad\quad = False$$
$$eq\ (l, k, r)\ \varnothing \qquad\quad = False$$
$$eq\ (l_1, k_1, r_1)\ (l_2, k_2, r_2) = k_1 = k_2\ \textbf{and}\ (eq\ l_1\ l_2)\ \textbf{and}\ (eq\ r_1\ r_2)$$

Answer of Exercise 3.2

3.2.1 Do all the even numbers form a group under addition?
Yes. For closure, even + even gives even; + is associative; the identity is 0; the inverse of a number is its negate.

3.2.2 Can we find a subset of integers that forms a group under multiplication?
The subset $\{-1, 1\}$ forms a group under multiplication. The identity is 1; each is the reverse of itself.

3.2.3 Do all the positive real numbers form a group under multiplication?
Yes, positive real numbers are close under multiplication; the identity is 1; for each r, the reverse is $1/r$.

3.2.4 Do integers form a group under subtraction?
No, because subtraction is not associative, e.g., $(3 - 2) - 1 = 0$, while $3 - (2 - 1) = 2$.

3.2.5 Find a group with only two elements.
As shown in Exercise 3.2.2, $\{-1, 1\}$ forms a group under multiplication. For another example, Boolean values $\{T, F\}$ form a group under logic exclusive or (xor). xor is closed and associative. F is the identity because every element xor F gives itself; each is the reverse of itself.

3.2.6 What is the identity element for Rubik cube group? What is the inverse of F?
It is the identity transform, which keeps any state unchanged. The reverse of F is F', which rotates the face side 90 degree counterclockwise.

Answer of Exercise 3.3

3.3.1 The set of Boolean values {True, False} forms a monoid under the logic-or (∨). It is called the "Any" logic monoid. What is the identity element for this monoid?

False

3.3.2 The set of Boolean values {True, False} forms a monoid under the logic-and (∧). It is called the "All" logic monoid. What is the identity element for this monoid?

True

3.3.3 For the comparable type, when comparing two elements, there can be four different results. We abstract them as $\{<, =, >, ?\}$. The first three are deterministic; the last "?" means undetermined. For this set, we can define a binary operation to make it a monoid. What is the identity element for this monoid?

For the binary operation (·), define below Cayley table:

·	<	=	>	?
<	<	<	?	?
=	<	=	>	?
>	?	>	>	?
?	?	?	?	?

These four relations form a monoid; the identity element is $=$.

3.3.4 Show that the power operation for group, monoid, and semigroup is commutative: $x^m x^n = x^n x^m$.

In order to show the power is commutative, we first prove a lemma: $x^n x = x x^n$ by induction on n. For group and monoid, when $n = 0$:

$$x^0 x = ex = x = xe = x x^0$$

As there is no identity in semigroup, we start from $n = 1$:

$$x^1 x = xx = xx^1$$

Assume $x^n x = xx^n$ holds for n, for $n + 1$:

$$
\begin{aligned}
x^{n+1}x &= (xx^n)x && \text{recursive definition of power} \\
&= x(x^n x) && \text{associativity} \\
&= x(xx^n) && \text{induction assumption} \\
&= xx^{n+1} && \text{recursive definition of power}
\end{aligned}
$$

With this lemma, we apply mathematical induction again to show the commutativity for power. For group and monoid, when $n = 0$:

$$x^m x^0 = x^m e \qquad\qquad \text{definition of 0-th power}$$
$$= x^m \qquad\qquad \text{identity axiom}$$
$$= e x^m \qquad\qquad \text{identity axiom}$$
$$= x^0 x^m \qquad\qquad \text{definition of 0-th power}$$

Because semigroup does not have identity, we start from $n = 1$:

$$x^m x^1 = x^m x \qquad\qquad \text{definition of 1-st power}$$
$$= x x^m \qquad\qquad \text{the lemma}$$
$$= x^1 x^m \qquad\qquad \text{definition of 1-st power}$$

Assume $x^m x^n = x^n x^m$ holds for n, for $n + 1$:

$$x^m x^{n+1} = x^m (x x^n) \qquad\qquad \text{recursive definition of power}$$
$$= (x^m x) x^n \qquad\qquad \text{associativity}$$
$$= x x^m x^n \qquad\qquad \text{the lemma}$$
$$= x (x^m x^n) \qquad\qquad \text{associativity}$$
$$= x (x^n x^m) \qquad\qquad \text{induction assumption}$$
$$= (x x^n) x^m \qquad\qquad \text{associativity}$$
$$= x^{n+1} x^m \qquad\qquad \text{recursive definition of power}$$

□

Answer of Exercise 3.4

3.4.1 Is the odd–even test function a homomorphism from the additive group of integers $(\mathbb{Z}, +)$ to the logic-and group of Boolean $(Bool, \wedge)$? What is about the multiplicative group of integers without zero?
The odd–even test function maps integers to true or false. We need to verify whether the homomorphism property $f(a) f(b) = f(a \cdot b)$ holds. When both a and b are odd: $odd(a) = odd(b) = True$; their sum is even;

hence $odd(a + b) = False$. However, the corresponding logic-and is true: $odd(a) \wedge odd(b) = True \neq False = odd(a + b)$; hence it is not a homomorphism.

There is a homomorphism from the multiplicative group of integers without zero to the logic-and group. We verify all the three cases as below:

1. a, b are both odd: $odd(a) = odd(b) = True$. Their product ab is also odd: $odd(ab) = True$; $odd(a) \wedge odd(b) = odd(ab)$.
2. a, b are both even: $odd(a) = odd(b) = False$. Their product ab is also even: $odd(ab) = False$; $odd(a) \wedge odd(b) = odd(ab)$.
3. a, b are old and even; let $odd(a) = True, odd(b) = False$. Their product ab is even: $odd(ab) = False$; $odd(a) \wedge odd(b) = odd(ab)$.

3.4.2 There is a homomorphism f from groups G to G', mapping $a \to a'$; does the order of a equal to the order of a'?

Their orders are the same. To prove it, let e' denote the image of the identity e and n be the order of a, i.e., $a^n = e$.

$$f(a^n) = f(e) = e'$$

Because f is a homomorphism:

$$f(a^n) = f(a) * f(a) \cdots f(a) \qquad\qquad \text{total } n$$
$$= a' * a' \cdots a' \qquad\qquad \text{total } n$$
$$= (a')^n$$

Therefore, $(a')^n = e'$; the orders of a and a' are both equal to n. □

3.4.3 Show that the identity element of transformation group is the identity transformation.

Assume the transformation $\epsilon' : a \to a^{\epsilon'} = \epsilon'(a)$ is the identity element. For any transformation τ, the identity axiom $\epsilon'\tau = \tau$ holds.

$$\tau : a \to a^\tau$$
$$\epsilon'\tau : a \to (a^{\epsilon'})^\tau$$

Because the transformations are bijective (Proposition 3.2.4), $a^{\epsilon'} = a = \epsilon'(a)$ for every a; ϵ' is the identity transformation. □

Answer of Exercise 3.5

3.5.1 Show that if the transform σ sends the i-th element to the j-th, i.e., $\sigma(i) = j$, then the k-cycle permutation is $(\sigma(i_1) \, \sigma(\sigma(i_1)) \, \ldots \, \sigma^k(i_1))$.

From the k-cycle definition:

$$i_1 \rightarrow i_2 \rightarrow i_3 \rightarrow \ldots \rightarrow i_{k-1} \rightarrow i_k \rightarrow i_1$$

expand to:

$$i_2 = \sigma(i_1)$$

$$i_3 = \sigma(i_2) = \sigma(\sigma(i_1)) = \sigma^2(i_1)$$

$$\ldots$$

$$i_k = \sigma(i_{k-1}) = \sigma(\sigma(i_{k-2})) = \sigma^{k-1}(i_1)$$

$$i_1 = \sigma(i_k) = \sigma^k(i_1)$$

□

3.5.2 List all the elements in S_4.

(1);
(12), (34), (13), (24), (14), (23);
(123), (132), (134), (143), (124), (142), (234), (243);
(1234), (1243), (1324), (1342), (1423), (1432);
(12)(34), (13)(24), (14)(23).
total $4! = 24$ elements.

3.5.3 Write a program to convert the product of k-cycles back to permutation.
We need to input a list of k-cycles and n, the number of elements. n helps to deal with the ambiguity case, like (1).

```
1: function PERMUTE(C, n)
2:     π ← [1, 2, ..., n]
3:     for c ∈ C do
4:         j ← c[1]
5:         m ← |c|
6:         for i ← 2 to m do
7:             π[j] ← c[i]
8:             j ← c[i]
9:         π[c[m]] ← c[1]
10:    return π
```

Answer of Exercise 3.6

3.6.1 Analogous to $S3$ and equilateral triangle, what symmetry does S_4 describe?
S_4 describes the symmetry of a tetrahedron: (1) Proper congruence: As shown in Fig. A.3, a tetrahedron has four vertexes (corners), each with an axis.

Fig. A.3 The rotation and
reflection axes

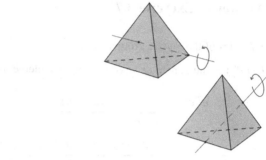

Fig. A.4 Improper
congruence

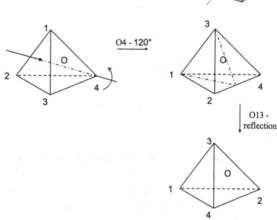

Rotations around these axes by $120°$ and $240°$ are symmetric. Connect the middle point of every two opposite edges and it also gives an axis. There are total 3 such axes, and each corresponds to a reflective symmetry. Including the identity transformation, there are total $2 \times 4 + 3 + 1 = 12$ symmetries. They form a group called A_4, an alternating group (Sect. 3.2.7).

For the tetrahedron 1234 in Fig. A.4, A_4 transforms it into:

3124, 2314, 1423, 1342, 4213, 3241, 4132, 2431, 2143, 3412, 4321, 1234

(2) Improper congruence: Every improper congruence corresponds to a proper one. For example in Fig. A.4, first rotate $120°$ against the axis; the tetrahedron transforms into 3124 on the right (proper); then reflect against plane $O13$; it transforms into the bottom-right 3142 (improper).

The reflexive corresponding to A_4 is one to one, in total 12:

3142, 2341, 1432, 1324, 4231, 3214, 4123, 2413, 2134, 3421, 4312, 1243

Although there are other two reflection planes, they do not generate new transformations. As there are $4! = 24$ permutations of four points, all the 24 transformations form symmetry group S_4 under composition.

Answer of Exercise 3.7

3.7.1 Proof that cyclic groups are abelian.

Proof Let the generator be a. For any two elements of a^p, a^q:

$$a^p a^q = \underbrace{(aa \cdots a)}_{p} \underbrace{(aa \cdots a)}_{q}$$

$$= \underbrace{(aa \cdots a)}_{q} \underbrace{(aa \cdots a)}_{p} \qquad \text{associativity, regroup}$$

$$= a^q a^p$$

or in a compact way, integers adding is commutative: $p + q = q + p$.

$$a^p a^q = a^{p+q} = a^{q+p} = a^q a^p$$

shows the group abelian. $\square$

3.7.2 Write a program for polynomial long division.
Represent polynomial $a(x) = a_0 x^n + a_1 x^{n-1} + \cdots + a_n x + a_{n+1}$ as a list
of coefficients: $[a_0, a_1, \ldots, a_{n+1}]$, where the leading $a_0 \neq 0$, and we pad 0
for any missing coefficient after a_0. For integer coefficients, the polynomial
division means

$$a(x) = q(x)b(x) + r(x)$$

where $q(x)$ is the quotient, $r(x)$ is the remainder, and the degree of $r(x)$ is
less than the degree of $b(x)$. Assume the degree of $a(x)$ is greater than the
degree of $b(x)$. Below is the example program in Haskell:

```
polyndiv as bs = pdiv as (length as - length bs) [] where
   pdiv as i qs = if i == 0 then (reverse (q:qs), as')
                     else pdiv as' (i - 1) (q:qs) where
     q = head as 'div' head bs
     as' = dropWhile (== 0) [a - b*q |
                           (a, b) <- zip as (bs ++ [0,
                             0..])]
```

3.7.3 Factor $x^{12} - 1$ to cyclotomic polynomials.

$$x^{12} - 1 = \Phi_1(x)\Phi_2(x)\Phi_3(x)\Phi_4(x)\Phi_6(x)\Phi_{12}(x)$$

$$= (x - 1)(x + 1)(x^2 + x + 1)(x^2 + 1)(x^2 - x + 1)(x^4 - x + 1)$$

3.7.4 Solve the quintic cyclotomic equation in radicals. Hint: The equation $x^4 +$ $x^3 + x^2 + x + 1 = 0$ changes to symmetric after being divided by x^2 as $x^2 + x + 1 + x^{-1} + x^{-2} = 0$.

Let $y = x + \frac{1}{x}$, and the original equation converts into:

$$y^2 + y - 1 = 0$$

Solving this quadratic equation gives:

$$y = \frac{-1 \pm \sqrt{5}}{2}$$

Then solve the quadratic equation of x:

$$x = \frac{\sqrt{5} - 1 \pm \sqrt{-2\sqrt{5} - 10}}{4}, \frac{-\sqrt{5} - 1 \pm \sqrt{2\sqrt{5} - 10}}{4}$$

Answer of Exercise 3.8

3.8.1 Prove that a nonempty subset H of group G forms a subgroup if and only if:

(a) For any a, b in H, the product $ab \in H$.
(b) For every a in H, its inverse $a^{-1} \in H$.

Proof H is closed because of (a). The associativity holds in G, hence also holds in H. Since H is not empty, there is an element $a \in H$, and (b) asserts a^{-1} is in H. Further (a) asserts their products $aa^{-1} = e \in H$. H satisfies all group axioms. Conversely, if H is a subgroup, (a) is true obviously. For (b), H contains the identity element e as a group; for every a in H, the inverse axiom asserts $a^{-1} \in H$. □

3.8.2 Show that if a nonempty subset H of a finite group G is a subgroup if and only if it is closed: i.e., for all $a, b \in H$, $ab \in H$.

Proof H as a subgroup is nonempty since it contains the identity element at least; the group axiom asserts closeness.
Conversely, assume H is nonempty; for any $a \in H$, a^n is in H for any positive n because H is closed. Since G is finite, the order of a is finite; there is some m, such that $a^m = e$; hence the identity element is in H; because $aa^{m-1} = e = a^{m-1}a$, the inverse $a^{-1} = a^{m-1}$ is in H; H satisfies all group axioms; hence it is a subgroup. □

3.8.3 Verify A_n is a subgroup of S_n.

With above exercise, we can easily show A_n is a subgroup of S_n. For any two even permutations, their product is even. A_n is closed. □

Answer of Exercise 3.9

3.9.1 Today is Sunday, what day will it be after 2^{100} days?

There are seven days in a week. By Fermat's little theorem $2^{7-1} \equiv 2^6 \equiv 1$ (mod 7).

$$2^{100} = 2^{16 \times 6 + 4} \equiv 1 \times 2^4 \ (\text{mod } 7)$$

$$\equiv 16 \equiv 2 \ (\text{mod } 7)$$

It will be Tuesday.

3.9.2 Given two strings (or list), write a program to test if they form the same necklace.

Let S_1, S_2 be two strings of the same length. Duplicate S_1 and append to itself, and then check whether S_2 is the sub-string of $S_1 S_1$. If yes, then they form the same necklace.

$$eqiv(S_1, S_2) = S_2 \subset (S_1 + S_1)$$

3.9.3 Weil thought Fermat might use the binomial theorem to derive the little theorem [27]. Prove the Fermat's little theorem in this way.

Proof Induction on a. When $a = 1$, p divides $a^p - a = 0$; assume p divides $(a^p - a)$, for $a + 1$:

$$(a + 1)^p - (a + 1)$$

$$= a^p + \binom{p}{1} a^{p-1} + \cdots + \binom{p}{p-1} a + 1 - a - 1 \quad \text{binomial theorem}$$

$$\equiv a^p - a \qquad\qquad\qquad\qquad\qquad\qquad\qquad p \text{ divides } \binom{p}{i}$$

$$\equiv 0 \qquad\qquad\qquad\qquad\qquad\qquad\qquad\qquad \text{induction hypothesis}$$

□

3.9.4 Write a program to realize Eratosthenes sieve.

Start from 2, pick the next as a prime, and then remove all its multiples repeatedly. Below is an example program in Haskell:

```
primes = sieve [2..] where
sieve (x:xs) = x : sieve [y | y ← xs, y 'mod' x > 0]
```

Below example programs in Python and Java take the first 100 prime numbers:

```python
def odds():
    i = 3
    while True:
        yield i
        i = i + 2

class prime_filter(object):
    def __init__(self, p):
        self.p = p
        self.curr = p

    def __call__(self, x):
        while x > self.curr:
            self.curr += self.p
        return self.curr != x

def sieve():
    yield 2
    iter = odds()
    while True:
        p = next(iter)
        yield p
        iter = filter(prime_filter(p), iter)

list(islice(sieve(), 100))
```

```java
public class Prime {
    private static LongPredicate sieves = x -> true; // init
sieve
    public final static long[] PRIMES = LongStream
        .iterate(2, i -> i + 1)
        .filter(i -> sieves.test(i))
        .peek(i -> sieves = sieves.and(v -> v % i != 0)) //
update
        .limit(100)                        // take first 100
        .toArray();
}
```

3.9.5 Extend the idea of Eratosthenes sieve algorithm, and write a program to generate Euler ϕ function values for all numbers from 2 to 100.
When running sieve of Eratosthenes within n, we use every prime number to update the list of Euler's ϕ values. The list is initialized as $[1, 1, \ldots, 1]$

of length n. For every prime p, its totient $\phi(p) = p(1 - \frac{1}{p}) = p - 1$. We multiply this value to all multiples of p. However, it is not enough as $\phi(p^2) = p^2(1 - \frac{1}{p}) = p\phi(p)$. We need to go on multiplying p to all values that are multiples of p^2 and repeat for multiples of p^3 and so on until $p^m > n$. Below algorithm implements this idea.

```
1: function EULER-TOTIENT(n)
2:     φ ← [1, 1, ..., 1]                          ▷ 1 to n
3:     P ← [2, 3, ..., n]                          ▷ sieve input
4:     while P ≠ [ ] do                            ▷ not empty
5:         p ← P[0]                                ▷ pick the next prime
6:         P ← [x|x ∈ P[1...], 0 ≠ x mod p]        ▷ apply sieve
7:         p' ← p
8:         repeat
9:             for i ← from p' to n step p' do
10:                if p' = p then
11:                    φ[i] ← φ[i] × (p − 1)
12:                else
13:                    φ[i] ← φ[i] × p
14:            p' ← p' × p
15:        until p' > n
16:    return φ
```

3.9.6 From Euler's theorem, how many n-th primitive roots of unity are?
$\phi(n)$, if n is prime, then all the roots except 1 are primitive.

3.9.7 Write a program to realize fast modulo multiplication and Fermat's primality test.
Analogous to fast power algorithm below:

$$x^y = \begin{cases} y \text{ is even}: & x^{\lfloor y/2 \rfloor} \\ y \text{ is odd}: & x \cdot x^{\lfloor y/2 \rfloor} \end{cases}$$

Here is the modulo multiplication:

```
1: function MOD-EXP(x, y, n)
2:     if y = 0 then
3:         return 1
4:     z ← MOD-EXP(x, ⌊y/2⌋, n)
5:     if y is even then
6:         return z² mod n
7:     else
8:         return x · z² mod n
```

Fermat's primality test:

```
1: function PRIMALITY(n)
2:     random select k positive numbers a₁, a₂, ..., aₖ < n as A
3:     for a in A do
4:         if MOD-EXP(a, n − 1, n) ≠ 1 then
5:             return composite
6:     return prime
```

Answer of Exercise 3.10

3.10.1 Prove Proposition 3.3.1 that the cancellation law holds in a nonzero ring (ring without zero divisor).

Proof Assuming R has no zero divisor, if $a \neq 0$ and $ab = ac$, then:

$$ab = ac$$
$$a(b - c) = 0 \qquad \text{both sides - } ac \text{, and apply distribution law}$$
$$b - c = 0 \qquad a \neq 0 \text{ and no zero divisor}$$
$$b = c$$

symmetrically we can show $ba = ca \Rightarrow b = c$; both cancellation laws hold in R. Conversely, assume the first cancellation law holds in R. If $a \neq 0$ and $ab = 0$, then:

$$ab = 0 = a0 \qquad \text{cancel } a \text{ as } a \neq 0$$
$$b = 0$$

Shows R has no zero divisor. We can cancel from right for the second cancellation law. □

3.10.2 Prove that all real numbers in the form of $a + b\sqrt{2}$, where a, b are integers form an integral domain under the normal addition and multiplication.

Proof We need to verify three things:

(1) $\times$ is commutative:

$$(a + b\sqrt{2})(c + d\sqrt{2}) = ac + 2bd + (ad + bc)\sqrt{2}$$
$$= (c + d\sqrt{2})(a + b\sqrt{2})$$

(2) The multiplicative identity exists:

$$1(a + b\sqrt{2}) = (a + b\sqrt{2})1 = a + b\sqrt{2}$$

(3) No zero divisor:

$$(a + b\sqrt{2})(c + d\sqrt{2}) = 0 \Rightarrow a = b = 0 \text{ or } c = d = 0$$

□

Answer of Exercise 3.11

3.11.1 Prove that $\mathbb{Q}(a, b) = \mathbb{Q}(a)(b)$, where $\mathbb{Q}(a, b)$ contains all the expressions combined with a and b, such as $2ab$, $a + a^2b$, etc.
For example: $\mathbb{Q}(\sqrt{2}, \sqrt{3}) = \{a + b\sqrt{2} + c\sqrt{3} + d\sqrt{6}\}$, where a, b, c, d in $\mathbb{Q}$

$$\mathbb{Q}(\sqrt{2})(\sqrt{3}) = \{a + b\sqrt{3}\} \qquad\qquad\qquad \text{where } a, b \in \mathbb{Q}(\sqrt{2})$$
$$= \{a' + b'\sqrt{2} + (c + d\sqrt{2})\sqrt{3}\} \qquad \text{where } a', b', c, d \in \mathbb{Q}$$
$$= \{a' + (b' + d)\sqrt{2} + c\sqrt{3} + d\sqrt{6}\} \quad \text{where } a', b', c, d \in \mathbb{Q}$$

Proof

$$\mathbb{Q}(a)(b) = \{x_0 + x_1 b + x_2 b^2 + \cdots + x_n b^n\} \quad \text{where } x_i \in \mathbb{Q}(a)$$

n is the minimum integer that polynomial $p(b) = 0$ exists. Substitute x_i with the expressions in field $\mathbb{Q}(a)$.

$$\mathbb{Q}(a)(b) = \{y_{0,0} + y_{0,1}a + y_{0,2}a^2 + \cdots + y_{0,m}a^m +$$
$$(y_{1,0} + y_{1,1}a + y_{1,2}a^2 + \cdots + y_{1,m}a^m)b +$$
$$\cdots +$$
$$(y_{n,0} + y_{n,1}a + y_{n,2}a^2 + \cdots + y_{n,m}a^m)b^n\}$$

where $y_{i,j} \in \mathbb{Q}$, and m is the minimum integer that polynomial $p(a) = 0$ exists. Assume $m < n$ (otherwise, let $m' = min(m, n)$, $n' = max(m, n)$ to swap), and further convert it into:

$$\mathbb{Q}(a)(b) = \{y_{0,0} + y_{0,1}a + y_{1,0}b + y_{0,2}a^2 + y_{1,1}ab + y_{2,0}b^2 + \cdots +$$
$$y_{0,m}a^m + y_{1,m-1}a^{m-1}b + \cdots + y_{m,0}b^m +$$

$$y_{1,m}a^m b + y_{2,m-1}a^{m-1}b^2 + \cdots + y_{m,1}b^{m+1} + \cdots +$$
$$y_{n,m}a^m b^n\}$$

This field contains all expressions of a, b.   □

Answer of Exercise 3.12

3.12.1 Prove Eq. (3.11), and give the relation between Lagrange resolvent and the coefficients of the original equation.

Proof Expand $(L + R)(L + \omega R)(L + \omega^2 R)$, and by $1 + \omega + \omega^2 = 0$:

$$(L + R)(L + \omega R)(L + \omega^2 R) = L^3 + R^3 \tag{A.2}$$

by Eq. (3.13):

$$\begin{cases} L + R & = 3r_3 \\ L + \omega R & = 3\omega^2 r_2 \\ L + \omega^2 R & = 3\omega r_1 \end{cases}$$

Substitute in Eq. (A.2):

$$\begin{aligned} L^3 + R^3 &= (L + R)(L + \omega R)(L + \omega^2 R) & \text{Eq. (A.2)} \\ &= (3r_3)(3\omega^2 r_2)(3\omega r_1) \\ &= 27r_1 r_2 r_3 & \omega^3 = 1 \\ &= -27q & \text{Vieta's theorem} \end{aligned}$$

Next evaluate LR:

$$\begin{aligned} LR &= (r_1 + \omega r_2 + \omega^2 r_3)(r_1 + \omega^2 r_2 + \omega r_3) & \text{definition of } L, R \\ &= (r_1^2 + r_2^2 + r_3^2) + (\omega + \omega^2)(r_1 r_2 + r_2 r_3 + r_1 r_3) & \text{expand} \\ &= (r_1^2 + r_2^2 + r_3^2) - (r_1 r_2 + r_2 r_3 + r_1 r_3) & 1 + \omega + \omega^2 = 0 \\ &= -2p - p & \text{Newton's notes} \\ &= -3p \end{aligned}$$

Therefore: $L^3 R^3 = -27p^3$.   □

3.12.2 Verify Galois group of the equation $x^3 - x^2 - 2x + 2 = 0$ is $\{(1), (23)\}$.
We skip the identity (1) and only verify permutation (23). It is easy to factor
the polynomial: $x^3 - x^2 - 2x + 2 = x^2(x - 1) - 2(x - 1) = (x - 1)(x^2 - 2)$.
It has three roots of $1, \pm\sqrt{2}$. From Sect. 3.4.1, all rational expressions of the
three roots are in the form of $r = a + b\sqrt{2}$, where a, b are rationals. If r is
known, then $b = 0$; $r = a$ is unchanged under the transpose of (23)(swap
$\pm\sqrt{2}$).
Conversely, permutation (23) swaps $\pm\sqrt{2}$, and it flips the sign of $\sqrt{2}$ in
expression $a + b\sqrt{2}$. If unchanged, then $a + b\sqrt{2} = a - b\sqrt{2}$; hence $b = 0$.
The rational number a is known.   □

Answer of Exercise 3.13

3.13.1 Prove that for any polynomial $p(x)$ with rational coefficients, $E/\mathbb{Q}$ is a
field extension, f is a $\mathbb{Q}$-automorphism of E, and then equation $f(p(x)) =
p(f(x))$ holds.

Proof Let $p(x) = a_0 + a_1x + \ldots + a_nx^n$, where a_i are rationals.

$$f(p(x)) = f(a_0 + a_1x + \cdots + a_nx^n)$$

$$= f(a_0) + f(a_1x) + \cdots + f(a_nx^n) \qquad f(x + y) = f(x) + f(y)$$

$$= f(a_0) + f(a_1)f(x) + \cdots + f(a_n)f(x)^n \quad f(ax) = f(a)f(x)$$

$$= a_0 + a_1f(x) + \cdots + a_nf(x)^n \qquad f(x) = x, \forall x \in \mathbb{Q}$$

$$= p(f(x))$$

□

3.13.2 What is the splitting field for polynomial $p(x) = x^4 - 1$, is it $\mathbb{C}$? What are
its $\mathbb{Q}$-automorphisms?
$x^4 - 1$ has four roots: $\pm 1, \pm i$; factor the polynomial to $p(x) = (x + 1)(x -
1)(x + i)(x - i)$. Actually, the splitting field of $p(x)$ is not $\mathbb{C}$; it is overly
big. The splitting field is $\mathbb{Q}(i)$.
There are two $\mathbb{Q}$-automorphisms: $f(a + bi) = a - bi$ and the identity
$g(x) = x$.
3.13.3 What is the Galois group for quadratic equation $x^2 - bx + c = 0$?
There are two roots:

$$x_1, x_2 = \frac{b \pm \sqrt{b^2 - 4c}}{2}$$

There are three cases: (1) There exists a rational number r, such that $b^2 - 4c = r^2$. The equation has two rational roots (including duplicated ones); (2) no such rational number; the equation is not solvable in rational field, but there are real roots; (3) the discriminant is negative; the equation is not solvable in real field; but there are complex roots. The corresponding Galois groups for these three cases are:

(1) Two rational roots: $\frac{b \pm r}{2}$. Its Galois group contains the identity automorphism $f(x) = x$ only.

(2) Two irrational roots: $\frac{b \pm \sqrt{d}}{2}$. Its Galois group has two automorphisms: the identity transformation and $f(p + q\sqrt{d}) = p - q\sqrt{d}$, where p, q are rationals.

(3) Two complex roots $\frac{b \pm i\sqrt{d}}{2}$. Its Galois group has two automorphisms: the identity and $f(p + qi) = p - qi$, where p, q are real numbers.

Actually, Galois groups in cases 2 and 3 are isomorphic (only the splitting fields are different). Because $f(f(x)) = x$, $f^2 = e$ has order of 2; they are also isomorphic to the group of two elements 0, 1 under the addition modulo 2 and isomorphic to the cyclic group C_2 or $\mathbb{Z}/2\mathbb{Z}$ (remind the notation of quotation group $\mathbb{Z}$ and its subgroup of even numbers $2\mathbb{Z}$).

3.13.4 Show that if p is a prime number, Galois group of equation $x^p - 1$ is cyclic group C_{p-1}.

Proof The p roots of $x^p - 1$ are the points along the unit circle in complex plane: $1, \omega, \omega^2, \ldots, \omega^{p-1}$; they can be expressed in $e^{2\pi ki/p}$. The splitting field is $\mathbb{Q}(\omega)$. Let automorphism f be an element in Galois group $Gal(\mathbb{Q}(\omega)/\mathbb{Q})$; by definition:

$$f(\omega)^k = f(\omega^k) = 1 \iff \omega^k = 1$$

It means $f(\omega)$ is also a p-th root of unity (a root of equation $x^p - 1 = 0^1$). Denote:

$$f(\omega) = h_i(\omega) = \omega^i$$

If $f(\omega)$ is an i-th root of unity, we name it as h_i, where $1 \le i \le p - 1$ (see below frame for why i cannot be 0). In this way, we establish a bijective (one-to-one) mapping from Galois group to cyclic group C_{p-1}:

$$Gal(\mathbb{Q}(\omega)/\mathbb{Q}) \xrightarrow{\sigma} C_{p-1} : \sigma(h_i) = i$$

---

[1] If p is not a prime number, but an integer n greater than 1, then the k-th power of the m-th root is $\zeta_m^k = e^{2\pi mki/n}$. There may exist some $k < n$, such that $\zeta_m^k = 1$. Actually, it holds as far as n divides mk. However, if n is a prime number p, then k cannot be less than p but must be some multiple of p.

where its Galois group contains $p-1$-automorphisms $\{h_1, h_2, \ldots, h_{p-1}\}$, the same order as the cyclic group C_{p-1}. We show this is a group isomorphism next:

$$(h_i \cdot h_j)(\omega) = h_i(h_j(\omega)) = h_i(\omega^j) = \omega^{ji} = \omega^{ij}$$

hence

$$\sigma(h_i \cdot h_j) = ij = \sigma(h_i) \cdot \sigma(h_j)$$

and $h_1(\omega)$ is the generator of this cyclic Galois group. □

Do not be confused with the two groups: The first one is the additive group of integers modulo n. It is a cyclic group, containing the residue classes modulo n of $\{0, 1, 2, \ldots, n-1\}$, total n elements. This group is often denoted as $\mathbb{Z}/n\mathbb{Z}$. It is isomorphic to the group of n roots of equation $x^n - 1 = 0$. The elements are the n-th root of unity $\{1 = \zeta_n^0, \zeta_n^1, \zeta_n^2, \ldots, \zeta_n^{n-1}\}$; the binary operation is multiplication among roots. The isomorphism is $i \mapsto \zeta_n^i$, maps 0 to 1, modulo addition to multiplication.

The second group is the multiplicative group of integers modulo n. The group elements are *not* all the residues from 0 to $n-1$, but those coprime to n. The binary operation is modulo multiplication, denoted as $(\mathbb{Z}/n\mathbb{Z})^\times$. There are total $\phi(n)$ elements, where ϕ is Euler's totient function. When n is a prime p, they are $\{1, 2, \ldots, p-1\}$, total $p-1$ elements. However, the multiplicative group modulo n is not necessary cyclic. Luckily it is cyclic when n is prime, and $(\mathbb{Z}/p\mathbb{Z})^\times$ is isomorphic to the additive group $\mathbb{Z}/(p-1)\mathbb{Z}$.

This exercise shows: For Galois group in rational field extension, if it is generated by n-th root of unity, then this group is isomorphic to the multiplicative group of integers modulo n, i.e., $(\mathbb{Z}/n\mathbb{Z})^\times$. For example, the cubic equation $x^3 - 1 = 0$ has three roots $\{1, \frac{-1 \pm i\sqrt{3}}{2}\}$. Its Galois group contains two automorphisms: the identity $f(x) = x$, which corresponds to $h_1(\omega) = \omega^1$, and $g(a + bi) = a - bi$, which corresponds to $h_2(\omega) = \omega^2$. The effect of h_2 is to transform three roots labeled from 1, 2, 3 to 1, 3, 2.

$$1 \mapsto 1^2 = 1$$

$$\omega \mapsto \omega^2$$

$$\omega^2 \mapsto (\omega^2)^2 = \omega^3 \omega = \omega$$

3.13.5 Let α be a root of equation $x^3 + x^2 - 4x + 1$, and verify that $2 - 2\alpha - \alpha^2$ is also a root. What is Galois group of this equation in rational field?

Let $f(x) = x^3 + x^2 - 4x + 1$; then $f(\alpha) = 0$. It is easy to verify $f(2 - 2\alpha - 2\alpha^2) = 0$. Next let $g(\alpha) = 2 - 2\alpha - 2\alpha^2$, and we verify the following:

$$g^2(\alpha) = g(2 - 2\alpha - 2\alpha^2)$$

$$= \alpha^3 + \alpha - 3 \qquad\qquad f(\alpha) = 0$$

$$g^3(\alpha) = g(\alpha^3 + \alpha - 3)$$

$$= \alpha \qquad\qquad f(\alpha) = 0$$

Therefore, $g^3 = e$, $G = \{e, g, g^2\}$, the cyclic group C_3.

Answer of Exercise 3.14

3.14.1 The quintic cyclotomic equation $x^5 - 1 = 0$ is radical solvable. What is its Galois group and what are the solvable group series?

The five roots along the unit circle in the complex plane are $\{1, \zeta, \zeta^2, \zeta^3, \zeta^4\}$, where $\zeta = e^{2\pi i/5} = \frac{\sqrt{5}-1+i\sqrt{10+2\sqrt{5}}}{4}$. Its Galois group in rational field is a cyclic group of order 4: $G(\mathbb{Q}(\zeta)/\mathbb{Q}) = C_4$. It is isomorphic to the multiplicative group modulo 5: $(\mathbb{Z}/5\mathbb{Z})^\times = \{1, 2, 3, 4\}_5$. There is no intermediate field extension. The splitting field is $\mathbb{Q}(\zeta)$. Its Galois group in splitting field is $\{e\}$. The group series are:

$$C_4 \rhd \{e\}$$

The quotient group $C_4/\{e\}$ is cyclic too. As cyclic group is abelian (by Exercise 3.7.1), the equation is radical solvable.

3.14.2 Show that, when $n \geq 5$, assume the subgroup $G \subset S_n$ contains all 3-cycles like $(a\ b\ c)$; if N is a normal subgroup of G, and the quotient group G/N is abelian, then N contains all 3-cycles too.

Proof Since the quotient group G/N is abelian, $g_1 N \cdot g_2 N = g_1 g_2 N = g_2 g_1 N$ for any $g_1, g_2 \in G$; multiply the inverse to the left: $g_1^{-1} g_2^{-1} g_1 g_2 N = N$; hence $g_1^{-1} g_2^{-1} g_1 g_2 \in N$.

For $n \geq 5$, there are at least five labels; for any 3-cycle $(a\ b\ c)$, choose another two different labels: d, f; let $g_1 = (d\ b\ a)$, $g_2 = (a\ f\ c) \in G$.

$$g_1^{-1} g_2^{-1} g_1 g_2 = (a\ b\ d)(c\ f\ a)(d\ b\ a)(a\ f\ c) = (a\ b\ c) \in N$$

shows N contains all 3-cycles too. $\qquad\qquad\square$

Answer of Exercise 4.1

4.1.1 Show that the identity arrow is unique for each object.

Proof Besides id_A, assume there exists another identity arrow id'_A, self-pointed:
$A \xrightarrow{id'_A} A$. For any arrow from A to B: $A \xrightarrow{f} B$, from the definition of identity arrow: $f \circ id_A = f$. Substitute B with A and f with id'_A:

$$id'_A \circ id_A = id'_A$$

Similarly, for any arrow from B to A: $B \xrightarrow{g} A$, by definition of identity arrow, $id'_A \circ g = g$ holds. Substitute B with A and g with id_A:

$$id'_A \circ id_A = id_A$$

Combine above two: $id_A = id'_A$; the identity arrow is unique for each object. $\square$

1.2.2 Verify the monoids $(S, \cup, \varnothing)$ (the elements are sets, the binary operation is set union, and the identity element is empty set) and $(\mathbb{N}, +, 0)$ (elements are natural numbers, the binary operation is add, and the identity element is zero) are all categories with only one object.

Every monoid is a category with only one object, while it is a bit difficult to answer: What is that object for each category? In fact, it does not matter what the object is; the object is not necessarily any element, monoid, or any subset. To avoid confusing with concrete object, we give it notation of ★.

For the monoid of sets under union, every set s as an element of S defines an arrow:

Note there is not any inner structures (of the monoid) involved. The arrow composition is exactly set union.

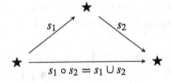

Since set union is associative, so as arrows composition. The empty set $\varnothing$ is the identity of this monoid; it defines the identity arrow, since the union of any set with the empty set equals to that set itself. $\varnothing$ serves as the identity arrow. The monoid of sets under union does form a category with one object.

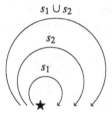

For the additive monoid for natural numbers, every number n defines an arrow:

$$\bigstar \xrightarrow{\ n\ } \bigstar$$

The arrow composition is $+$. Since $+$ is associative, so as arrow composition. 0 defines the identity arrow because the sum of 0 and any number equals to this number itself. The additive monoid of natural numbers does form a category.

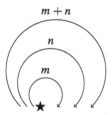

1.2.3 Chapter 1 introduces Peano's axioms for natural numbers and things with the same structure of nature numbers, like linked-list. They can be described in categories. This was found by Dedekind although the category theory was not established by his time. We named this category as Peano category, denoted as **Pno** nowadays. The objects in this category are (A, f, z), where A is a set, for example, natural numbers $\mathbb{N}$; $f : A \to A$ is a successor function. It is *succ* for natural numbers; $z \in A$ is the starting element, and it is zero for natural numbers. Given any two Peano objects (A, f, z) and (B, g, c), define morphism from A to B as:

$$A \xrightarrow{\ \phi\ } B$$

It satisfies:

$$\phi \circ f = g \circ \phi \quad \text{and} \quad \phi(z) = c$$

Verify that **Pno** is a category.

An object in Peano category is a tuple of (A, f, z). An arrow is a map ϕ that preserves the tuple structure. Arrow composition is function composition:

$$A \xrightarrow{\phi} B \xrightarrow{\psi} C$$

$$A \xrightarrow{\psi \circ \phi} C$$

Since function composition is associative, so as arrow composition. The identity arrow $A \xrightarrow{id_A} A$ satisfies $id_A(z) = z$ and $id_A \circ f = f \circ id_A$.   □
Obviously, the tuple $(\mathbb{N}, succ, 0)$ is an object in Peano category. Interestingly, for every object (A, f, z), there is a unique arrow:

$$(\mathbb{N}, succ, 0) \xrightarrow{\sigma} (A, f, z)$$

where:

$$\sigma(n) = f^n(z)$$

which maps any natural number n by applying f to z for n times.

Answer of Exercise 4.2

4.2.1 For the list functor, define the arrow map with *foldr*.
      It is about to define list map with *foldr* essentially:

$$fmap\ f = foldr\ (x\ ys \mapsto f(x):ys)\ [\ ]$$
$$= foldr\ ((:) \circ f)\ [\ ] \qquad\qquad \eta\text{-reduction}$$

4.2.2 Verify that the functor composition of **Maybe** ∘ **List** and **List** ∘ **Maybe** are all functors.

Proof We only prove **Maybe** ∘ **List** is a functor. The other proof is similar. It sends any object A to **Maybe**(**List** A). For arrows, we first show it sends identity to identity:

(**Maybe** ∘ **List**) id = **Maybe**(**List** id)	composition
= **Maybe** id	identity arrow for list functor
= id	identity arrow for maybe functor

When applies to composited arrows:

$(\textbf{Maybe} \circ \textbf{List})\ (f \circ g)$

$= \textbf{Maybe}(\textbf{List}\ (f \circ g))$ composition

$= \textbf{Maybe}((\textbf{List}\ f) \circ (\textbf{List}\ g))$ composition of list functor

$= (\textbf{Maybe}\ (\textbf{List}\ f)) \circ (\textbf{Maybe}\ (\textbf{List}\ g))$ composition of maybe functor

$= ((\textbf{Maybe} \circ \textbf{List})\ f) \circ ((\textbf{Maybe} \circ \textbf{List})\ g)$ composition

$\square$

4.2.3 Show that the composition for any functors $\textbf{G} \circ \textbf{F}$ is still a functor.

Proof Similar to previous exercise, we need to show the composed functor preserves identity and composition. First for identity arrow:

$(\textbf{G} \circ \textbf{F})\ id = \textbf{G}(\textbf{F}\ id)$ composition

$= \textbf{G}\ id$ $\textbf{F}$ preserves identity arrow

$= id$ $\textbf{G}$ preserves identity arrow

Second for arrow composition:

$(\textbf{G} \circ \textbf{F})\ (\phi \circ \psi) = \textbf{G}(\textbf{F}\ (\phi \circ \psi))$ composition

$= \textbf{G}((\textbf{F}\ \phi) \circ (\textbf{F}\ \psi))$ $\textbf{F}$ preserves arrow composition

$= (\textbf{G}\ (\textbf{F}\ \phi)) \circ (\textbf{G}\ (\textbf{F}\ \psi))$ $\textbf{G}$ preserves arrow composition

$= ((\textbf{G} \circ \textbf{F})\ \phi) \circ ((\textbf{G} \circ \textbf{F})\ \psi)$ composition

$\square$

4.2.4 Give an example functor for preset.

The functor for preset category is a monotone function. Remind a preset as a category has its elements as objects; there is an arrow from i to j if $i \leq j$. A monotone function f sends element (object) i to $f(i)$. If $i \leq j$, then $f(i) \leq f(j)$ because f is monotone; hence it sends the arrow $i \rightarrow j$ to $f(i) \rightarrow f(j)$. It is easy to verify the identity and composition conditions: (1) $i \leq i$ implies $f(i) \leq f(i)$; (2) if $i \leq j \leq k$, then $f(i) \leq f(j) \leq f(k)$; hence $f(i) \leq f(k)$. The monotone function f therefore is a functor.

4.2.5 Define functor for binary tree (see Sect. 2.6).

For any object A in *Set* (category of sets—total functions), the binary tree functor sends it to *Tree A* as below:

```
data Tree A = Empty | Branch (Tree A) A (Tree A)
```

For any arrow $A \xrightarrow{f} B$, the binary tree functor sends it to $Tree\ A \to Tree\ B$ as:

$$\begin{aligned} fmap\ f\ Empty &= Empty \\ fmap\ f\ (Branch\ l\ x\ r) &= Branch\ (fmap\ f\ l)\ (f\ x)\ (fmap\ f\ r) \end{aligned}$$

or use the *mapt* defined in Eq. (2.15):

$$fmap = mapt$$

Answer of Exercise 4.3

4.3.1 What is the product of two objects in a poset? What is the coproduct?
In a poset as category, objects are elements. There is at most one arrow between two objects a and b (if $a \leq b$ or vice versa). If they have arrows to both the upstream or downstream, then:

$$\text{meet } a \wedge b \qquad \text{join } a \vee b$$

is the

$$\text{product} \qquad \text{coproduct}$$

for this pair of objects.
Whereas meet is the least upper bound of a and b, join is the greatest lower bound of them. In general, join and meet of a subset of the poset need not exist; hence the product or coproduct of the poset need not exist too.

4.3.2 Prove the absorption law for coproduct, and verify the coproduct functor satisfies the composition condition.
The absorption law for coproduct states:

$$[p, q] \circ (f + g) = [p \circ f, q \circ g]$$

Proof

$$\begin{aligned} &[p, q] \circ (f + g) \\ =&[p, q] \circ [left \circ f, right \circ g] && \text{definition of } + \\ =&[[p, q] \circ (left \circ f), [p, q] \circ (right \circ g)] && \text{fusion law} \end{aligned}$$

$$=[[p, q] \circ left \circ f, [p, q] \circ right \circ g] \qquad \text{associative}$$

$$=[p \circ f, q \circ g] \qquad \text{cancellation law}$$

□

The composition condition for coproduct states:

$$(f + g) \circ (f' + g') = f \circ f' + g \circ g'$$

Proof Let $p = left \circ f$ and $q = right \circ g$.

$$(f + g) \circ (f' + g')$$

$$=[left \circ f, right \circ g] \circ (f' + g') \qquad \text{definition of } +$$

$$=[p, q] \circ (f' + g') \qquad \text{substitute with } p, q$$

$$=[p \circ f', q \circ g'] \qquad \text{absorption law}$$

$$=[left \circ f \circ f', right \circ g \circ g'] \qquad \text{substitute } p, q \text{ back}$$

$$=[left \circ (f \circ f'), right \circ (g \circ g')] \qquad \text{associative law}$$

$$=f \circ f' + g \circ g' \qquad \text{reverse of } +$$

□

Answer of Exercise 4.4

4.4.1 Prove $swap$ satisfies the natural transformation condition $(g \times f) \circ swap = swap \circ (f \times g)$.

For $A \xrightarrow{f} C$ and $B \xrightarrow{g} D$, we need to show below diagram commutes.

$$(A, B) \xrightarrow{swap_{A,B}} (B, A)$$
$$f \times g \downarrow \qquad\qquad \downarrow g \times f = swap\ f \times g$$
$$(C, D) \xrightarrow{swap_{C,D}} (D, C)$$

Proof

$$((g \times f) \circ swap)\ (A, B)$$

$$=(g \times f)\ (swap\ (A, B)) \qquad \text{definition of composition}$$

$$=(g \times f) \circ (B, A) \qquad \text{definition of } swap$$

$$=(g \ B, f \ A) \qquad\qquad \text{product of arrows}$$

$$=(D, C) \qquad\qquad\quad \text{definition of } g, f$$

$$=swap \ (C, D) \qquad\quad \text{reverse of } swap$$

$$=swap \ (f \ A, g \ B) \qquad \text{reverse of } f, g$$

$$=swap \ ((f \times g) \ (A, B)) \qquad \text{product of arrows}$$

$$=(swap \circ (f \times g)) \ (A, B) \qquad \text{reverse of composition}$$

□

4.4.2 Prove that the polymorphic function *length* defined below is a natural transformation.

$$length : [A] \to Int$$
$$length \ [\ ] = 0$$
$$length \ (x : xs) = 1 + length \ xs$$

Proof For any object A, the arrow *length* indexed by A is:

$$[A] \xrightarrow{length_A} \mathbf{K}_{Int} \ A$$

where $\mathbf{K}_{Int}$ is the constant functor (see Example 4.2.2). It sends any object to Int and sends every arrow to identity arrow id of Int. For arrow $A \xrightarrow{f} B$, we need to show below diagram commutes.

$$
\begin{array}{ccc}
A & [A] \xrightarrow{\ length_A\ } \mathbf{K}_{Int} \ A \\
\Big\downarrow f & \mathbf{List}(f)\Big\downarrow \qquad\qquad \Big\downarrow \mathbf{K}_{Int}(f) \\
B & [B] \xrightarrow[\ length_B\ ]{} \mathbf{K}_{Int} \ B
\end{array}
$$

From the definition of constant functor, this diagram is equivalent to:

$$
\begin{array}{ccc}
A & [A] \xrightarrow{\ length_A\ } Int \\
\Big\downarrow f & \mathbf{List}(f)\Big\downarrow \qquad\qquad \Big\downarrow id \\
B & [B] \xrightarrow[\ length_B\ ]{} Int
\end{array}
$$

We are going to show:

$$id \circ length_A = length_B \circ \mathbf{List}(f)$$

which is equivalent to $length_A = length_B \circ \mathbf{List}(f)$. Use mathematical induction on list. First for empty list:

$length_B \circ \mathbf{List}(f)[\,]$

$= length_B [\,]$ definition of list functor

$= 0$ definition of $length$

$= length_A [\,]$ reverse of $length$

Assume $length_B \circ \mathbf{List}(f)\ as = length_A\ as$ is true, for $(a{:}as)$:

$length_B \circ \mathbf{List}(f)(a{:}as)$

$= length_B\ (f(a) : \mathbf{List}(f)\ as)$ definition of list functor

$= 1 + length_B\ (\mathbf{List}(f)\ as)$ definition of $length$

$= 1 + length_B \circ \mathbf{List}(f)\ as$ arrow composition

$= 1 + length_A\ as$ induction assumption

$= length_A\ (a{:}as)$ reverse of $length$

$\square$

4.4.3 Natural transformation is composable. Consider two natural transformations $\mathbf{F} \xrightarrow{\phi} \mathbf{G}$ and $\mathbf{G} \xrightarrow{\psi} \mathbf{H}$. For any arrow $A \xrightarrow{f} B$, draw the diagram for their composition, and list the commutative condition.

The commutative condition is:

$$\mathbf{H}(f) \circ (\psi_A \circ \phi_A) = (\psi_B \circ \phi_B) \circ \mathbf{F}(f)$$

Answer of Exercise 4.5

4.5.1 In the example of poset, if there exists the minimum (or the maximum) element, then the minimum (or the maximum) is the initial object (or the final object). Consider the category *Poset* of all posets, if there exists the initial object, what is it? If there exists the final object, what is it?

For the *Poset* category, objects are posets; arrows are monotone functions.

For two posets P and Q, arrow $P \xrightarrow{h} Q$ means for any two ordered elements $a \leq b$ in P, $h(a) \leq h(b)$ holds in Q.

Analogue to *Set*, the initial object in this category is the empty poset $\varnothing$. There is unique arrow from it to any poset P:

$$\varnothing \longrightarrow P$$

The final object is the singleton poset $\{\bigstar\}$, and the only ordering is $\bigstar \leq \bigstar$. From any poset P, there is unique arrow:

$$P \xrightarrow{f} \{\bigstar\}$$

f sends $p \rightsquigarrow \bigstar$ for every p in P. If $a \leq b$ holds in P, then $f(a) = \bigstar \leq \bigstar = f(b)$ holds in $\{\bigstar\}$.

4.5.2 In Peano category *Pno* (see Exercise 4.1.3), what is the initial object in the form of (A, f, z)? What is the final object?

The initial object is $(\mathbf{N}, succ, 0)$. There is unique arrow from it to any object:

$$(\mathbf{N}, succ, 0) \xrightarrow{\sigma} (A, f, z) : \sigma(n) = f^n(z)$$

The final object is a singleton $1 = (\{\bigstar\}, \bigstar, id)$. There is unique arrow from any object (A, f, z) to it:

$$(A, f, z) \xrightarrow{\sigma} 1 : \sigma(a) = \bigstar$$

Answer of Exercise 4.6

4.6.1 Verify that *Exp* is a category. What is the *id* arrow and how to compose arrows?

Proof We show that the *id* arrow is $h \xrightarrow{id} h$, such that the following diagram commutes:

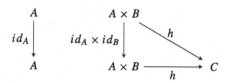

The composition of $h \xrightarrow{i} k$ and $k \xrightarrow{j} m$ is $j \circ i$ such that below diagram commutes:

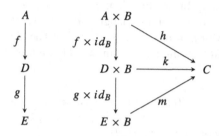

For arrow $h \xrightarrow{j} k$, it means $id_k \circ j = j = j \circ id_h$ holds. The associative law holds for any three arrows.   □

4.6.2 What is the subscript of *id* in the reflection law $curry\ apply = id_?$? Find another way to prove the reflection law.

The identity axiom of category asserts id_X for each object X. The subscript of the *id* in the reflection law corresponds to the object (a binary function) of $A \times B \to C$.

Proof For arbitrary binary function $f : (a, b) \rightsquigarrow c$.

$$curry \circ apply\ f\ a\ b = curry\ (apply\ f)\ a\ b \qquad \text{composition}$$
$$= (apply\ f)\ (a, b) \qquad \text{reverse of } curry$$
$$= f\ (a, b) \qquad \text{definition of } apply$$
$$= (id_{A \times B \to C} \circ f)\ (a, b)$$

□

4.6.3 Define

$$(curry\ f) \circ g = curry(f \circ (g \times id))$$

as the fusion law for Currying. Prove this law and give its diagram.

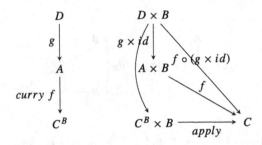

Proof Let f be arbitrary binary function: $f : A \times B \to C$ and g be a function from D to A. In the above diagram, triangle with vertexes $D \times B$, $A \times B$, and C commutes. The arrow of $D \times B \to C$ is $f \circ (g \times id)$. The big triangle with vertexes $D \times B$, $C^B \times B$, and C commutes. From the definition of exponential transpose arrow (Eq. (4.23)):

$$apply \circ (((curry\ f) \circ g) \times id) = f \circ (g \times id)$$

By universal property of *curry* and *apply*:

$$(curry\ f) \circ g = curry(f \circ (g \times id))$$

$\square$

Answer of Exercise 4.7

4.7.1 Draw the diagram to illustrate the group axiom of inverse element. Formalize the axiom of inverse element as: $m \circ (id, i) = m \circ (i, id) = e$

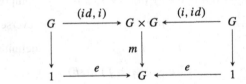

4.7.2 Let p be a prime number. Use the F-algebra to define the α arrow for the multiplicative group for integers modulo p (refer to Chap. 3 for the definition of this group).
According to the α arrow of group:

$$FA \xrightarrow{\ \alpha = e + m + i\ } A$$

Define the multiplicative group for integers modulo p as:

$$e\,() = 1 \qquad\qquad \text{1 is the identity element}$$
$$m(a, b) = ab \bmod p \qquad \text{multiplication modulo } p$$
$$i(a) = a^{p-2} \bmod p \qquad \text{Fermat's little theorem } a^{p-1} \equiv 1 \ (\bmod\ p)$$

4.7.3 Define F-algebra for ring (see Sect. 3.3.1 for the definition of ring).
The algebraic structure of ring contains three parts:

(1) Carrier object R, the set that carries the algebraic structure of ring.
(2) Polynomial functor $FR = 1 + 1 + R \times R + R \times R + R$.
(3) Arrow $FR \xrightarrow{\ \alpha = z + e + p + m + n\ } R$ consists of the identity element
of addition z, the identity element of multiplication e, addition p,
multiplication m, and negation n.

These define the F-algebra (R, α) for ring. When the carrier object is
integers, for example, the ring is defined as below under normal arithmetic:

$$z\,() = 0$$
$$e\,() = 1$$
$$p\,(a, b) = a + b$$
$$m\,(a, b) = ab$$
$$n\,a = -a$$

4.7.4 What is the id arrow for F-algebra category? How to compose arrows?
The id arrow is the automorphism from F-algebra (A, α) to itself.
When composing two F-morphisms, the arrows between carrier objects
$A \xrightarrow{f} B \xrightarrow{g} C$ make the following diagram commute:

where:

$$g \circ f \circ \alpha = \gamma \circ \mathbf{F}(g) \circ \mathbf{F}(f) = \gamma \circ \mathbf{F}(g \circ f)$$

Answer of Exercise 4.8

4.8.1 Someone write the natural number like functor as the below recursive form. What do you think about it?

data NatF A = ZeroF | SuccF (NatF A)

No, it does not work. Consider carrier object A, functor **NatF** is recursive, and it does not send A to a determined object. In fact, we expect to map it to an object in Peano category (A, f, z).

4.8.2 We can define an α arrow for **NatF** $Int \rightarrow Int$, named $eval$:

$$eval : \textbf{NatF} \; Int \rightarrow Int$$
$$eval \; ZeroF \quad = 0$$
$$eval \; (SuccF \; n) = n + 1$$

We can recursively substitute $A' = \textbf{NatF} \; A$ into functor **NatF** by n times. We denote the functor obtained as $\textbf{NatF}^n \; A$. Can we define the following α arrow?

$$eval : \textbf{NatF}^n \; Int \rightarrow Int$$

Note $ZeroF$ is an object of $\textbf{NatF}^i Int$, where $i = 1, 2, \ldots, n$.

$$eval : \textbf{NatF}^n \; Int \rightarrow Int$$
$$eval \; ZeroF = 0$$
$$eval \; (SuccF \; ZeroF) = 1$$
$$eval \; (SuccF \; (SuccF \; ZeroF)) = 2$$
$$\cdots$$
$$eval \; (SuccF^{n-1} \; ZeroF) = n - 1$$
$$eval \; (SuccF^n \; m) = m + n$$

Answer of Exercise 4.9

4.9.1 For the binary tree functor **TreeF** $A \; B$, fix A and use the fixed point to show that (**Tree** A, $[nil, branch]$) is the initial algebra.

Proof Let $B' = \textbf{TreeF} \; A \; B$. Recursively apply the functor to itself, and denote this result by **Fix** (**TreeF** A).

Fix (TreeF A**)**

=**TreeF** A (**Fix** (**TreeF** A))	fixed point
=**TreeF** A (**TreeF** $A(\ldots)$)	expand
=$NilF|BrF$ A (**TreeF** A $(\ldots)$) (**TreeF** A $(\ldots)$)	binary tree functor
=$NilF|BrF$ A (**Fix** (**TreeF** A)) (**Fix** (**TreeF** A))	reverse of fixed point

Compare with the definition of $TreeA$:

```
data Tree A = Nil | Br A (Tree A) (Tree A)
```
Hence **Tree** A = **Fix** (**TreeF** A). The initial algebra is (**Tree** A, $[nil, br]$). □

Answer of Exercise 5.1

5.1.1 Verify the definition of *foldl* in *foldr*:

$$foldl\ f\ z\ xs = foldr\ (b\ g\ a \mapsto g\ (f\ a\ b))\ id\ xs\ z$$

Proof Extract the λ expression to a *step* function, and we are going to show:

$$foldl\ f\ z\ xs = foldr\ step\ id\ xs\ z$$
$$\text{where } step\ x\ g\ a = g\ (f\ a\ x)$$

$(foldr\ step\ id\ [x_1, x_2, \ldots, x_n])\ z$

$=(step\ x_1(step\ x_2(\ldots (step\ x_n\ id)))\ldots)\ z$

$=(step\ x_1(step\ x_2(\ldots (a_n \mapsto id\ (f\ a_n\ x_n))))\ldots)\ z$

$=(step\ x_1(\ldots (a_{n-1} \mapsto (a_n \mapsto id\ (f\ a_n\ x_n))\ (f\ a_{n-1}\ x_{n-1})))\ldots)\ z$

$=(a_1 \mapsto (\ldots (a_n \mapsto id\ (f\ a_n\ x_n))\ (f\ a_{n-1}\ x_{n-1}))\ldots (f\ a_1\ x_1))\ z$

$=(a_1 \mapsto (a_2 \mapsto (\ldots (a_n \mapsto f\ a_n\ x_n)\ (f\ a_{n-1}\ x_{n-1}))\ldots)\ (f\ a_1\ x_1))\ z$

$=(a_1 \mapsto (a_2 \mapsto (\ldots (a_{n-1} \mapsto f\ (f\ a_{n-1}\ x_{n-1})\ x_n)\ \ldots))\ (f\ a_1\ x_1))\ z$

$=(a_1 \mapsto f\ (f\ (\ldots (f\ a_1\ x_1)\ x_2)\ \ldots)\ x_n)\ z$

$=f\ (f\ (\ldots (f\ z\ x_1)\ x_2)\ \ldots)\ x_n$

$=foldl\ f\ z\ [x_1, x_2, \ldots, x_n]$

□

We often write f as an infix of $\oplus$ to highlight the difference between *foldl* and *foldr*:

$$foldl \ \oplus \ f \ z = ((\ldots (z \oplus x_1) \ \oplus x_2)\ldots) \ \oplus x_n$$

5.1.2 Verify below *build*...*foldr* forms hold:

$$concat \ xss = build \ (f \ z \mapsto foldr \ (xs \ x \mapsto foldr \ f \ x \ xs) \ z \ xss)$$

$$map \ f \ xs = build \ (\oplus z \mapsto foldr \ (y \ ys \mapsto (f \ y) \oplus ys) \ z \ xs)$$

$$filter \ p \ xs = build \ (\oplus z \mapsto foldr$$

$$\left(x \ xs' \mapsto \begin{cases} p(x): & x \oplus xs' \\ \text{otherwise}: & xs' \end{cases} \right) z \ xs)$$

$$repeat \ x = build \ (\oplus z \mapsto let \ xs = x \oplus xs \ in \ xs)$$

Proof

(1) *Concat*:

$$build \ (f \ z \mapsto foldr \ (xs \ x \mapsto foldr \ f \ x \ xs) \ z \ xss)$$

$=(f \ z \mapsto foldr \ (xs \ x \mapsto foldr \ f \ x \ xs) \ z \ xss) \ (:) \ [\ ]$	definition of *build*
$=foldr \ (xs \ x \mapsto foldr \ (:) \ x \ xs) \ [\ ] \ xss$	β-reduction
$=foldr \ +\!\!+ \ [\ ] \ xss$	concatenate two
$=concat \ xss$	concatenate multiple

(2) *Map*:

$$build \ (\oplus z \mapsto foldr \ (y \ ys \mapsto (f \ y) \oplus ys) \ z \ xs)$$

$=(\oplus z \mapsto foldr \ (y \ ys \mapsto (f \ y) \oplus ys) \ z \ xs) \ (:) \ []$	definition of *build*
$=foldr \ (y \ ys \mapsto f(y): ys) \ [] \ xs$	β-reduction
$=foldr \ (x \ ys \mapsto f(x): ys) \ [] \ xs$	α-conversion, rename
$=map \ f \ xs$	list map

(3) *Filter*:

$$build \ (\oplus z \mapsto foldr \ (x \ xs' \mapsto$$

$$\begin{cases} p(x): & x \oplus xs' \\ \text{otherwise}: & xs' \end{cases}) z \ xs)$$

$$=(\oplus \, z \mapsto foldr \,(x \, xs' \mapsto$$

$$\begin{cases} p(x): & x \oplus xs' \\ \text{otherwise}: & xs' \end{cases}) \, z \, xs)\,(:)\,[\,] \qquad\qquad \text{expand } build$$

$$=foldr \,(x \, xs' \mapsto \begin{cases} p(x): & x:xs' \\ \text{otherwise}: & xs' \end{cases})\,[\,]\,xs \qquad\qquad \beta\text{-reduction}$$

$$=filter \, p \, xs \qquad\qquad\qquad\qquad \text{by Eq.}(5.3)$$

(4) *Repeat*:

$$build \,(\oplus \, z \mapsto let \, xs = x \oplus xs \, in \, xs)$$

$$=(\oplus \, z \mapsto let \, xs = x \oplus xs \, in \, xs)\,(:)\,[] \qquad \text{definition of } build$$

$$=(let \, xs = x:xs \, in \, xs) \qquad\qquad\qquad \beta\text{-reduction}$$

$$=repeat \, x \qquad\qquad\qquad\qquad\qquad \text{by Eq.}(5.12)$$

□

5.1.3 Simplify the quick sort algorithm.

$$qsort \,[\,] \quad = [\,]$$
$$qsort \,(x:xs)= qsort \,[a|a \in xs, a \le x] +\!\!+ [x] +\!\!+ qsort \,[a|a \in xs, x < a]$$

Transform the ZF-expression to *filter*, and combine the two rounds of filtering into one pass:

$$qsort \,[\,] \quad = [\,]$$
$$qsort \,(x:xs)= qsort \, as +\!\!+ [x] +\!\!+ qsort \, bs$$

where:

$$(as, bs) = foldr \, h \,([\,],[\,])\,xs$$

$$h \, y \,(as', bs') = \begin{cases} y \le x: & (y:as', bs') \\ \text{otherwise}: & (as', y:bs') \end{cases}$$

Next simplify list concatenation:

$$qsort \, as +\!\!+ [x] +\!\!+ qsort \, bs$$
$$=qsort \, as +\!\!+ (x : qsort \, bs)$$
$$=foldr \,(:)\,(x : qsort \, bs)(qsort \, as)$$

5.1.4   Verify the type constraint of fusion law by category theory.
As shown in below diagram:

$$
\begin{array}{ccc}
\textbf{ListF}A\ [A] & \xrightarrow{\ (:)\ +\ []\ } & [A] \\[2pt]
\textbf{ListF}A(h) \Big\downarrow & & \Big\downarrow h = (\!|\alpha|\!) \\[2pt]
\textbf{ListF}A\ B & \xrightarrow[\ \alpha\ =\ f\ +\ z\ ]{} & B
\end{array}
$$

The catamorphism $(\!|\alpha|\!)$ is abstracted to build some algebraic structure from α, i.e., $g\ \alpha$, where g accepts the α arrow of the F-algebra and generates result of B. The α arrow is the coproduct of $f : A \to B \to B$ and $z : 1 \to B$, that the type is:

$$ g : \forall A.(\forall B.(A \to B \to B) \to B \to B) $$

The definition of build is $build(g) = g\ (:)\ []$. It applies g to the arrow of the initial algebra and builds the carrier object of the initial algebra, which is a list of $[A]$.

$$ build : \forall A.(\forall B.(A \to B \to B) \to B \to B) \to \textbf{List}\ A $$

Answer of Exercise 5.2

5.2.1   Use the fusion law to optimize the expression evaluation function:

$$ eval = sum \circ map\ (product \circ (map\ dec)) $$

$eval\ es$

$= sum(map\ (product \circ (map\ dec))\ es)$

$= sum(map\ f\ es)$ let $f = product \circ$

                                     $(map\ dec)$

$= \textbf{foldr}\ (+)\ 0\ (\textbf{build}\ (\oplus z \mapsto$

     $foldr\ (t\ ts \mapsto (f\ t) \oplus ts)\ z\ es))$ sum in fold, map in build

$= (\oplus z \mapsto foldr\ (t\ ts \mapsto (f\ t) \oplus ts)\ z\ es)\ (+)\ 0$ fusion law

$= foldr\ (t\ ts \mapsto (f\ t) + ts)\ 0\ es$ β-reduction

Written in point-free form:

$$eval = foldr\ (t\ ts \mapsto (f\ t) + ts)\ 0$$
$$= foldr\ ((+) \circ f)\ 0 \qquad\qquad \text{(A.3)}$$

Next simplify the *product* $\circ$ (*map dec*) part

$$(product \circ (map\ dec))\ t$$
$$= product\ (map\ dec\ t)$$
$$= \boldsymbol{foldr}\ (\times)\ 1\ (\boldsymbol{build}\ (\oplus z \mapsto \qquad\qquad\qquad product\ \text{in}\ foldr$$
$$\qquad foldr\ (d\ ds \mapsto (dec\ d) \oplus ds)\ z\ t)) \qquad\qquad map\ \text{in}\ build$$
$$= (\oplus z \mapsto foldr(d\ ds \mapsto (dec\ d) \oplus ds)\ z\ t)\ (\times)\ 1 \qquad \text{fusion law}$$
$$= foldr\ (d\ ds \mapsto (dec\ d) \times ds)\ 1\ t \qquad\qquad \beta\text{-reduction}$$
$$= foldr\ ((\times) \circ fork\ (dec, id))\ 1\ t \qquad\qquad fork(f, g)$$
$$\qquad\qquad\qquad\qquad\qquad\qquad\qquad\qquad\qquad x = (f\ x, g\ x)$$

Substituting this for f in Eq. (A.3), we obtain the final simplified result:

$$eval = foldr\ ((+) \circ (foldr\ ((\times) \circ fork\ (dec, id))\ 1\ t))\ 0$$

5.2.2 How to expand all expressions from left?

There are three options for every digit d when expanding from left to right:

(1) Insert nothing. Append d to the last factor of the last term of e_i. Combine $f_n +\!\!\!+ [d]$ as a new factor. For example e_i is $1+2$, d is 3, write 3 after $1+2$ without inserting any symbols, and we obtain a new expression $1 + 23$.

(2) Insert $\times$. Create a new factor $[d]$, and then append it to the last term of e_i. Combine $t_m +\!\!\!+ [[d]]$ as a new term. For the same example of $1 + 2$, write 3 after it, put a $\times$ between 2 and 3, and the new term is $1 + 2 \times 3$.

(3) Insert +. Create a new term $[[d]]$, and then append it to e_i to generate a new expression $e_i +\!\!\!+ [[[d]]]$. For the same example of $1 + 2$, write 3 after it, put a + between 2 and 3, and the new expression is $1 + 2 + 3$.

Define *append* function, which adds element to the end of a list:

$$append\ x = foldr\ (:)\ [x]$$

Next define function $onLast(f)$, which applies f to the last element of a list:

$$onLast\ f = foldr\ h\ [\ ]$$
$$\text{where:}$$
$$h\ x\ [\ ] \ = [(f\ x)]$$
$$h\ x\ xs \ = x : xs$$

Then implement the above three options as below:

```
add d exp = [((append d) 'onLast') 'onLast' exp,
             (append [d]) 'onLast' exp,
             (append [[d]]) exp]
```

5.2.3 The following definition converts expression into string:

$$str = (join\ \text{"+"}) \circ (map\ ((join\ \text{" $\times$ "}) \circ (map\ (show \circ dec))))$$

where $show$ converts number into string. Function $join(c, s)$ concatenates multiple strings s with delimiter c. For example:

$$join(\text{"#"}, [\text{"abc"}, \text{"def"}]) = \text{"abc#def"}$$

Use the fusion law to optimize str.
$join$ in Eq. (5.14) inserts space between every two strings. Extract the space as a parameter to define $join(c, s)$:

$$join\ c = foldr\ (w\ b \mapsto foldr\ (:)\ (c : b)\ w)\ [\ ]$$

The definition of str contains embedded $(join\ c) \circ (map\ f)$:

$$str = (join\ c) \circ (map\ f) \quad \text{where:}\ f = (join\ d) \circ (map\ g)$$

where $c = \text{"+,"}$ $d = \text{"}\times\text{,"}$ and $g = show \circ dec$. Simplify $(join\ c) \circ (map\ f)$ as below.

$(join\ c) \circ (map\ f)\ es$

$= \textbf{foldr}\ (w\ b \mapsto foldr\ (:)(c : b)\ w)\ [\ ]\ ($ *join* in *foldr*

     $\textbf{build}\ (\oplus z \mapsto foldr\ (y\ ys \mapsto (f\ y) \oplus ys)\ z\ es))$ *map* in *build*

$= (\oplus z \mapsto foldr\ (y\ ys \mapsto$

     $(f\ y) \oplus ys)\ z\ es))\ (w\ b \mapsto foldr\ (:)\ (c : b)\ w)\ [\ ]$ fusion law

$= foldr\ (y\ ys \mapsto foldr\ (:)\ (c : ys)\ (f\ y))\ [\ ]\ es$ β-reduction

Substitute $c =' +', d =' \times'$, and $g = show \circ dec$ into the result; it simplifies to the following:

$$str = foldr\ (x\ xs \mapsto foldr\ (:)\ (\text{'+'}:xs)($$
$$foldr(y\ ys \mapsto foldr\ (:)\ (\text{'}\times\text{'}:ys)\ (show \circ dec\ y))\ [\ ])\ [\ ]$$

Answer of Exercise 6.1

6.1.1 Equation (1.12) in Chap. 1 implements Fibonacci numbers by folding. How to define Fibonacci numbers as potential infinity with *iterate*?

$$F = (fst \circ unzip)\ (iterate\ ((m, n) \mapsto (n, m + n))\ (1, 1)).$$

For example, *take* 100 *F* gives the first 100 Fibonacci numbers.

6.1.2 Define *iterate* by folding.

For the infinite stream *iterate f x*, if applying *f* to each element followed by prepending *x* as the first, we obtain the same infinite stream back.

$$iterate\ f\ x = x : foldr\ (y\ ys \mapsto (f\ y):ys)\ [\ ]\ (iterate\ f\ x)$$
$$= x : foldr\ ((:) \circ f)\ [\ ]\ (iterate\ f\ x)$$

For example:

```
take 10 $ iterate (+1) 0 = [0,1,2,3,4,5,6,7,8,9]
```

Answer of Exercise 6.2

6.2.1 Show **Stream** is the fixed point of **StreamF**.

Proof Let $A' = $ **StreamF** $E\ A$; apply it to itself repeatedly. Let this result be **Fix (StreamF** E).

Fix (StreamF E) $=$ **StreamF** E **(Fix (StreamF** E))	fixed point
$=$ **StreamF** E **(StreamF** E (...))	expand recursively
$=$ **Stream** E **(Stream** E (...))	rename
$=$ **Stream** E	reverse of *Stream*

Shows **Stream** is the fixed point of **StreamF**. □

6.2.1 Define *unfold*.

We use *Maybe* to define the terminate condition:

```
unfold :: (b → Maybe (a, b)) → (b → [a])
unfold f b = case f b of
                 Just (a, b') → a : unfold f b'
                 Nothing → []
```

6.2.2 The fundamental theorem of arithmetic states that any integer greater than 1 can be uniquely factored as the product of prime numbers. Given a text T and a string W, does any permutation of W exist in T? Solve this programming puzzle with this theorem and the stream of primes.

Our idea is to map every distinct character to a prime number, for example, a $\rightsquigarrow$ 2, b $\rightsquigarrow$ 3, c $\rightsquigarrow$ 5, ... We can convert any string W, no matter it contains repeated characters or not, to a product of prime numbers:

$$F = \prod_{c \in W} p_c$$

We call F the number theory finger print of string W. When W is empty, we define its finger print as 1. Because multiplication of integers is commutative, the finger print is the same for all permutations of W, and according to the fundamental theorem of arithmetic, the finger print is unique. We develop an elegant solution based on this: First, calculate F of string W, and then slide a window of length $|W|$ along T from left to right. When starting, we compute the finger print within this window of T and compare it with F. If they are equal, it means T contains some permutation of W. Otherwise, slide the window to the right by a character. We can easily compute the updated finger print value for this new window position: Divide the product by the prime number of the character slides out, and multiply the prime number of the character slides in. Whenever the finger print equals to F, we find a permutation. In order to map different characters to prime numbers, we use sieve of Eratosthenes to generate a series of prime numbers. Below is the example algorithm accordingly.

1: **function** CONTAINS?(W, T)
2:    $P \leftarrow$ *ana era* $[2, 3, \ldots]$ ▷ prime numbers
3:    **if** $W = \phi$ **then**
4:       **return** True
5:    **if** $|T| < |W|$ **then**
6:       **return** False
7:    $m \leftarrow \prod P_c, c \in W$
8:    $m' \leftarrow \prod P_c, c \in T[1 \ldots |W|]$
9:    **for** $i \leftarrow |W| + 1$ to $|T|$ **do**
10:      **if** $m = m'$ **then**

11:                    **return** True
12:              $m' \leftarrow m' \times P_{T_i}/P_{T_{i-|W|}}$
13:        **return** $m = m'$

Answer of Exercise 6.3

6.3.1 In the 1-to-1 correspondence between the rooms and guests with the numbering scheme shown in Fig. 6.10, which room number should be assigned for guest i of group j? Which guest of which group will live in room k?
Use the convention to count from zero, and use pair (i, j) to denote the j-th guest of the i-th group. List first several guests and their rooms:

(i, j)	$(0, 0)$	$(0, 1)$	$(1, 0)$	$(2, 0)$	$(1, 1)$	$(0, 2)$	$(0, 3)$	$(1, 2)$	$(2, 1)$	$(3, 0)$	...
k	0	1	2	3	4	5	6	7	8	9	...
$i + j$	0	1	1	2	2	2	3	3	3	3	...

Writing down the values of $i + j$, there is a pattern: There are one instance of number 0, two instances of number 1, three instances of number 2, four instances of number 3, ... They are exactly the triangle numbers found by Pythagoreans. Let $m = i + j$; there are total $\frac{m(m+1)}{2}$ grid points along the diagonals on the left-bottom side of a given grid.
For the diagonal where this point belongs to, if m is odd, then the room number increases along the left up direction, i increases, and j decreases; if m is even, then the direction is right-bottom. Summarizing the two cases gives the following result:

$$k = \frac{m(m+1)}{2} + \begin{cases} m - j & m \text{ is odd} \\ j & m \text{ is even} \end{cases}$$

Further, we use $(-1)^m$ to simplify the conditions:

$$k = \frac{m(m+2) + (-1)^m(2j - m)}{2}$$

6.3.2 For Hilbert's Grand hotel, there are alternative solutions for the problem on the third day. Figure 6.11 is the cover page of the book *Proof without word*. Can you give a numbering scheme based on this figure?
As shown in Fig. A.5, count along the gnomon shaped path. There are an odd number of grid points along every gnomon.

Fig. A.5 Alternative
numbering scheme for
infinity of infinity

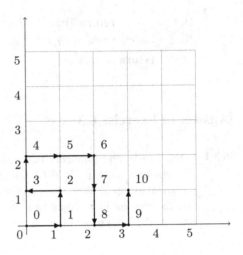

Answer of Exercise 6.4

6.4.1 Let $x = 0.9999\cdots$; then $10x = 9.9999\cdots$. Subtracting them gives $10x - x = 9$. Solving this equation gives $x = 1$. Hence $1 = 0.9999\cdots$. Is this reasoning correct?

Yes, it is correct.

Answer of Exercise 6.5

6.5.1 Light a candle between two mirrors, what image can you see? Is it potential or actual infinity?

The candle image reflects between the two mirrors endlessly and generates infinitely many images. If we consider the speed of light is limited, then it is potential infinity from physics viewpoint.

Answer of Exercise 7.1

7.1.1 We can define numbers in natural language. For example "the maximum of two digits number" defines 99. Define a set containing all numbers that cannot be described within 20 words. Consider such an element: The minimum number that cannot be described within 20 words. Is it a member of this set? This is an instance of Russell's paradox. Whether it is a member, all lead to contradiction.

7.1.2 "The only constant is change" said by Heraclitus. Is this Russell's paradox?
Yes, this is an instance of Russell's paradox.

7.1.3 Is the quote saying by Socrates (the beginning of this chapter) Russell's paradox?
Yes, it is an instance of Russell's paradox.

Answer of Exercise 7.2

7.2.1 Translate Fermat's last theorem into a TNT string.
Define power operation first.

$$\begin{cases} \forall a : e(a, 0) = S0 & \text{0-th power is 1} \\ \forall a : \forall b : e(a, Sb) = a \cdot e(a, b) & \text{recursion} \end{cases}$$

Next define Fermat's last theorem atop it.

$$\forall d : \neg \exists a : \exists b : \exists c : \neg (d = 0 \vee d = S0 \vee d = SS0) \rightarrow e(a, d)$$
$$+ e(b, d) = e(c, d)$$

7.2.2 Prove the associative law of addition with TNT reasoning rules.
Surprisingly, we can prove every theorem below:

$$\begin{aligned} a + b + 0 &= a + (b + 0) \\ a + b + S0 &= a + (b + S0) \\ a + b + SS0 &= a + (b + SS0) \end{aligned}$$

$$\cdots$$

For example:

$$a + b + 0 = a + b = a + (b + 0)$$

and:

$$\begin{aligned} a + b + SS0 &= SS(a + b + 0) \\ &= SS(a + b) \\ &= a + SSb \\ &= a + (b + SS0) \end{aligned}$$

However, we cannot prove: $\forall c : a + b + c = a + (b + c)$.
To do that, we need to introduce mathematical induction.

Answer of Exercise 7.3

7.3.1 Show $\forall a : (0 + a) = a$ with the induction rule.

Proof By induction on a, first for 0:

$$0 + 0 = 0$$

Next assuming $(0 + a) = a$, we are going to show Sa holds.

$$(0 + Sa) = S(0 + a) \qquad\qquad \text{axiom (3)}$$
$$= Sa \qquad\qquad \text{induction assumption}$$

By rule of induction, it proves $\forall a : (0 + a) = a$. $\qquad\qquad\qquad\qquad\qquad$ □

Appendix B
Proof of Commutativity of Addition

To prove the commutativity of addition $a + b = b + a$, we prepare three things. The first states that, for any natural number a, we have:

$$0 + a = a \qquad \text{(B.1)}$$

It means the zero on the left side can be cancelled for addition. When $a = 0$, according to the first rule of addition, it holds:

$$0 + 0 = 0$$

As the induction, suppose $0 + a = a$ holds, and we are going to show that $0 + a' = a'$.

$$
\begin{aligned}
0 + a' &= (0 + a)' && \text{second rule of addition} \\
&= a' && \text{induction assumption}
\end{aligned}
$$

Next we define the successor of 0 be 1 and prove the second fact:

$$a' = a + 1 \qquad \text{(B.2)}$$

It means the successor of any natural number is this number plus 1. This is because:

$$
\begin{aligned}
a' &= (a + 0)' && \text{first rule of addition} \\
&= a + 0' && \text{second rule of addition} \\
&= a + 1 && \text{1 is defined as the successor of 0}
\end{aligned}
$$

© China Machine Press 2024
X. Liu, *Mathematics in Programming*,
https://doi.org/10.1007/978-981-97-2432-1

The third thing we are going to prove is the starting case:

$$a + 1 = 1 + a \tag{B.3}$$

When $a = 0$, we have:

$$0 + 1 = 1 \qquad \text{We proved the left 0 can be cancelled}$$
$$= 1 + 0 \qquad \text{first rule of addition}$$

For induction case, suppose $a + 1 = 1 + a$ holds, and we are going to show $a' + 1 = 1 + a'$.

$$a' + 1 = a' + 0' \qquad \text{1 is the successor of 0}$$
$$= (a' + 0)' \qquad \text{first rule of addition}$$
$$= ((a + 1) + 0)' \qquad \text{second result we proved. Eq. (B.2)}$$
$$= (a + 1)' \qquad \text{second rule of addition}$$
$$= (1 + a)' \qquad \text{induction assumption}$$
$$= 1 + a' \qquad \text{second rule of addition}$$

On top of these three results, we can prove the commutative law of addition. We first show that when $b = 0$ it holds. According to the first rule of addition, we have $a + 0 = a$; while from the first result we proved, $0 + a = a$ holds too. Hence $a + 0 = 0 + a$. Then we prove the induction case. Suppose $a + b = b + a$ holds, and we are going to show $a + b' = b' + a$.

$$a + b' = (a + b)' \qquad \text{second rule of addition}$$
$$= (b + a)' \qquad \text{induction assumption}$$
$$= b + a' \qquad \text{second rule of addition}$$
$$= b + a + 1 \qquad \text{second result we proved. Eq. (B.2)}$$
$$= b + 1 + a \qquad \text{third result we proved. Eq. (B.3)}$$
$$= (b + 1) + a \qquad \text{associative law proved in Chap. 1}$$
$$= b' + a \qquad \text{third result we proved. Eq. (B.3)}$$

Therefore we proved commutative law of addition with Peano's axioms ([9], p147–148).

Appendix C
Uniqueness of Product and Coproduct

The following proposition asserts product and coproduct are unique.

Proposition C.0.1 *For any pair of object A, B of category* **C**, *let the objects and arrows in below diagram*

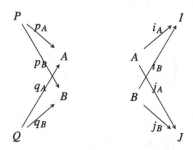

be a pair of

product   coproduct

wedges, and then

$$P, Q \qquad I, J$$

are isomorphic wedges. There are unique arrows:

$$
\begin{array}{cc}
P & I \\
f\downarrow \uparrow g & f\downarrow \uparrow g \\
Q & J
\end{array}
$$

© China Machine Press 2024
X. Liu, *Mathematics in Programming*,
https://doi.org/10.1007/978-981-97-2432-1

Such that:

$$\begin{cases} p_A = q_A \circ f & p_B = q_B \circ f \\ q_A = p_A \circ g & q_B = p_B \circ g \end{cases} \quad \begin{cases} i_A = g \circ j_A & i_B = g \circ j_B \\ j_A = f \circ i_A & j_B = f \circ i_B \end{cases}$$

where f and g are inverse pair of isomorphisms.

Proof We only prove the left side part. The right side can be proved in a similar way. Given A, B, Object Q and the pair q_A, q_B form a product wedge. Object P and the pair p_A, p_B form another wedge. From the definition of product, there is a unique mediator f satisfying:

$$p_A = q_A \circ f \quad \text{and} \quad p_B = q_B \circ f$$

By reversing the roles of P and Q (let P be product and Q be arbitrary wedge), we have:

$$q_A = p_A \circ g \quad \text{and} \quad q_B = p_B \circ g$$

Hence we have:

$$\begin{cases} p_A \circ g \circ f = q_A \circ f = p_A \\ p_B \circ g \circ f = q_B \circ f = p_B \end{cases}$$

and:

$$\begin{cases} q_A \circ f \circ g = p_A \circ g = q_A \\ q_B \circ f \circ g = p_B \circ g = q_B \end{cases}$$

Therefore:

$$g \circ f = id_P \qquad f \circ g = id_Q$$

$\square$

This proved if two objects have product (or coproduct), then it is unique.

Appendix D
Product and Coproduct of Set

Proof We prove this by construction. The Cartesian product $A \times B$ contains all combinations from the two sets.

$$\{(a, b) | a \in A, b \in B\}$$

We define two special arrows (functions) as p_A and p_B:

$$\begin{cases} fst\ (a, b) = a \\ snd\ (a, b) = b \end{cases}$$

Consider an arbitrary wedge

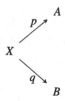

where $p\ x = a, q\ x = b, x \in X$. For example, let X be *Int*, A be *Int*, and B be *Bool*, and the two functions p and q are defined as

$$\begin{cases} p(x) = -x \\ q(x) = even(x) \end{cases}$$

© China Machine Press 2024
X. Liu, *Mathematics in Programming*,
https://doi.org/10.1007/978-981-97-2432-1

Such that p negates an integer, and q examines if it is even. We define the function $X \xrightarrow{m} A \times B$ as below:

$$m(x) = (a, b)$$

For this example, we have:

$$m(x) = (-x, even(x))$$

Such that the following diagram commutes:

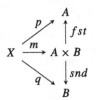

Let us verify it:

$$
\begin{aligned}
(fst \circ m)(x) &= fst\ m(x) & \text{function composition} \\
&= fst\ (a, b) & \text{definition of } m \\
&= a & \text{definition of } fst \\
&= p(x) & \text{reverse of } p
\end{aligned}
$$

and

$$
\begin{aligned}
(snd \circ m)(x) &= snd\ m(x) & \text{function composition} \\
&= snd\ (a, b) & \text{definition of } m \\
&= b & \text{definition of } snd \\
&= q(x) & \text{reverse of } q
\end{aligned}
$$

For the above example, we have:

$$
\begin{cases}
(fst \circ m)(x) = fst\ (-x, even(x)) = -x = p(x) \\
(snd \circ m)(x) = snd\ (-x, even(x)) = even(x) = q(x)
\end{cases}
$$

We also need to prove the uniqueness of m. Suppose there exists another function $x \xrightarrow{h} A \times B$, such that the diagram also commutes:

$$fst \circ h = p \quad \text{and} \quad snd \circ h = q$$

We have:

$$
\begin{aligned}
(a, b) &= (p(x),\ q(x)) && \text{definition of } p, q \\
&= ((fst \circ h)(x),\ (snd \circ h)(x)) && \text{commute} \\
&= (fst\ h(x),\ snd\ h(x)) && \text{function composition} \\
&= (fst\ (a, b),\ snd\ (a, b)) && \text{reverse of } fst, snd
\end{aligned}
$$

Hence $h(x) = (a, b) = m(x)$, which proves the uniqueness of m.

Next we prove the coproduct part. The elements in the disjoint union $A + B$ have two types. One comes from A as $(a, 0)$, and the other comes from B as $(b, 1)$. We can define two special arrows (functions) as i_A and i_B:

$$
\begin{cases}
left(a) = (a, 0) \\
right(b) = (b, 1)
\end{cases}
$$

Consider the wedge of an arbitrary set X:

We define arrow $A + B \xrightarrow{m} X$ as below:

$$
\begin{cases}
m\ (a, 0) = p(a) \\
m\ (b, 1) = q(b)
\end{cases}
$$

Such that the following diagram commutes:

$$
\begin{array}{ccc}
& A & \\
left \downarrow & \searrow{\scriptstyle p} & \\
A + B & \xrightarrow{m} & X \\
right \uparrow & \nearrow{\scriptstyle q} & \\
& B &
\end{array}
$$

Next we need to prove the uniqueness of m. Suppose there exists another arrow $A + B \xrightarrow{h} X$ and also makes the diagram commutes:

$$
h \circ left = p \quad \text{and} \quad h \circ right = q
$$

Taking any $a \in A, b \in B$, we have:

$$\begin{cases} h\,(a,0) = h(left(a)) = (h \circ left)(a) = p(a) = m\,(a,0) \\ h\,(b,1) = h(right(a)) = (h \circ right)(b) = q(b) = m\,(b,1) \end{cases}$$

Hence $h = m$, which proves the uniqueness of m.   □

References

1. Clawson, C.C.: The Mathematical Traveler, Exploring the Grand History of Numbers. Springer, Springer (1994). ISBN: 9780306446450
2. Wikipedia: Babylonian numerals. https://en.wikipedia.org/wiki/Babylonian_numerals
3. Kline, M.: Mathematics: The Loss of Certainty. Oxford University Press, Oxford (1980). ISBN: 0-19-502754-X
4. Hofstadter, D.R.: Gödel, Escher, Bach: An Eternal Golden Braid. Basic Books; Anniversary edition (1999). ISBN: 978-0465026562
5. Bird, R., de Moor, O.: Algebra of Programming. University of Oxford, Prentice Hall Europe (1997). ISBN: 0-13-507245-X
6. Gu, S.: "浴缸里的惊叹". People's Postal Press (2014). ISBN: 9787115355744
7. Han, X.: "数学悖论与三次数学危机". People's Postal Press (2016). ISBN: 9787115430434
8. Klein, M.: Mathematical thought from Ancient to Modern Times, vol. 1. Oxford University Press, Oxford (1972). ISBN: 9780195061352
9. Stepanov, A.A., Rose, D.E.: From Mathematics to Generic Programming, 1st edn. Addison-Wesley Professional (2014). ISBN-13: 978-0321942043
10. Euclid, Thomas Heath (Translator): Euclid's Elements (The Thirteen Books). Digireads.com Publishing (2017). ISBN-13: 978-1420956474
11. Han, X.: "好的数学——"下金蛋"的数学问题". Hunan Science and Technology Press (2009). ISBN: 9787535756725
12. Liu, X.: "Elementary Functional Algorithms (算法新解)". People's Postal Press (2017). ISBN: 9787115440358. https://github.com/liuxinyu95/AlgoXY
13. Wikipedia: Alan Turing. https://en.wikipedia.org/wiki/Alan_Turing
14. Abiteboul, S., Dowek, G., Nelson, K.-R.: The Age of Algorithms. Cambridge University Press, Cambridge (2019). ISBN-13: 978-1108745420
15. Peyton Jones, S.L.: The Implementation of Functional Programming Language. Prentice Hall, London (1987). ISBN: 013453333X
16. Curry, H.B.: The Paradox of Kleene and Rosser. Trans. Am. Math. Soc. **50**(3), 454–516 (1941)
17. Han, X.: "好的数学——方程的故事". Hunan Science and Technology Press (2012). ISBN: 9787535770066
18. Wikipedia: Galois theory. https://en.wikipedia.org/wiki/Galois_theory
19. Wikipedia: Évariste Galois. https://en.wikipedia.org/wiki/Evariste_Galois
20. Singh, A.R.: The Last Mathematical Testament of Galois, pp. 93–100. Resonance (1999)
21. Sawilowsky, S.S., Cuzzocrea, J.L.: Joseph Liouville's 'Mathematical Works of Évariste Galois'. J. Mod. Appl. Stat. Methods **5**(2), 32. https://doi.org/10.22237/jmasm/1162355460
22. Wikipedia: Rubik's Cube group. https://en.wikipedia.org/wiki/Rubik's_Cube_group

© China Machine Press 2024
X. Liu, *Mathematics in Programming*,
https://doi.org/10.1007/978-981-97-2432-1

23. Zhang, H.: "近世代数基础". Higher Education Press (1978). ISBN: 9787040012224
24. Armstrong, M.A.: Groups and Symmetry. Springer, Berlin (1988). ISBN: 0387966757
25. Wikipedia: Joseph-Louis Lagrange. https://en.wikipedia.org/wiki/Joseph-Louis_Lagrange
26. Wikipedia: Proofs of Fermat's little theorem. https://en.wikipedia.org/wiki/Proofs_of_Fermat's_little_theorem
27. Weil, A.: Number Theory, An Approach Through History from Hammurapi to Legendre". Birkhäuser, Boston (2006). ISBN: 978-0-8176-4565-6
28. Wikipedia: Leonhard Euler. https://en.wikipedia.org/wiki/Leonhard_Euler
29. Wikipedia: Carmichaeal number. https://en.wikipedia.org/wiki/Carmichael_number
30. Dsgupta, S., Papadimitriou, C., Vazirani, U.: Algorithms. McGraw-Hill Education, New York (2006). ISBN: 9780073523408
31. Wikipedia: Miller-Rabin primality test. https://en.wikipedia.org/wiki/Miller-Rabin_primality_test
32. Wikipedia: Emmy Noether. https://en.wikipedia.org/wiki/Emmy_Noether
33. Zhang, P.: "伽罗瓦理论：天才的激情". Higher Education Press (2013). ISBN: 9787040372526
34. Edwards, H.M.: Galois Theory, 3rd edn. Springer, Berlin (1984). ISBN: 978-0387909806
35. Goodman, D.: An Introduction to Galois Theory. https://nrich.maths.org/1422
36. Weyl, H.: Symmetry. Princeton University Press, Princeton; Reprint edition (2016). ISBN: 978-0691173252
37. Bell, E.T.: Men of Mathematics. Touchstone; Reissue edition (1986). ISBN: 978-0671628185
38. Feng, C.: "从一元一次方程到伽罗瓦理论", 2nd edn. East China Normal University Press (2019). ISBN: 9787567587380
39. Yuki, H.: Mathematical Girls 5: Galois Theory (Chinese Edition). People's Postal Press (2021). ISBN: 9787115559623
40. Dieudonne, J: Mathematics—The Music of Reason. Springer, Berlin (1998). ISBN: 9783540533467
41. Wiki, H.: Monad. https://wiki.haskell.org/Monad
42. Wikipedia: Samuel Eilenberg. https://en.wikipedia.org/wiki/Samuel_Eilenberg
43. Wikipedia: Saunders Mac Lane. https://en.wikipedia.org/wiki/Saunders_Mac_Lane
44. Simmons, H.: An introduction to Category Theory, 1st edn. Cambridge University Press, Cambridge (2011). ISBN: 9780521283045
45. Wikipedia: Tony Hoare. https://en.wikipedia.org/wiki/Tony_Hoare
46. Philip, W.: Theorems for free! In: Functional Programming Languages and Computer Architecture, pp. 347–359. Association for Computing Machinery (1989)
47. Milewski, B.: Category Theory for Programmers. https://bartoszmilewski.com/2014/10/28/category-theory-for-programmers-the-preface/
48. Smith, P.: Category Theory—A Gentle Introduction. http://www.academia.edu/21694792/A_Gentle_Introduction_to_Category_Theory_Jan_2018_version_
49. Wikipedia: Exponential Object. https://en.wikipedia.org/wiki/Exponential_object
50. Manes, E.G., Arbib, M.A.: Algebraic Approaches to Program Semantics. Texts and Monographs in Computer Science. Springer, Berlin (1986)
51. Lambek, J.: A fixpoint theorem for complete categories. Math. Z. **103**, 151–161 (1968)
52. Wikibooks: Haskell/Foldable. https://en.wikibooks.org/wiki/Haskell/Foldable
53. Gill, A., Launchbury, J., Peyton Jones, S.L.: A Short Cut to Deforestation. Functional Programming Languages and Computer Architecture, pp. 223–232 (1993)
54. Bird, R.: Pearls of Functional Algorithm Design, 1st edn. Cambridge University Press, Cambridge (2010). ISBN: 978-0521513388
55. Hinze, R., Harper, T., James, D.W.H.: Theory and Practice of Fusion. In: 22nd International Symposium of IFL. Implementation and Application of Functional Languages, pp.19–37 (2010)
56. Takano, A., Meijer, E.: Shortcut Deforestation in Calculational Form. In: Functional Programming Languages and Computer Architecture, pp. 306–313 (1995)

57. Knuth, D.: The Art of Computer Programming, vol. 4, Fascicle 4: Generating All Trees. Addison-Wesley, Reading (2006). ISBN: 978-0321637130

58. Luminet, J.-P., Lachièze-Rey, M.: De l'infini—Horizons cosmiques, multivers et vide quantique. DUNOD (2016). ISBN: 9782100738380

59. Noguti, T.: "数学的センスが身につく練習帳". SoftBank Creative (2007). ISBN: 9784797339314

60. Wikipedia: Googol. https://en.wikipedia.org/wiki/Googol

61. Wikipedia: Zeno's Paradoxes. https://en.wikipedia.org/wiki/Zeno's_paradoxes

62. Zhang, J.., Wang, X.: Continuum Hypothesis. Liaoning Education Press (1988). ISBN: 7-5382-0436-9/G.445

63. Courant, R., Robbins, H., Stewart, I.: What Is Mathematics? An Elementary Approach to Ideas and Methods, 2nd edn. Oxford University Press, Oxford (1996). ISBN: 978-0195105193

64. Poincaré, H.: Science and Hypothesis. Franklin Classics (2018). ISBN: 9780342349418

65. Gatys, L.A., Ecker, A.S., Bethge, M.: A Neural Algorithm of Artistic Style. In: 2015 IEEE Conference on Computer Vision and Pattern Recognition (CVPR) (2017). arXiv:1508.06576 [cs.CV]

66. Gu, S.: "思考的乐趣——Matrix67数学笔记". People's Postal Press (2012). ISBN: 9787115275868

67. Abelson, H., Sussman, G.J., Sussman, J.: Structure and Interpretation of Computer Programs. MIT Press, Cambridge (1984). ISBN 0-262-01077-1

68. Poincaré, H.: The Value of Science. Modern Library (2001). ISBN: 978-0375758485

69. Ried, C.: Hilbert, 1st edn. Springer, Berlin (1996). ISBN: 978-0387946740

70. Lockhart, P.: Measurement. Belknap Press: An Imprint of Harvard University Press; Reprint edition 2014. ISBN: 978-0674284388

Index

© China Machine Press 2024
X. Liu, *Mathematics in Programming*,
https://doi.org/10.1007/978-981-97-2432-1

Printed in the United States
by Baker & Taylor Publisher Services